Emerging Energy Materials

Emerging Energy Materials: Applications and Challenges guides the reader through materials used in progressive energy systems.

It tackles their use in energy storage across solar, bio, geothermal, wind, fossil, hydrogen, nuclear, and thermal energy. Specific chapters are dedicated to energy-reaping systems currently in development. This book contributes to the current literature by highlighting concerns that are frequently overlooked in energy materials textbooks. Awareness of these challenges and contemplation of possible solutions is critical for advancing the field of energy material technologies.

Key features:

- Provides up-to-date information on the synthesis, characterization, and a range of applications using various physical and chemical methods.
- Presents the latest advances in future energy materials and technologies subjected to specific applications.
- Includes applied illustrations, references, and advances in order to explain the challenges and trade-offs in the field of energy material research and development.
- Includes coverage of solar cell and photovoltaic, hydro power, nuclear energy, fuel cell, battery electrode, supercapacitor and hydrogen storage applications.

This book is a timely reference for researchers looking to improve their understanding of emerging energy materials, as well as for postgraduate students considering a career within materials science, renewable energy, and materials chemistry.

Govind B. Nair is a Post-doctoral researcher at University of the Free State, Bloemfontein, South Africa.

H. Nagabhushana is Chair of the Department of Physics at Tumkur University, India.

Nirupama S. Dhoble is a Professor of chemistry at Sevadal Mahila Mahavidhyalaya, India.

Sanjay J. Dhoble is a Professor of physics at Rashtrasant Tukadoji Maharaj Nagpur University, Nagpur, India.

Series in Materials Science and Engineering

The series publishes cutting edge monographs and foundational textbooks for interdisciplinary materials science and engineering.

Its purpose is to address the connections between properties, structure, synthesis, processing, characterization, and performance of materials. The subject matter of individual volumes spans fundamental theory, computational modeling, and experimental methods used for design, modeling, and practical applications. The series encompasses thin films, surfaces, and interfaces, and the full spectrum of material types, including biomaterials, energy materials, metals, semiconductors, optoelectronic materials, ceramics, magnetic materials, superconductors, nanomaterials, composites, and polymers.

It is aimed at undergraduate and graduate level students, as well as practicing scientists and engineers.

Proposals for new volumes in the series may be directed to Carolina Antunes, Commissioning Editor at CRC Press, Taylor & Francis Group (Carolina.Antunes@tandf.co.uk).

2D Materials for Infrared and Terahertz Detectors
Antoni Rogalski

Fundamentals of Fibre Reinforced Composite Materials
A. R Bunsell. S. Joannes, A. Thionnet

Fundamentals of Low Dimensional Magnets
Ram K Gupta, Sanjay R Mishra, Tuan Anh Nguyen, Eds.

Emerging Applications of Low Dimensional Magnets
Ram K Gupta, Sanjay R Mishra, Tuan Anh Nguyen, Eds.

Handbook of Silicon Carbide Materials and Devices
Zhe Chuan Feng, Ed.

Bioelectronics: Materials, Technologies, and Emerging Applications
Ram K. Gupta, Anuj Kumar, Eds.

Advances in 3D Bioprinting
Roger J. Narayan, Ed.

Hydrogels: Fundamentals to Advanced Energy Applications
Ram K. Gupta, Anuj Kumar, Eds.

Emerging Energy Materials: Applications and Challenges
Govind B. Nair, H. Nagabhushana, Nirupama S. Dhoble, Sanjay J. Dhoble, Eds.

For more information about this series, please visit: https://www.routledge.com/Series-in-Materials-Science-and-Engineering/book-series/TFMATSCIENG

Emerging Energy Materials

Applications and Challenges

Edited by

Govind B. Nair

H. Nagabhushana

Nirupama S. Dhoble

Sanjay J. Dhoble

CRC Press is an imprint of the
Taylor & Francis Group, an **informa** business

Front cover image: Leo Matyushkin/Shutterstock

First edition published 2024
by CRC Press
2385 NW Executive Center Drive, Suite 320, Boca Raton FL 33431

and by CRC Press
4 Park Square, Milton Park, Abingdon, Oxon, OX14 4RN

CRC Press is an imprint of Taylor & Francis Group, LLC

ISBN: 978-1-032-31209-5 (hbk)
ISBN: 978-1-032-32483-8 (pbk)
ISBN: 978-1-003-31526-1 (ebk)

DOI: 10.1201/9781003315261

Typeset in Minion
by SPi Technologies India Pvt Ltd (Straive)

Contents

List of Contributors

Fuad Ameen
King Saud University
Riyadh, Saudia Arabia

Shahid Bashir
Higher Institution Centre of Excellence, UM Power Energy Dedicated Advanced Centre
University of Malaya
Kuala Lumpur, Malaysia

Ashok Bera
Department of Physics
Indian Institute of Technology
Jammu, India

Ramesh Chandra
Nanoscience Laboratory, Institute Instrumentation Centre
IIT Roorkee
Roorkee, India

Vijay Chaudhari
S. N. Mor College of Arts & Commerce & Smt. G. D. Sarda Science College
Tumsar, India

Mohd Fakhruddin
Centre for Ionics, Department of Physics
University of Malaya
Kuala Lumpur, Malaysia

Durvesh Gautam
Department of Physics
Ch. Charam Singh University
Meerut, India

Yogendra K. Gautam
Department of Physics
Ch. Charam Singh University
Meerut, India

Abhijeet R. Kadam
Department of Physics
R. T. M Nagpur University
Nagpur, India

Tania Kalsi
Department of Nanoscience and Materials
Central University of Jammu
Jammu, India

Ashwani Kumar
Nanoscience Laboratory, Institute Instrumentation Centre
IIT Roorkee
Roorkee, India

Pragati Kumar
Department of Nanoscience and Materials
Central University of Jammu
Jammu, India

Chaitali M. Mehare
Department of Physics
R.T.M Nagpur University
Nagpur, India

M. D. Mehare
Department of Applied Physics
Priyadarshini College of Engineering
Nagpur, India

Marta Michalska-Domanska
Institute of Optoelectronics
Military University of Technology
Warszawa, Poland

Amol Nande
Guru Nanak College of Science
Ballarpur, India

Sarojini Jeeva Panchu
Department of Physics
University of the Free State
Bloemfontein, South Africa

Yatish R. Parauha
Department of Physics
Shri Ramdeobaba College of Engineering and Management
Nagpur, India

M. Pershaanaa
Centre for Ionics, Department of Physics
University of Malaya
Kuala Lumpur, Malaysia

J. D. Punde
S S Girls College
Gondia, India

K. Ramesh
Centre for Ionics, Department of Physics
University of Malaya
Kuala Lumpur, Malaysia

S. Ramesh
Centre for Ionics, Department of Physics
University of Malaya
Kuala Lumpur, Malaysia

Nupur Saxena
Department of Physics
Indian Institute of Technology
Jammu, India

Vijay Singh
Department of Chemical Engineering
Konkuk University
Seoul, Republic of Korea

Hendrik C. Swart
Department of Physics
University of the Free State
Bloemfontein, South Africa

Sumedha Tamboli
Department of Physics
University of the Free State
Bloemfontein, South Africa

B. Vengadaesvaran
UM Power Energy Dedicated Advanced Centre (UMPEDAC)
University of Malaya
Kuala Lumpur, Malaysia

Ariff Zahiruddin
Centre for Ionics, Department of Physics
University of Malaya
Kuala Lumpur, Malaysia

I

Energy Storage Devices and Energy Conversion Devices

CHAPTER 1

Basics and Design of the Supercapattery

An Energy Storage Device

M. Pershaanaa, Ariff Zahiruddin, and Mohd Fakhruddin

Universiti Malaya, Kuala Lumpur, Malaysia

Shahid Bashir

UM Power Energy Dedicated Advanced Centre, Universiti Malaya, Kuala Lumpur, Malaysia

S. Ramesh and K. Ramesh

Universiti Malaya, Kuala Lumpur, Malaysia

1.1 INTRODUCTION

Over decades, the earth has faced a periodic upsurge in its temperature due to excessive emission of greenhouse gases, especially carbon dioxide (CO_2), that have caused severe climate change and ecological disruption. To minimize greenhouse gas emissions and restore the environment, several initiatives have been taken under the Paris Agreement focusing on energy, industries, infrastructure, land, and transportation [1]. Among these, the transition from fossil fuel to renewable energy as the energy source was the major initiative imposed to produce a net-zero carbon environment. Renewable energy has numerous advantages, namely, diminishing greenhouse gas emissions and ensuring environmental protection due to reduced pollution from power plants for electricity generation [2]. However, the major pitfall of renewable energy implementation is the uncertainty of the energy source supply. For instance, if solar energy is used as the energy source, then the electricity supply will be in excess during daytime but insufficient during night-time. To counter this issue, energy storage devices such as supercapacitors are embedded into the energy storage system to provide energy during peak demands and when the energy supply is low. Supercapacitors are often selected as the backup energy storage device due to the

DOI: 10.1201/9781003315261-2

high-power density of the supercapacitor that enables sudden and rapid release of energy over a short period. Howbeit, supercapacitors have limited energy storage capacity which is ill-suited to large-scale power plants. Batteries are a persuasive alternative, as they have outstanding energy storage capacity suitable for large-scale power plants, but they suffer from low power density making then unable to release large amount of energy in a short time. A hybrid energy storage device of a battery/supercapacitor combination known as a 'supercapattery' was introduced to extract and utilize the strengths of supercapacitors and batteries, concomitantly suppressing their weaknesses. This chapter includes a brief and fundamental discussion of the supercapattery and of the building blocks of the supercapattery, namely, batteries and supercapacitor. A general in design process of a supercapattery and its electrode materials are also discussed for better understanding of the device and its charge storage mechanism.

1.2 WHAT IS A SUPERCAPATTERY?

A supercapattery as the name suggests, is a combination of batteries and a supercapacitor. The supercapattery is being explored as creative hybrid electrochemical energy storage (EES) device that combines the advantages of rechargeable batteries and supercapacitors in a single device. The word 'supercapattery' was first coined in a 2007 industrial EES study and, since then, academics have aggressively advocated its usage, resulting in its progressive acceptance by the EES community [3]. Fundamentally, secondary batteries or rechargeable batteries such as LIBs have high energy density, up to 200 Wh kg^{-1} as they store charges via redox reactions but suffer from low power density, as low as 100 W kg^{-1}. On the other hand, supercapacitors exhibit a high power density of an average of 10 kW kg^{-1} and a low energy density of 10 Wh kg^{-1} due to their fast and reversible non-faradaic charge storage mechanism. To bridge the merits and demerits of both devices, a hybrid device was presented by using the battery-type materials as positive terminal and carbonaceous materials as negative terminal. In this design, the faradaic component (positive electrode) is responsible for specific energy, whereas the non-faradaic (carbonaceous materials) component controls specific power [4]. Ragone plot in Figure 1.1 shows how a supercapattery compensates for an electric double-layer capacitor's (EDLC) poor energy density and for the batteries' low power density [5].

1.3 BUILDING BLOCKS OF A SUPERCAPATTERY

As mentioned in the previous section, a supercapattery consists of two electrochemical energy storage devices, namely, a battery and supercapacitor that act as the key components in the hybrid device. Therefore, to understand the overall mechanism and fabrication of the hybrid device it is crucial to study single electrochemical energy storage devices separately. Thus, in the upcoming subsections, a brief discussion on batteries and supercapacitors is included.

1.3.1 Battery (Li-ion Battery)

Arguably, the most popular and well-known electrochemical technology in use today is the battery. The huge energy storage capacity of batteries is a result of their high energy

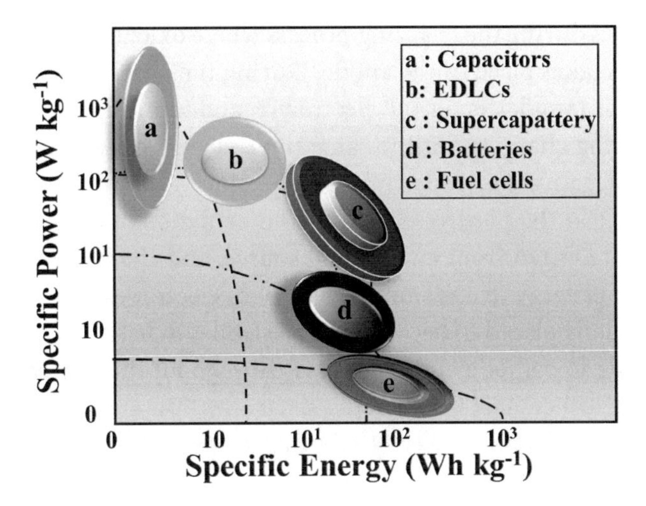

FIGURE 1.1 Supercapattery performance compared to EDLCs and batteries. (Reproduced with permission from Ref. [5], © 2021 Elsevier Ltd.)

density. Every battery has three fundamental components: an electrolyte, a negative side (known as anode), and a positive side (known as cathode). These electrodes are frequently separated by separators and immersed in an electrolyte, which acts as a conducting path for ions to travel between them (Figure 1.2 [6]).

As soon as the battery (for example Li-ion battery) is connected to an application (discharging process), positive ions (Li^+) form between the negative electrode (anode) and the electrolyte, due to the oxidation process, and deliver electrons into the positive electrode (cathode) via the external circuit (resistors, connecting wires, capacitors, and so on). These electrons are then accepted by the positive electrode and undergo a reduction process resulting in an electrical difference between the negative electrode and positive electrode, which causes an electric current to flow to operate the device [6, 7]. The reverse occurs in

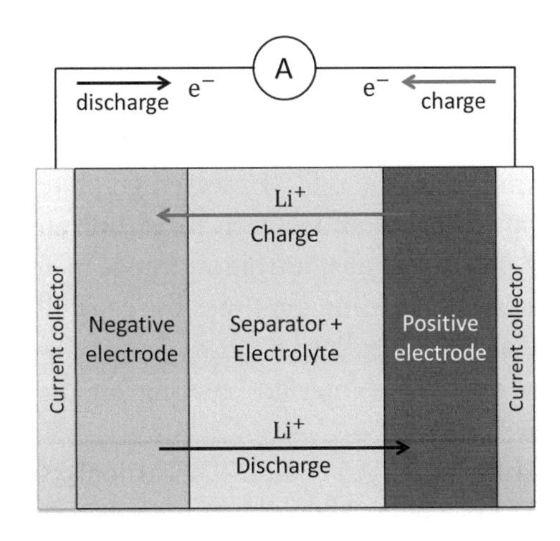

FIGURE 1.2 Schematic diagram of a rechargeable lithium-ion battery. (Reproduced with permission from Ref. [6], © 2019 Elsevier Ltd.)

rechargeable batteries during the charging process where oxidation takes place in the cathode while reduction takes place in the anode. During the charging process, Li^+ ions leave the positive electrode (anode), enter the electrolyte, and intercalate into the negative electrode (cathode) during charging. At the same time, electrons leave the positive electrode and flow into the negative electrode through the outer electric circuit. Upon charging, reduction takes place at the positive electrode and oxidation at the negative electrode. In this case, energy is required from an external source, and the battery converts electrical energy into chemical energy that is stored inside both electrodes.

A separator is usually placed in between the electrodes in batteries to allow ions to readily move from anode to cathode and vice versa throughout the charge/discharge process and to prevent electrical contact between the electrodes. The active components of batteries are frequently encased with a cover system that keeps air from entering and electrolyte solvent from escaping while supporting the assembly to prevent an explosion. Certain cells, such as lithium-ion batteries, may experience thermal runaway in some failure situations due to the production of flammable gases, which put the cells in danger from fire and explosion [8].

1.3.1.1 Electrode Material for Batteries

The active material distinguishes the anode from the cathode in Li-ion batteries. Silicon and/or carbonaceous materials are commonly used in the anode because silicon serves as the active element and offers high capacity, while carbon boosts conductivity and reduces silicon expansion. Lithium compounds are used in the cathode because lithium has the highest electropositive element. The ability of an element to contribute electrons to form positive ions is measured by its electro-positivity. In other terms, it quantifies how readily an element may generate energy. Lithium quickly loses electrons, which implies it readily generates a large amount of energy. Graphitic carbon, hard carbon, synthetic graphite, tin-based alloys, and silicon-based materials are commonly utilized in the fabrication of the anode. On the other hand, cathode materials include lithium manganese oxide, lithium cobalt oxide, lithium nickel cobalt manganese oxide, lithium-ion phosphate, and electrical conducting polymers [9]. Other components include a binder, gel precursor, and electrolyte solvent [10].

Given its high conductivity, good reversibility, and low cost, graphite is utilized commercially as an electrode material in anodes. However, graphite has a low energy density since it can only retain one lithium atom per each six carbon atoms. Silicon, on the other hand, is widely considered one of the most intriguing anode materials as each silicon atom can attach to four lithium ions, providing ten times the capability of a graphite electrode [11]. However, the volume of the silicon lattice varies drastically during repetitive charge/discharge cycles, and the electrolyte interface coating on the anode surface becomes unstable.

According to a recent study by W. An et al. [12], a carbon-coated porous silicon anode has a high capacity of 1,271 mAh g^{-1} at 2,100 mA g^{-1} with 90% capacity retention for 1,000 cycles and a lower electrode swelling of 17.8% at such a great areal capacity of 5.1 mAh cm^{-2}. After 400 charge/discharge cycles, the complete cell, with pre-lithiated silicon anode as well as $Li\left(Ni_{1/3}Co_{1/3}Mn_{1/3}\right)O_2$ cathode, has a high energy density of 502 Wh kg^{-1} and an 84% capacity retention.

Besides, a study by Z. Li et al. shows the synergistic combination of these characteristics enables the synthesized $SiO_x - TiO_2@C$ nanocomposite for a Li-ion battery anode to have good electrochemical properties, namely, high specific capacity, great rate capability, and with stable long-term cyclability. After 200 cycles at 0.1 A g^{-1}, a steady specific capacity of 910 mAh g^{-1} was established, and after 600 cycles at 1 A g^{-1}, a stable specific capacity of 700 mAh g^{-1} was achieved [13].

Despite the benefits of high specific energy, the limitations due to low power density and short cycle life restrict commercialization. Non-monotonic energy consumption and fast changes throughout a battery's discharge process, which are harmful to the battery, are the main causes [14]. One viable solution is to combine a battery with a supercapacitor to form a hybrid device, which has a similar construction to a battery but a longer life cycle and higher energy density.

1.3.2 Types of Supercapacitors

A supercapacitor is an energy storage device that is categorized as an electrochemical energy storage device that stores and releases energy at the interfaces between the electrolyte and electrode by adsorption and desorption of ions. The charge storage mechanism of supercapacitors is rapid and reversible due to the adsorption/desorption process occurring at the interface of the electrode-electrolyte. Consequently, supercapacitors can charge and discharge the charges in the device faster than batteries. There are three types of supercapacitors which are: (1) the Electric Double Layer Capacitor (EDLC), (2) the pseudocapacitor, and (3) the hybrid capacitor, as shown in Figure 1.3.

1.3.2.1 EDLC

The EDLC is the pioneer type of supercapacitor which uses non-faradaic reaction to store its charges. An EDLC is usually made up of symmetrical and highly porous electrodes that are made up of carbonaceous compound. At first the EDLC used the Helmholtz layer to store energy electrostatically where the electrolyte ions moved to the electrode to form an electrical double layer (EDL) that was divided by one atomic distance [16]. The model was then modified by Gouy and Chapman where, in their diffuse model, the anions and cations were treated as a continuous distribution in the electrolyte. Gouy and Chapman also stated that the ion concentration in the solution near the surface follows the Boltzmann distribution and forms the diffuse layer due to thermal motion drive [16]. However, the Gouy-Chapman model of EDL was lacking in terms of capacitance estimation and thus, to counter the issue, the model was revised by Stern. Stern merged Helmholtz's theory with Gouy-Chapman's theory and demonstrated the current theory of the EDL where the ions undergo electrostatic adsorption without any charge transfer to store the charges in the interface of the electrode and electrolyte [16]. Since the charge storage is held in the EDL, the thickness of the EDL plays a vital role in the value capacitance of EDLC. Apart from the EDL, the contact surface area of the electrode is also important in controlling the capacitance of the EDLC

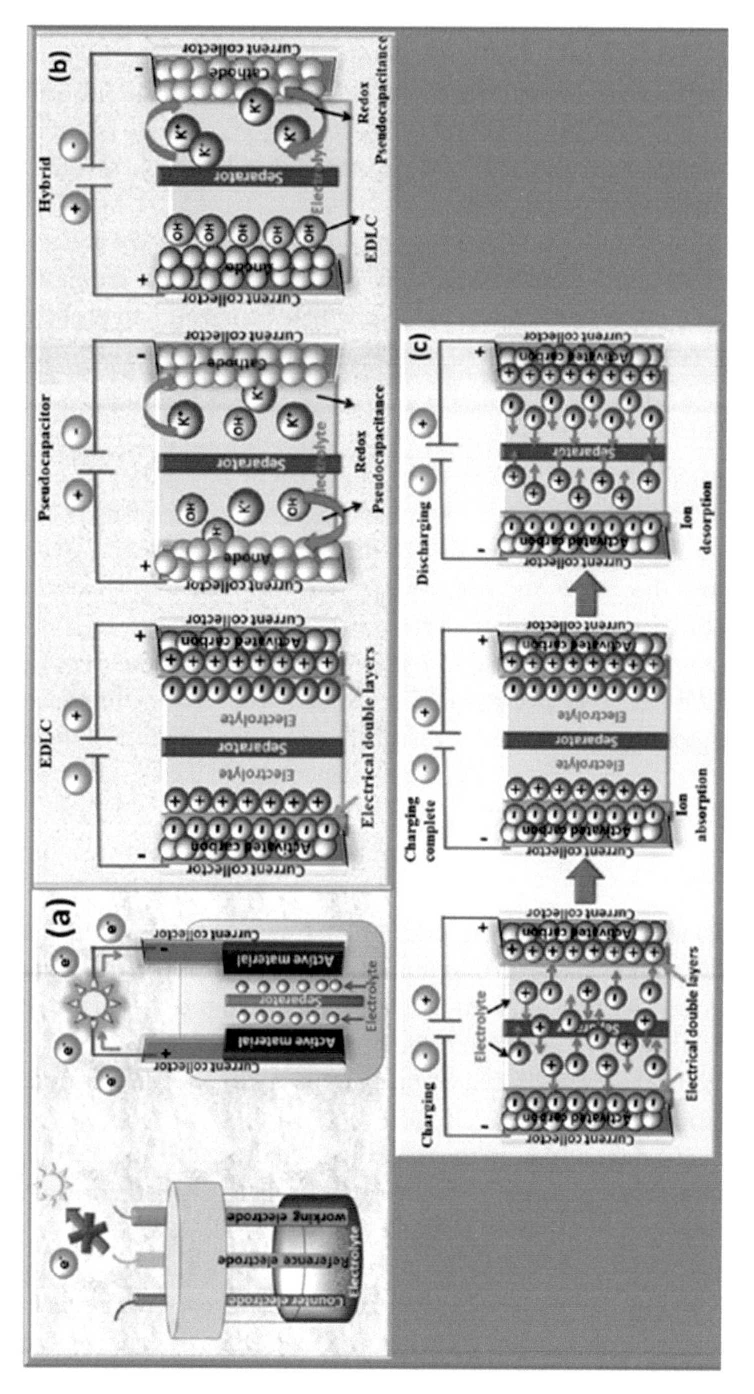

FIGURE 1.3 Schematic diagram of supercapacitor types: (a) EDLC, (b) pseudocapacitor, (c) hybrid capacitor. (Reproduced under the Creative Commons Attribution license from Ref. [15] © 2022 S. H. Nagarajarao et al.)

because the surface area is directly proportional to the capacitance. Meanwhile, the thickness of the EDL is conversely proportional to the capacitance of the EDLC. This relation was demonstrated in the formula of capacitance as given in Eqn. (1.1) [16, 17].

$$\text{Capacitance, } C = \frac{\varepsilon_r \varepsilon_o}{d} A \tag{1.1}$$

Where A represents the effective surface area of the electrode interacting with the electrolyte, d represents the thickness of the EDL formed, while ε_r and ε_o represent the relative permittivity and permittivity in vacuum respectively. The charge transfer-free mechanism of the EDLC is rapid and reversible which contributes to greater power densities than standard batteries, making them ideal for power-quality applications. Instantly, EDLCs suffer from low energy density due to the nature of the charge storage mechanism.

1.3.2.1.1 Electrode Materials for EDLCs Non-metallic and inert carbonaceous compounds such as activated carbons (AC), graphene, porous carbon (PC), carbon nanotubes (CNTs), carbide-derived carbon (CDC), carbon aerogels, and graphdiyne are widely explored as the active material for EDLCs [16–24]. This is because carbonaceous compounds are known for their high abundance, easy manufacturure, and for having great electronic conductivity and mechanical properties [16]. On top of that, carbonaceous materials also have low toxicity, can work in a wide range of temperature, have a larger surface area for ion interaction and, most importantly, are low-cost compared to metallic compounds, as some of these carbonaceous materials are derived from biomass [16].

Among these carbonaceous materials, AC was the first explored material for EDLCs due to its splendid specific surface area, as AC is often made up of different pore distributions, like macroporous structures (pore size > 500 Å), mesoporous structures (pore size in the range of 20–500 Å) and microporous structures (pore size < 20 Å) [24]. Apart from having a large surface area, AC is also known as an environmentally friendly material because it is derived from biomass products. For instance, Kanruethai et al. in their study fabricated nitrogen-doped AC that was derived from biowaste of rice straw using a two-step process of carbonization followed by potassium hydroxide (KOH) activation [23]. From the study, it was revealed that AC derived from rice straw, with and without nitrogen doping, exhibits a sizeable specific surface area of 2,651 m^2g^{-1} and 2,537 m^2g^{-1}, respectively. It was also perceived that AC doped with nitrogen exhibits excellent capacity retention of 95% even after 10,000 charge/discharge cycles with an energy density of 48.9 Wh kg^{-1} when the power density is 750 W kg^{-1}. In preparation of AC, the biomass selected for AC derivation is important because different biomass contributes to different electrochemical performances due to their different cellulose (C) to hemicellulose (HC) to lignin (L) compositions. For confirmation, Murugan and co-workers collected different types of biomass, namely, *Syzygium cumini* fruit shells (SCFS) and *Chrysopogin zizanioides* roots (CZR) for AC and fabricated both the sources using the same two-step process of carbonization followed by CO_2 activation (as shown in Figure 1.4) [22]. It was revealed that different source

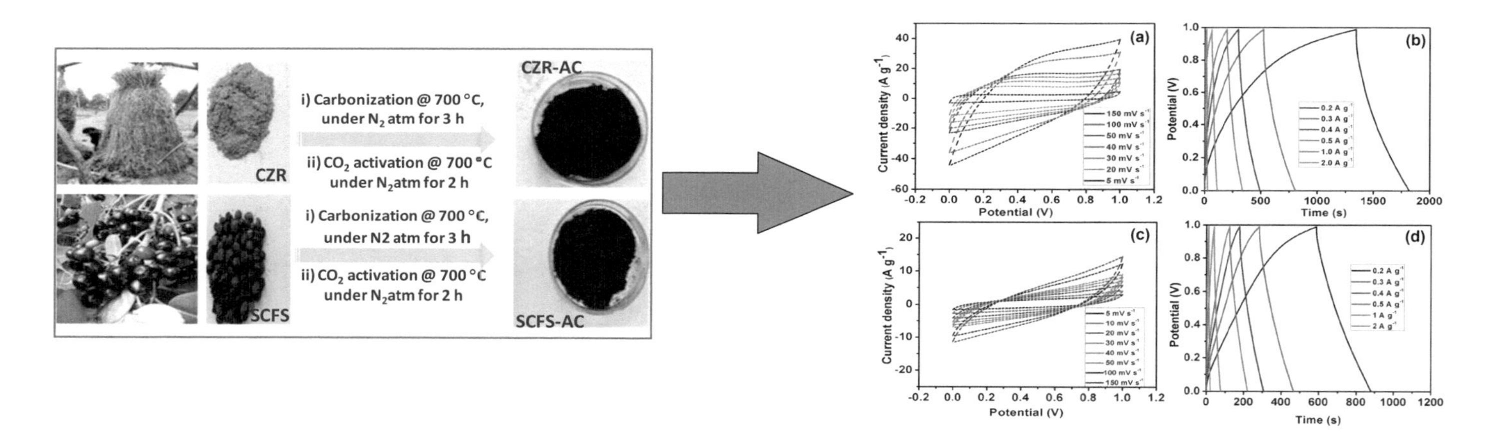

FIGURE 1.4 The illustration of the fabrication process and electrochemical analysis (Cyclic Voltammetry (CV) and Galvanostatic Charge Discharge (GCD)) of SCFS (a, b) and CZR (c, d) respectively. (Reproduced with permission from Ref. [22], © 2020 Elsevier Ltd.)

contributes to different morphology which directly affects the electrochemical performance of the supercapacitor. In this case, it was revealed that the ash content of CZR was higher than SCFS leading to the formation of mesopores with thin walls in CZR while micropores with thick walls in SCFS during the CO_2 activation process. As a result, the overall electrochemical performance of the SCFS device was better than CZR device where the specific capacitance of SCFS was 196 F g^{-1} 0.2 A g^{-1} with a maximum energy density of 27.22 Wh kg^{-1} at a power density of 200 W kg^{-1} while CZR exhibited 120 F g^{-1} at 0.2 A g^{-1} with an energy density of 16.72 Wh kg^{-1} at power density 200 W kg^{-1}. Even though the source of the AC differed yet the capacity retention of the ACs retained more than 90% over 5,000 charge/discharge cycles [22].

Besides AC, graphene-based materials and carbon nanotubes (CNTs) are commonly used materials for EDLCs. Graphene is a 2D carbon material that is made up of hexagonal-like arranged atoms to form a one-atom-thick sheet with outstanding mechanical and electrochemical properties [7, 21]. Graphene is also known as nanocomposite paper owning to its flexibility with an elastic modulus of 1.3 TPa and mechanical strength higher than steel making it suitable for wearable supercapacitors [7]. The properties of graphene are often enhanced by the fabrication of graphene derivatives such as graphene oxide, reduced graphene oxide, graphene doped with heteroatoms, or dual atom/co-doped graphene as graphene derivation counters the restacking and aggregation issue of graphene and concurrently increases the electrochemical properties of graphene. For instance, Debabrata et al. fabricated a multi-heteroatom self-doped graphene nanosheet derived from petroleum coke and revealed a satisfactory overall performance of specific capacitance around 170 F g^{-1} at 0.5 A g^{-1} for a single electrode and maximum energy density and power density of 8.8 Wh kg^{-1} and 800 W kg^{-1} respectively for EDLC [25]. In another research, duplicate laser pyrolysis was used to synthesize and incorporate nitrogen (N) and boron (B) co-doped heteroatoms into laser-induced graphene to fabricate a solid-state flexible supercapacitor. It was reported that the addition of N/B co-doped heteroatoms has remarkably increased compared to graphene with only N heteroatom and without doping of any heteroatoms by 2.5 and 12 times, respectively [26]. The combination of different heteroatoms is an effective way to enhance the electrochemical properties of graphene because it creates a synergistic effect between them and together increases the interlayer spacing supplying more active sites for ion adsorption/desorption processes.

Graphene is also utilized in the formation of CNTs as layers of nanosized graphene sheet are rolled into a cylindrical shape to form CNT with half of the fullerene molecule closing the ends of the cylinder (Figure 1.5) [7]. The combination of nanoscale carbons and the interconnected mesoporous structure of CNTs spikes the effective surface area of the CNT and consequently increases its electrical conductivity leading to higher specific capacitance compared to ACs. There are three types of CNTs: (1) single-walled carbon nanotubes (SWCNT), (2) double-walled carbon nanotubes (DWCNT), and (3) multi-walled carbon nanotubes (MWCNT), depending on the layers of the graphene sheet, as seen in the Figure 1.5. Among them, SWCNT and MWCNT are commonly used CNTs in supercapacitor application. Pure CNTs often exhibit both EDLC and pseudo-capacitance characteristics as the carbon compounds of the CNT undergo a non-faradaic reaction, while the

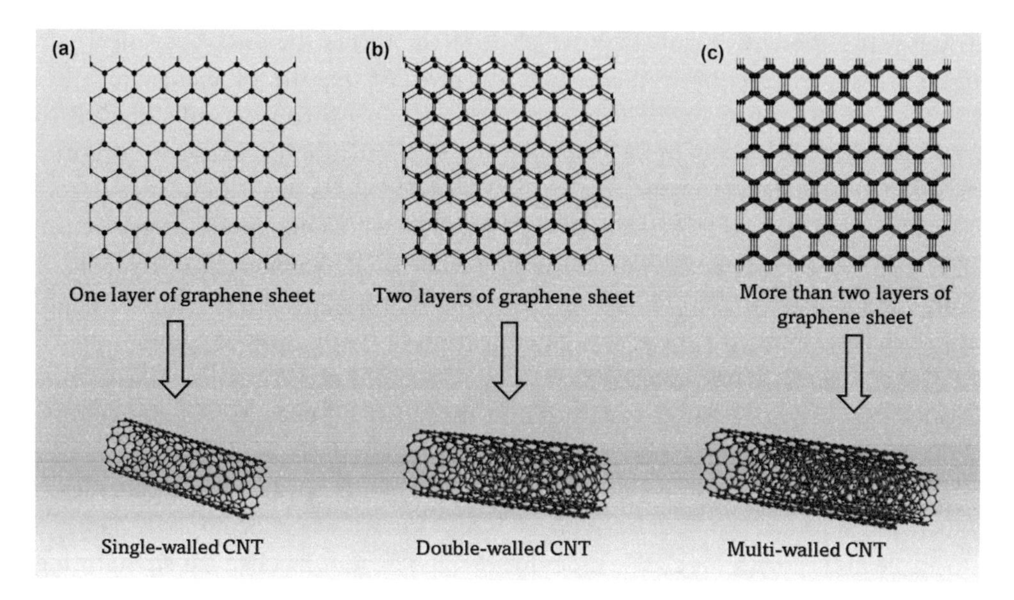

FIGURE 1.5 Types of CNTs. (Reproduced with permission from Ref. [7], © 2022 Elsevier Ltd.)

impurities on CNT undergo a faradaic reaction at the electrode-electrolyte interface [27]. According to Wang et al., the average specific capacitance exhibited by pure CNTs in aqueous electrolytes is from 20 F g^{-1} to 100 F g^{-1} [16]. Therefore, to optimize the performance of CNTs, graphene, nanomaterials, metallic compounds, or conducting polymers are incorporated with CNTs. In a recent review, Zakaria and co-workers reported a list of past research on graphene-CNT hybrid active material for supercapacitors and revealed that this hybrid overall exhibits magnifying specific capacitance, energy density, and capacity retention of the supercapacitor. For instance, Kshetri et al. fabricated a ternary hybrid structure composed of graphene, CNT, and carbon nanofibers and reported energy density up to 62 Wh kg^{-1} with a specific capacitance of 218 F g^{-1} at 1 A g^{-1} and capacity retention up to 91.7% over 10,000 charge/discharge cycles [28]. Meanwhile, Xiong et al. interconnected CNT with nanoporous reduced graphene oxide and polyaniline (PANI) and revealed significantly better results than the previous study, where this combination exhibited an outstanding specific capacitance of 741 F g^{-1} at 1 A g^{-1} with energy density and capacity retention up to 92.4 Wh kg^{-1} and 95% over 5,000 cycles, respectively [29].

1.3.2.2 Pseudocapacitor

The pseudocapacitor was developed when researchers were investigating other alternatives for porous, and high surface area materials like manganese oxide, and other oxides upon producing greater capacitance. The phenomenon was dubbed 'pseudocapacitance' after it was observed that reversible electrochemical reactions occurred between the aqueous electrolytes and these compounds [30]. The term pseudocapacitance was first introduced by David C. Grahame in 1941 referring to pseudocapacity that demonstrates the reversible capacity process [31]. Pseudocapacitors are also known as faradaic supercapacitors as the capacitance contributed to this supercapacitor was quick and had reversible redox processes at or near the electrode surface [32]. In contrast to EDLCs, pseudocapacitors have a relatively low power density and poor cycle stability due to mechanical degradation which originated from swelling/shrinkage of active material during the charging and discharging process [7, 20]. However, the charge storage mechanism of this type of supercapacitor has significantly improved energy density compared to EDLC apart from simultaneously increasing the specific capacitance [33].

According to the pseudocapacitor's operating principle, electrical energy is stored by transferring electron charge between the electrode and electrolyte through electrosorption, reduction-oxidation reactions, and intercalation processes [34]. Unlike EDLCs, charge transfer occurs in pseudocapacitors during the charging/discharging processes. Upon receiving electricity, similar to EDLCs, the electron and hole accumulation occurs on terminals due to potential differences leading the ions in the electrolyte to move towards respective electrodes. As the electrolyte ion reaches the surface of the electrode, the charge transfer process occurs at the electrode/electrolyte interface owing to the oxidation and reduction processes. This then creates a passage for Faradaic current to pass through the double layer and causes a spike in its energy density. These supercapacitor materials are called pseudocapacitive materials and there are three types of Faradaic mechanisms that occur to store its charges which are: (1) underpotential deposition or adsorption pseudocapacitance, (2) redox pseudocapacitance, and (3) intercalation pseudocapacitance (Figure 1.6). When metal ions form an adsorbed monolayer at a different metal surface and a more negative potential than the equilibrium potential for the reduction of the metal, this process is known as underpotential deposition. The deposition of lead on the surface of a gold electrode is a well-known illustration of underpotential deposition. The Langmuir-type electrosorption of H on a noble metal substrate, such as Pt, Rh, or Ru, is another classic example of underpotential deposition [35].

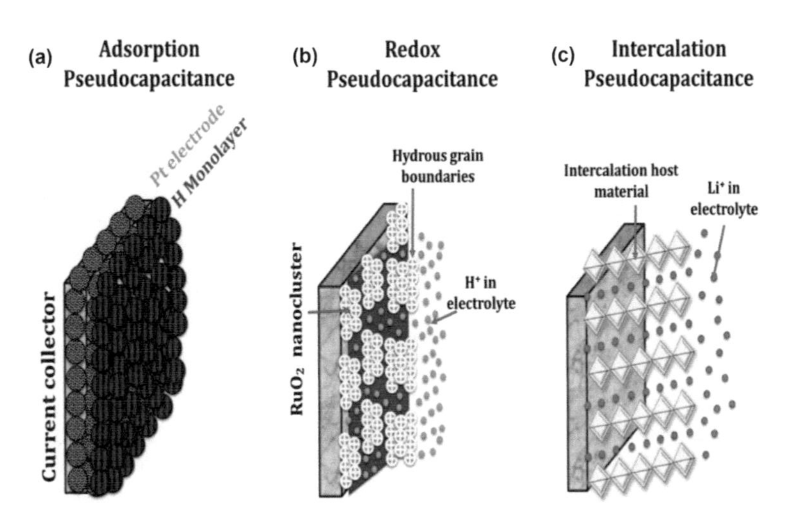

FIGURE 1.6 Different types of pseudocapacitance. (Reproduced with permission from Ref. [31], © 2021 Elsevier Ltd.)

When ions are electrochemically adsorbed onto a material's surface or close to its surface with a concurrent faradaic charge transfer, redox pseudocapacitance results. This is the common type of charge storage mechanism in pseudocapacitors as this occurs at the electrode/electrolyte interface and results in a fast and reversible redox reaction. Classical examples of redox pseudocapacitive materials are conducting polymers (polypyrrole and PANI) and metal oxides like RuO_2 and MnO_2 [34]. A brand-new electrical energy storage mechanism called intercalation pseudocapacitance appears to work similarly to an electrode in supercapacitors while storing energy in the electrode's bulk (fast reaction kinetics). Such intercalation pseudocapacitance can significantly reduce the gap in energy density and power density between supercapacitors and lithium-ion batteries, opening up new possibilities for the creation of sophisticated energy storage systems with high energy density and power density [34].

1.3.2.2.1 Electrode Materials for Pseudocapacitators

Intrinsic pseudocapacitor and extrinsic pseudocapacitor are well-known terms used in pseudocapacitors according to the behavior of the electrode materials towards the charge-storing mechanism. An intrinsic pseudocapacitor refers to a pseudocapacitor that stores charge using a surface-controlled redox reaction similar to the redox pseudocapacitance due to the nature of the material itself. Meanwhile, an extrinsic pseudocapacitor behaves like a battery type as the electrolyte ions intercalate into the bulk material similar to intercalation pseudocapacitance, due to the changes made in the material like doping or change in the particle size. The vast difference between the two pseudocapacitors is observed from

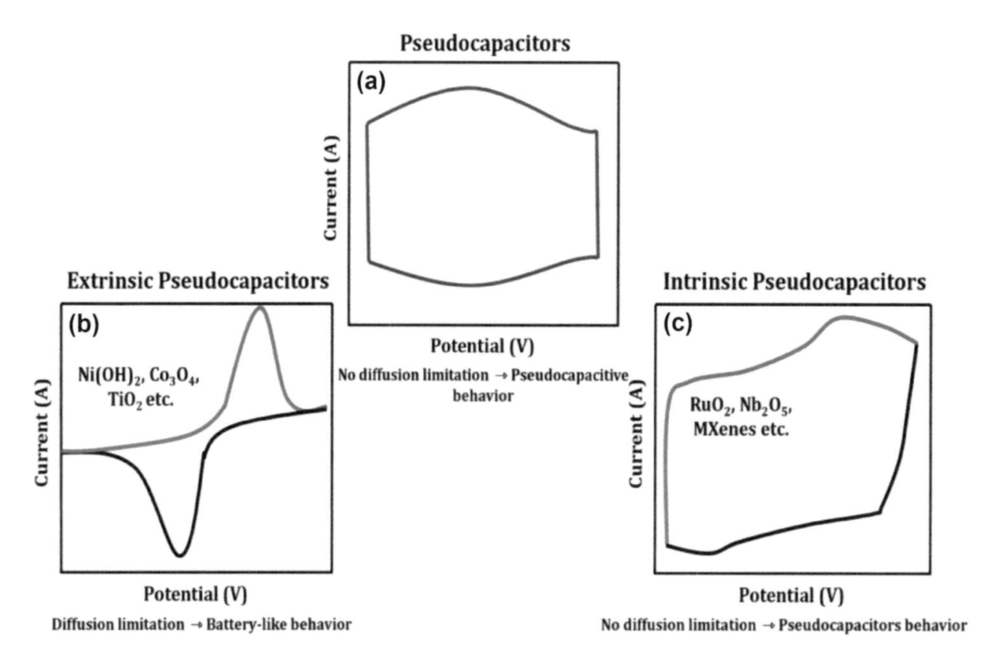

FIGURE 1.7 Cyclic voltammetry curves of intrinsic and extrinsic pseudocapacitors. (Reproduced with permission from Ref. [31], © 2021 Elsevier Ltd.)

the cyclic voltammetry (CV) curve in Figure 1.7 where the extrinsic pseudocapacitor curve exhibits a dominant oxidation and reduction peak mimicking the battery behavior while the intrinsic pseudocapacitor exhibits a nearly ideal electrochemical reversibility with a broad oxidation/reduction peak indicating the surface-controlled kinetics [31].

Transition metals-based materials are the key active material used in pseudocapacitors because it has multiple oxidation states for effective redox reaction at a wide range of potential. Over decades, transition metal oxides (TMOs) have been widely explored as active materials because of their high theoretical capacitance with lower internal resistance. However, TMO often suffers from low electrical conductivity due to the narrow bandgap of TMs which prevents the TMO from achieving its high capacitance. Thus, to counter this issue, TMO size is reduced to nano-size, doped with heteroatoms, and some of the TMOs are combined to form binary or ternary TMO as a reduction in the particle size leads to the increase of effective surface area by enlarging the intercalation channel size, meanwhile, the incorporation of different metals creates a synergistic effect to boost the specific capacitance. A spike in the specific capacitance directly causes a rise in the energy density of the supercapacitor as capacitance is directly proportional to the energy density. For instance, Jing et al. fabricated a ternary TMO nanoribbon composed of $NiO/V_2O_5/MnO_2$ to increase the interfaces of the interpenetrating channels and concurrently create a synergistic effect for better ion transport and electron conduction [36]. As predicted, the ternary TMOs exhibited 10 to 20 times larger energy storage capacity (energy density of 138 Wh kg^{-1} at power density 450 W kg^{-1}) compared to traditional V_2O_5/MnO_2 combination. TMO combination also exhibited fascinating specific capacitance and capacity retention of 788 F g^{-1} at 5 mV s^{-1} and 83.6% over 10,000 charge/discharge cycles, respectively. In another study, nanoparticles were prepared through a metal-organic framework (MOF)-74 to utilize $NiCo_2O_4$ and its uniform metal distribution to strengthen the cyclic stability and capacitance of the binary $NiCo_2O_4$ by reducing the lattice distortion of the binary TMO during fabrication process. As a result, the supercapacitor exhibited specific capacitance up to 684 F g^{-1} at 0.5 A g^{-1} with capacity retention of 86% over 3,000 cycles [37].

Besides TMOs, non-TMOs like metal sulfide/hydroxide/nitride/phosphate and MXene have also been widely explored as supercapacitor material in recent years as each type of non-TMOs has its unique feature. For example, transition metal nitrides (TMNs) are well known for their electrical conductivity (10^6 Ω^{-1} m^{-1}), electrochemical stability, and corrosion resistance [7]. Meanwhile, transition metal hydroxides (TMHs) are known for their structural and electrochemical features. According to Wan and Wang, TMHs exhibit their best performance in the form of ultrathin nanosheets due to their large effective surface that provides plenty of active sites for ion diffusion processes and simultaneously increases the ion transfer kinetics due to its short diffusion path length [38]. As an example, Xuan et al. fabricated a Co-Ni composite hydroxide ultrathin nanosheet that exhibits superlative electrochemical performance (specific capacitance of 1,608 F g^{-1} at 1 A g^{-1} with energy density of 55.9 Wh kg^{-1} at a power density of 750 W kg^{-1}) with rate capability of 90.6% even at a current density of 20 A g^{-1} [39]. On the flip side, Sangeetha et al. fabricated Co-Ni hydroxide particles as the active material and revealed a specific capacitance of 1,366 F g^{-1} at 1.5 A g^{-1} with a relatively lower rate capability (75% over current density 9 A g^{-1}) than

Xuan et al.'s [40]. Howbeit, Sangeetha, et al. reported an outstanding capacity retention of 96.26% over 2,000 charge/discharge cycles [40].

Recently, the transition metal carbide/nitride 2D family are being explored due to their hydrophilic tendencies, abundance of surface functionalities, and excellent metallic conductivity. This 2D family is known as MXene. Commonly investigated MXenes are Titanium Carbide ($Ti_3C_2T_x$, Ti_3C_2) MXene [41–43], Vanadium Carbide (V_2CT_x, V_2C, V_4C_3) MXene [44–46], Titanium Nitride (T_2NT_x, $Ti_4N_3T_x$) MXene [47, 48], Vanadium Nitride (V_2NT_x) MXene [49], and Niobium Carbide (Nb_2CT_x) MXene [50]. These MXenes are either used as an active material of the supercapacitor or act as a template for the incorporation of TMOs or non-TMOs or conducting polymer to enhance the electrical conductivity of the composite to uplift the capacity and cyclic stability of the electrode. For example, He and co-workers synthesized V_2CT_x MXene using NaF plus HCl etching method and used seawater as the electrolyte in the supercapacitor to enhance the cyclic stability of the MXene and simultaneously increase the specific capacitance [45]. As predicted, the V_2CT_x MXene exhibited a good specific capacitance of 181.1 F g^{-1} with splendid capacity retention of 89.1% over 5,000 charge/discharge cycles [45]. On the flip side, Wang et al. incorporated PANI onto V_2C MXene and reported higher specific capacitance of 337.5 F g^{-1} at 1 A g^{-1} with fair energy density and power density of 11.25 Wh kg^{-1}, 415.38 W kg^{-1} respectively [51]. Although, MXenes have superlative physiochemical properties yet they suffer from stability when exposed to humid or water. The degradation process of MXene is caused by oxidation reaction when exposed to water and past research has reported the oxidation rate varies according to the morphology, flake size, composition, etching method, and also storage conditions [42, 52]. Therefore, to reduce the oxidation rate, various approaches have been taken, namely, by addition of anti-oxidant, modification of etching method and stabilizing MXene morphology [42, 53]. For instance, Wu et al. fabricated $Ti_3C_2T_x$ MXene ink for solid-state micro-supercapacitor using an in situ etching method followed by capping of the MXene with antioxidant sodium ascorbate, and reported the stability of the MXene significantly improved even after storing for 80 days at ambient temperature and in exposure to air. To add on, it was revealed that the MXene exhibited fascinating areal capacitance of 108.1 mF cm^{-2} in absence of current collector [42].

Other known active materials for supercapacitors are conducting polymers, specifically PANI, polypyrrole (PPy), poly (3,4-ethylenedioxythiophene) (PEDOT), and polythiophene (PTs) due to their low cost, easy fabrication with splendid flexibility and conductivity. However, conducting polymers suffer from low cycling stability due to severe swelling of the polymers during the charge/discharge process and to enhance its stability it is often incorporated with TMO, non-TMOs, MXenes, or carbonaceous compounds [7]. Recently, PANI was incorporated with a rare metal oxide lanthanum oxide (La_2O_3) to prevent the electrochemical degradation of PANI and revealed that the incorporation of La_2O_3 boosted the efficiency up to 91.95% with an energy density of 53.8 Wh kg^{-1} and specific capacitance of 718 F g^{-1} at 5 mV s^{-1} compared to pure PANI (efficiency of 77.78% with energy density and specific capacitance of 7.8 Wh kg^{-1} and 686 F g^{-1} at 5 mV s^{-1}, respectively) [54]. In another study, two types of conducting polymers (PT and PPy) were incorporated into titanium oxide (TiO_2) to form a ternary composite to boost the conductivity of the supercapacitor. The study revealed that the mixture of two conducting polymers gives a better specific capacitance compared to utilizing single conducting polymer as Ppy-PT/TiO_2 exhibited a significantly higher specific capacitance (271.8 F g^{-1}) compared to PPy/TiO_2 (80.4 F g^{-1}) and PT/TiO_2 (109.6 F g^{-1}) [55].

1.3.2.3 Hybrid Capacitor

Hybrid capacitors are devices that combine EDL with pseudocapacitive electrodes to produce a device that exhibits both non-faradaic and faradaic reactions to store charges. They often provide specific advantages over conventional EDLC and pseudocapacitor technologies, such as increased power density, increased energy density, increased temperature stability, and enhanced electrical performance, because of their special structure. Hybrid wet tantalum and hybrid polymer aluminum are two types of hybrid capacitors. The highest working voltage of hybrid supercapacitors is 3.8 V and can provide energy density and capacitance up to ten times of symmetric supercapacitors [56]. Hybrid capacitors are small and compact. The size of capacitors is becoming increasingly significant because of the constant effort to miniaturize electrical devices. Since a hybrid capacitor combines an EDLC-type electrode with a pseudocapacitor-type electrode, when the voltage is supplied on the terminal it undergoes a redox reaction while in the other forms, EDL stores charges. Thus, the device contributes to high power density and cyclic stability from EDL while high energy density and higher capacitance from the redox reaction counter the drawbacks of EDLC and pseudocapacitors. There are two types of hybrid capacitors: (1) asymmetric supercapacitors, and (2) composite hybrid depending on the electrode material used. An asymmetric supercapacitor is the combination of two types of electrodes of a different nature or the same nature, but different material. A classic example of an asymmetric capacitator is when EDLC material is used as a negative electrode while pseudocapacitive is used as a positive electrode. Besides, is it also an asymmetric supercapacitor if both the electrodes are made up of either pseudocapacitor- or EDLC-type but comprise different material. For instance, Li et al. fabricated a nickel cobalt selenium ($NiCo_2Se_4$) nanotube as a positive electrode and combined it with AC (negative electrode) to form a typical asymmetric supercapacitor [57] while Wei et al. fabricated an all pseudocapacitive asymmetric supercapacitor by utilizing MnO_2 as the positive electrode active material with MXene as the negative electrode active material [58]. In line with that, M. Mandal fabricated CuO – MnO_2 composite pseudocapacitive electrode the same as the second list in his research [59]. First, the CuO nano-needle produced has a high specific capacitance of 2,519 F g^{-1} at a specific current of 8 A g^{-1}. However, the electrode's rate capability was compromised since the specific capacitance rapidly reduced when the specific current was raised to 30 and 50 A g^{-1}, respectively. The CuO nanoparticles were then hydrothermally modified with MnO_2, resulting in a CuO-MnO_2 (CMO_6) electrode with a high specific capacitance of 539 F g^{-1} even at a very high specific current density of 50 A g^{-1}.

On the flip side, hybrid composites are a single electrode that comprises both EDLC and pseudocapacitive material. For instance, Shang et al. incorporated ternary flower-like Ni-Co-Mn hydroxide into CNTs to fabricate a hybrid positive electrode for an asymmetric supercapacitor and revealed an outstanding specific capacitance of 2,136 F g^{-1} with excellent rate capability (approximately 90%) over 10 A g^{-1} with satisfactory performance as asymmetric device (retaining 53.8% of device maximum specific capacitance 150.88 F g^{-1} over 20 A g^{-1} current density) [60]. In another study, rGO, CNT, and PANI were combined to synthesize a ternary nanocomposite positive electrode that exhibited magnifying volumetric capacitance up to 1,038 F cm^{-3}, and upon combination $Ti_3C_2T_x$ MXene negative electrode, the hybrid supercapacitor exhibited splendid energy and power density of 70 Wh L^{-1} and 111 kW L^{-1}, respectively [61].

1.4 PREPARATION OF THE ELECTRODE MATERIALS

1.4.1 Hydrothermal Method

The method of fabricating the electrodes is crucial in defining the performance of the battery because it impacts the morphology and interface characteristics, which in turn affect metrics like porosity, pore size, and effective transport coefficient [8, 62, 63]. One of the most popular techniques for synthesizing nanomaterials is hydrothermal synthesis. Essentially, it uses a solution-reaction-based methodology. Nanomaterials are formed via hydrothermal synthesis over a wide range of temperatures, from very low to extremely high. Depending upon the vapor pressure of the primary components in the reaction, both low-pressure and high-pressure variables were utilized to regulate the morphology of materials that are synthesized. Powder materials have higher chemical purity because hydrothermal synthesis starts with high-quality precursors [64]. During the crystallization process, growing crystals and crystallites reject pollutants in the growth environment [64].

Preliminary research on how nickel ferrite ($NiFe_2O_4$) nanoparticles develop under hydrothermal conditions is presented in a study by Z. Rák and D. W. Brenner [65]. A model was created using a technique that incorporates first-principle calculation outcomes, an aqueous thermochemistry component, and empirical free energies of creation. Negative formation energy for the (111) surfaces while positive free energies for bulk production of nickel ferrite were expected based on calculations utilizing the model. Nickel ferrite nanoparticles between 30 and 150 nm in size are produced in the temperature range of 300 to 400 K under alkaline circumstances by combining the negative surface as well as positive bulk energies.

L. Ma et al. reported their work on the hydrothermal synthesis of different Co-doped $Zn_{1-x}Co_xMn_2O$ spinel nanocrystals [66]. The nanocrystals combine to produce hollow nanospheres. The effect of Co-doping concentration on sample structure, morphology, composition, and optical and photocatalytic capabilities was investigated. It was discovered that Co^{2+} ions substituted part of the Zn^{2+} lattice positions in $ZnMn_2O_4$ nanocrystals. The crystalline size reduced as Co doping increased as shown in Figure 1.8. The band gap of $Zn_{1-x}Co_xMn_2O$ is lower than that of $ZnMn_2O_4$, and the doped sample is red shifted.

In another research study, T. Wang used a one-step hydrothermal process to synthesize cobalt-tellurium incorporated into reduced graphene oxide (CoTe@rGO) as the active materials in supercapacitors [67]. The electrochemical properties of CoTe@rGO were greatly enhanced over that of pure CoTe. The ideal CoTe@rGO electrode material has a strikingly large specific capacitance at 810.6 F g^{-1} when current density is 1 A g^{-1}, according to the data. Yet after 5,000 charge/discharge cycles, the synthesized material preserved

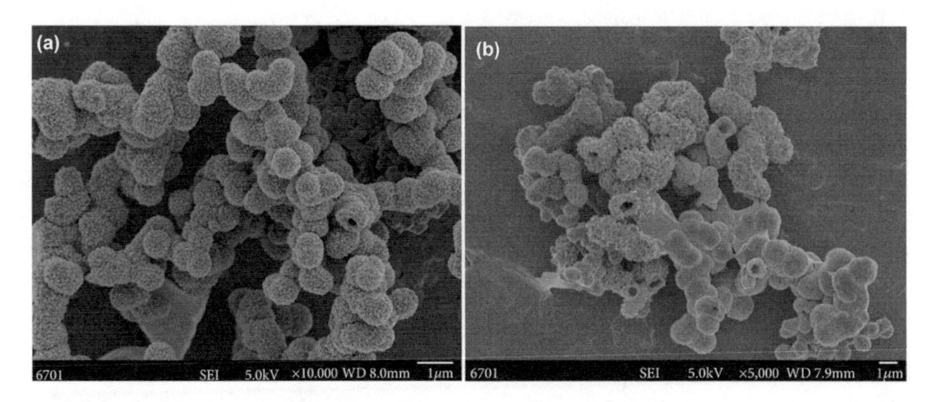

FIGURE 1.8 SEM images of (a) pure $ZnMn_2O_4$ and (b) co-doped $ZnMn_2O_4$ samples. (Reproduced under the Creative Commons Attribution license from Ref. [66], © 2019 Long Ma et al.)

77.2% of its original capacitance at 5 A g^{-1}, exhibiting high cycling stability. Furthermore, the composite electrode preserved 79.0% from its specific capacitance at 1 A g^{-1} despite a high current density of 20 A g^{-1}, indicating its superior rate performance.

1.4.2 Electrodeposition Method

As an alternative to immersion assembly, electrodeposition is a well-known technology for surface coatings using an applied voltage in electrochemical cells. This method may quickly combine ions, polymers, and colloids. Electrodeposition, often known as electroplating, is the controlled depositing of material onto conducting surfaces utilizing electric current from an ionic species-containing solution [68]. Electrodeposition methods are very suitable for single-element deposition. However, it is conceivable to do simultaneous depositions of multiple elements as well as the synthesis of very well-alternated layers of metals as well as oxides having a thickness of a few nanometers [69]. The method entails the breakdown of organic contaminants at the cathode while heavy metals are reduced and deposited on the cathode at the same time [70].

Mohammad H. Tahmasebi's work investigates the influence of deposition scan rate and Ni(II) to Mn(II) molar ratio in the depositing bath on the capacitive behavior of mixed oxide electrodes [71]. Increased nickel concentration in oxide thin films of the composition $Ni_xMn_{1-x}O_y$ (with x ranging from 0 to 0.17) leads to a rise in specific capacitance reaching a maximum of roughly 10% Ni. Through its impact on surface structure and layer morphology, the deposition scan rate influences the capacitive behavior of mixed oxide electrodes.

After 400 cycles, tin-based anodes, synthesized using a simple and classic electrodeposition process, exhibit high electrode capacity, and great rate performance, with excellent stable cycling stability, yielding extraordinary cycling stability of 728 mAh g^{-1} at a current density of 100 mA g^{-1} in research by Y. Liu [72]. The capacity of the tin-based anode was even retained at around 300 mAh g^{-1} after 250 cycles at a current density of 6 A g^{-1}.

In a study by Q. Hu, CoS and MnO_2 composite electrodes (MnO_{2-x} @ CoS) with a core-shell structure were created by using an easy electrodeposition process to coat CoS on the MnO_{2-x} arrays [73]. The metallic CoS coating and the oxygen vacancies of MnO_{2-x}

produced at a high temperature both enhance the electrical conductivity of composite electrode. The composite electrode's large surface area also offers the supercapattery a sufficient number of electrochemically active spots. 781.1 C g^{-1} of specific capacitance is present in the synthesized MnO_{2-x} @ CoS. A supercapattery (SCT) with a MnO_{2-x} @ CoS composite electrode has a large areal capacitance of 1,064 mF cm^{-2} as well as a high energy density of 34.72 Wh kg^{-1} at 597.24 W kg^{-1}. Furthermore, the supercapattery has high cycling stability, with 89.6% capacitance retention after 9,000 cycles.

1.4.3 Chemical Bath Deposition

Although chemical bath deposition is a widespread process, it is less recognized outside of the profession. A solid reaction product forming on a substrate as a result of a somewhat slow chemical reaction occurring in solution is a suitable way to conceptualize chemical bath deposition [74]. Along with the reaction product's deposition on the substrate, the reaction product frequently (but not always) precipitates in the deposition solution. Because the reaction to produce the material does not occur in solution, spin coating of a solution or a suspension of a ready-made substance is not regarded as chemical bath deposition.

A study by Y. Anil Kumar [75], obtained the greatest specific capacitance of the as-fabricated $MoNiO_4$ electrode, which was 1,140 F g^{-1} at 2 A g^{-1}, which is comparable to that of state-of-the-art $MoNiO_4$ electrodes. Furthermore, the $MoNiO_4$ material demonstrated an excellent energy density of 64.2 W h kg^{-1} at 2 A g^{-1}, an exceptional power density of 1,750 W kg^{-1}, and then an excellent electrochemical stability of 97.8% retention after 3,000 continuous charge-discharge cycles, even with a high current density of 4 A g^{-1}, that is comparable to that of the state-of-the-art $MoNiO_4$ flower-like structure (Figure 1.9).

A time-consuming chemical bath deposition approach was used to effectively manufacture $NiCo_xO_y$ in a study by F. Zhao [76]. The electrode material attained a specific capacitance of

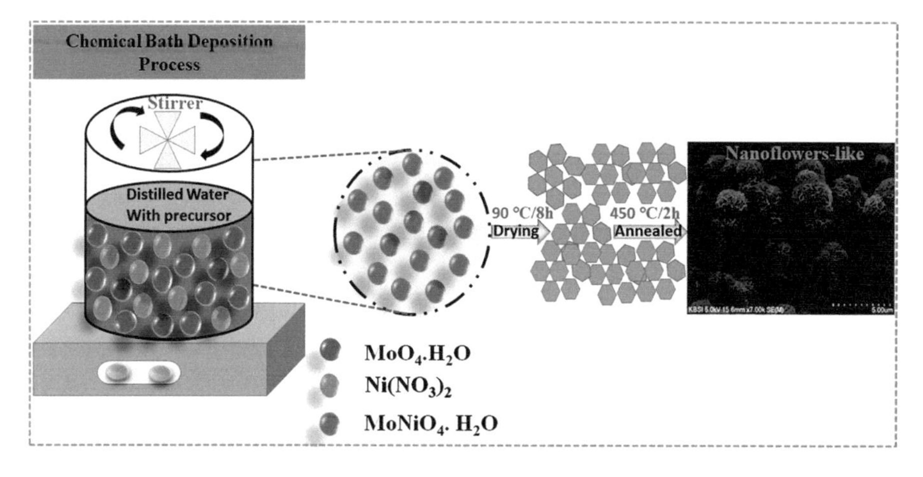

FIGURE 1.9 Chemical bath deposition process for the synthesis of $MoNiO_4$. (Reproduced with permission from Ref. [75], © 2020, The Royal Society of Chemistry and the Centre National de la Recherche Scientifique.)

TABLE 1.1 Advantages and Disadvantages of Synthesis Methods for Electrode Materials [77]

Method	Morphology	Advantages	Disadvantages
Hydrothermal Method	Nanostructured film	It takes less time to regulate morphology when synthesis factors like time, temperature, etc. are controlled.	Large-scale manufacturing is not suitable.
Electrodeposition	Nanostructured film and powder	Easy morphological control and used for large-scale manufacturing.	Time-consuming and operations at high temperatures.
Chemical Bath Deposition	Nanostructured film	Large-scale production, simple morphological control, and faster delivery than the hydrothermal method.	One can only synthesize a little amount of metal oxide.

507.3 F g^{-1} at a current density of at 1 A g^{-1} and excellent long-term cycle stability. Meanwhile, the fabricated $NiCo_xS_y$ can give excellent specific capacitance (1,196.1 F g^{-1} at 1 A g^{-1}), good cycle stability, and better rate performances (61.7% rate capability from 1 to 20 A g^{-1}) due to its flaky pores and channels and strong conductivity.

These electrode material production processes have several benefits as well as some drawbacks in some areas. This covers the amount of time required, whether the technology is commercialized, and, most critically, the morphology of the nanostructures created. Table 1.1 summarizes the benefits and drawbacks of the hydrothermal technique, electrodeposition, and chemical bath deposition [77].

1.5 FABRICATION OF SUPERCAPATTERY

With higher energy and power densities, the supercapattery bridges the gap between battery and supercapacitor. To attain the optimal energy density and power density, this device essentially combines both capacitive and Faradaic charge storage processes in a single device. The supercapattery uses highly efficient battery-type electrode active materials to overcome sluggish kinetics, poor rate performance, and poor cycle stability [78]. With a positive electrode that has a high energy density and a negative electrode made up of carbonaceous materials, a supercapattery combines the high energy capacity of battery-type electrode materials (redox-active materials that enable Faradaic reaction processes) and high power density [79]. The summarized concept is illustrated in Figure 1.10.

1.5.1 Types of Combination

There are five possible combinations for assembling a supercapattery device:

i) Pseudocapacitor that mimics battery + EDLC

ii) Pseudocapacitor that mimics battery + Pseudocapacitor

iii) EDLC + Battery

iv) Pseudocapacitor + Battery

v) Hybrid capacitor + Battery

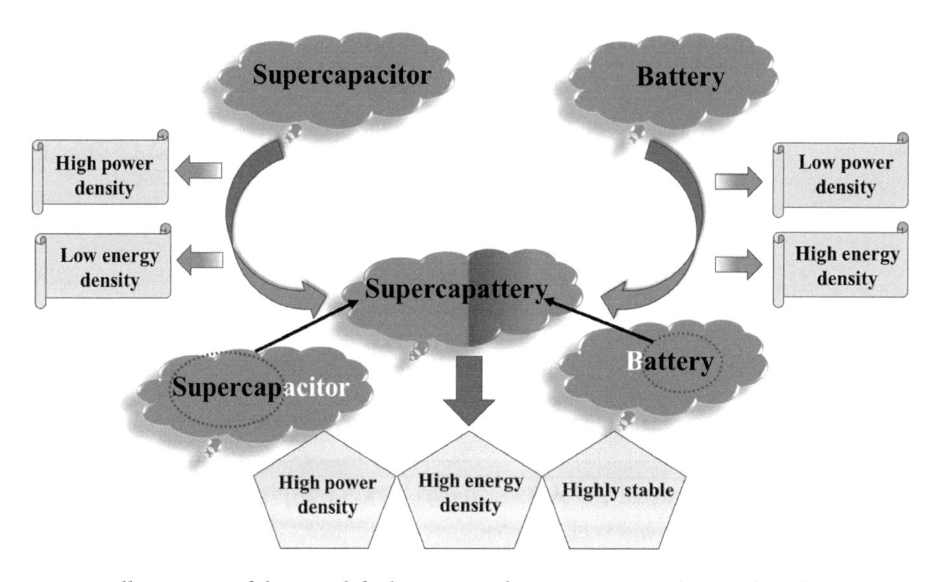

FIGURE 1.10 Illustration of the simplified concept of supercapattery. (Reproduced with permission from Ref. [5], © 2021 Elsevier Ltd.)

The base concept of supercapattery is a combination of battery-grade electrodes with supercapacitor-type material to significantly enhance the energy density of the device compared to what has been done by a hybrid capacitor. Thus, supercapatteries are assembled according to the five different combinations to achieve the optimized physiochemical properties. A pseudocapacitive that mimics battery behavior with a carbonaceous compound was the pioneer combination explored in supercapattery as a result of overlapping with a hybrid capacitor [3]. This is because the base concept of the charge storage mechanism of the hybrid capacitor is almost the same when the slight difference lies in the type of redox reaction that occurs in battery type (intercalation/de-intercalation into the bulk material) and that differs from the one that occurs in the pseudocapacitor (adsorption/de-adsorption that occurs at the surface of the active material). For instance, E. Agudosi researched an EDLC + battery combo using graphene-NiO battery-type electrodes and activated charcoal capacitive-type for supercapattery applications [78]. In a three-electrode cell, the G-NiO binder-free electrode has a specific capacity of roughly 243 C g^{-1} at 3 mV s^{-1}. Then, a two-electrode configuration of G-NiO/activated charcoal was used to create a supercapattery that powered in a steady potential range of 1.4 V. Energy density and power density were estimated to be 47.3 Wh kg^{-1} and 140 W kg^{-1}, respectively, at a current density of 0.2 A g^{-1}. Furthermore, the manufactured supercapattery demonstrated exceptional cyclic stability, with 98.7% specific capacity retention after 5,000 cycles.

According to S. Nayak's research [80], nickel phosphide nanoparticles are manufactured using a hydrothermal approach, whilst polyaniline is synthesized using an aniline polymerization approach to produce nickel phosphide-PANI composite for pseudocapacitive and EDLC combination. The supercapattery device uses the composite as the electrode material and chitosan biopolymer as the electrode binder. The supercapattery device made from polyaniline and nickel phosphide nanocomposite has an increased energy density of 21.9 Wh kg^{-1} with a maximum power density of 2,249 W kg^{-1}, and an amazing cyclic stability with a retention capacity of 89% even after 4,000 continuous GCD cycles. An asymmetric supercapacitor based on cobalt-molybdenum sulfide nanosheets with N-doped graphene nanosheet as the negatrode and Co-Mo-S NS as the positrode was described by Balamurugan et al. for use in a pseudocapacitive + battery type combination. Energy and power densities for this device were 89.6 Wh kg^{-1} and 20.07 kW kg^{-1}, respectively, and after 50,000 cycles, capacitance retention was 86.8%. Using a carbon cathode with S/P co-doping and a carbon cloth (CC) anode, F. Yu has developed an aqueous zinc bromine ($Zn - Br_2$) supercapattery [81]. The as-fabricated supercapattery had EDLC-type, pseudocapacitive, and battery-type charge storage mechanisms while effectively combining capacitive and battery-like characteristics in one device. As a result, the energy density was extraordinarily high at 270 W h kg^{-1} and the maximum power density was 9,300 W kg^{-1}. In the $Zn - Br_2$ supercapattery, peak power output may be produced via the capacitive and pseudocapacitive charge storage processes. Meanwhile, diffusion-controlled battery-type redox processes can provide the high energy demand.

Another famous type of battery grade materials explored is by utilizing and enhancing the famous battery compounds like lithium metal and orthosilicate nanomaterials in supercapattery. For example, Zhuang et al. fabricated lithium vanadium phosphate ($Li_3V_2(PO_4)_3$) composite coated with carbon using delayed sol-gel followed by a hydrothermal method to boost the low ionic diffusion co-efficient of $Li_3V_2(PO_4)_3$ from 10^{-12} to 1.4×10^{-7} cm^2 S^{-1} which resulted in the enhanced lithium-ion diffusion kinetics during charge/discharge process [82]. This cathode was then combined with AC anode with $LiPF_6$ as the electrolyte of the supercapattery and achieved exceptional energy density and power density of 53.22 Wh kg^{-1} and 3,010 W kg^{-1}, respectively [82]. In another research study, lithium-based orthosilicate nanocrystal (Li_2MnSiO_4) doped with 4% of zinc (Zn) revealed a specific capacity of 80.46 C g^{-1} even at high scan rate (20 mV s^{-1}) and upon a combination of AC as the anode material, the device exhibited splendid energy storage capacity (maximum 38.4 Wh kg^{-1} at 516 W kg^{-1}) and power density (maximum 14.429 kW kg^{-1} at 19.9 Wh kg^{-1}) [83].

1.5.2 General Procedure of Design-in Process

Since the primary goal of a supercapattery is to provide superlative structural and physiochemical characteristics that counter the pitfalls of both battery and supercapacitor, material selection plays a vital role in supercapattery fabrication. According to Balasubramaniam et al., the performance of a supercapattery is solely determined by the capacitance of the electrode, which is directly proportional to the active surface area, which contributes to quick ion and electron transport, low interfacial resistance, and the active material's meso-/microporous structure [84]. This is because capacitance directly and indirectly influences the energy density and power density of the device as shown in Equations (1.2)–(1.6) [5, 7].

$$\text{Capacitance, } C = \frac{\varepsilon A}{d} \text{ or } \frac{\Delta Q}{\Delta V}, \text{F} \tag{1.2}$$

$$\text{Specific Capacitance, } C_{sp} = \frac{I\Delta t}{\Delta V m}, \text{F g}^{-1} \quad (1.3)$$

$$\text{Specific Capacity, } Q_{sp} = \frac{I\Delta t}{m}, \text{C g}^{-1} \quad (1.4)$$

$$\text{Energy Density, } E_d = \frac{C_{sp} V^2}{2 \times 3.6}, \text{Wh kg}^{-1} \quad (1.5)$$

$$\text{Power Density, } P_d = \frac{E_d}{\Delta t} \text{ or } \frac{V^2}{R \times m}, \text{W kg}^{-1} \quad (1.6)$$

Where $\mathcal{E}$, A, d, Q, ΔV, m, R, and Δt represent the permittivity of the dielectric of the device, effective surface area (m^2), the thickness of a double layer formed, or the distance between two electrodes (m), charge capacity (C), potential window (V), active mass (g), resistance (Ω) of the device and discharge time(s), respectively. The formulas above on specific capacitance and specific capacity represent single electrode calculation where, for device calculation, multiplication of two is added, assuming the charges are distributed equally to both cathode and anode. Usually, for a symmetrical supercapacitor, the assumption is made because the same material with similar mass is used in the positive electrode (positrode) and negative electrode (negatrode) and when the same current and potential window is imposed on the electrodes, it will produce the same specific capacitance obeying Eqn. (1.3). However, the case is different for supercapattery fabrication because supercapattery brings different natures and types of materials into a single device. Thus, the possibilities of the charge on cathode and anode being equally distributed is extremely low. As a result of unbalanced charge distribution, the electrode with higher charge concentration would dominate the charge storage mechanism and result in unsymmetrical charge/discharge processes. To counter this issue, charge balancing is done as a crucial step during the fabrication of the supercapattery. Equation (1.7) represents the charge balancing equation for asymmetrical electrodes where "+" and "–" represents the positrode and negatrode terminal respectively. Equation (1.7) represents the charge storage in a supercapattery device:

$$Q_{\pm} = m_{\pm} C_{sp\pm} \Delta V_{\pm}, \quad (1.7)$$

Since the charge should be balanced for all the electrochemical energy storage devices according to the law of conservation of charge [3, 84], thus,

$$Q_{+} = Q_{-} \quad (1.8)$$

$$\frac{m_{+}}{m_{-}} = \frac{C_{sp-} \Delta V_{-}}{C_{sp+} \Delta V_{+}} \quad (1.9)$$

The mass ratio of the electrodes is derived from the Equation (1.9) according to the maximum charge achieved by the positrode and negatrode at respective potential windows.

Therefore, during the fabrication of a supercapattery device the compatible positrode and negatrode (that match the mass ratio) are selected and assumption on equal charge distribution is made.

During fabrication of a supercapattery device, material selection plays an important in selection of suitable electrode material and electrolyte. For electrode materials, several factors such as the morphology, effective surface area, pore-size distribution, doping, and hybridization are controlled and manipulated to achieve the optimized single electrode as all these factors greatly affect the capacitance. For example, different morphology of the same material can exhibit contrasting physiochemical behavior due to accessible active site, effective surface, and ion-diffusion kinetics. In a recent piece of research, cobalt carbonate hydroxide ($CO_2(CO_3)(OH)_2$) was fabricated in nanoflake and polyhedron flower morphology and revealed that nanoflake exhibited a better areal capacitance due to higher specific surface area (16 $m^2\ g^{-1}$) compared to polyhedron flower of specific surface area of 7 $m^2\ g^{-1}$ [85]. To add on, larger specific surface area has also increased the ion diffusion kinetic of the nanoflake $CO_2(CO_3)(OH)_2$ due to abundance of active sites exposure. As shown in the Figure 1.11, nanoflake morphology demonstrated a better specific capacitance, power density and energy

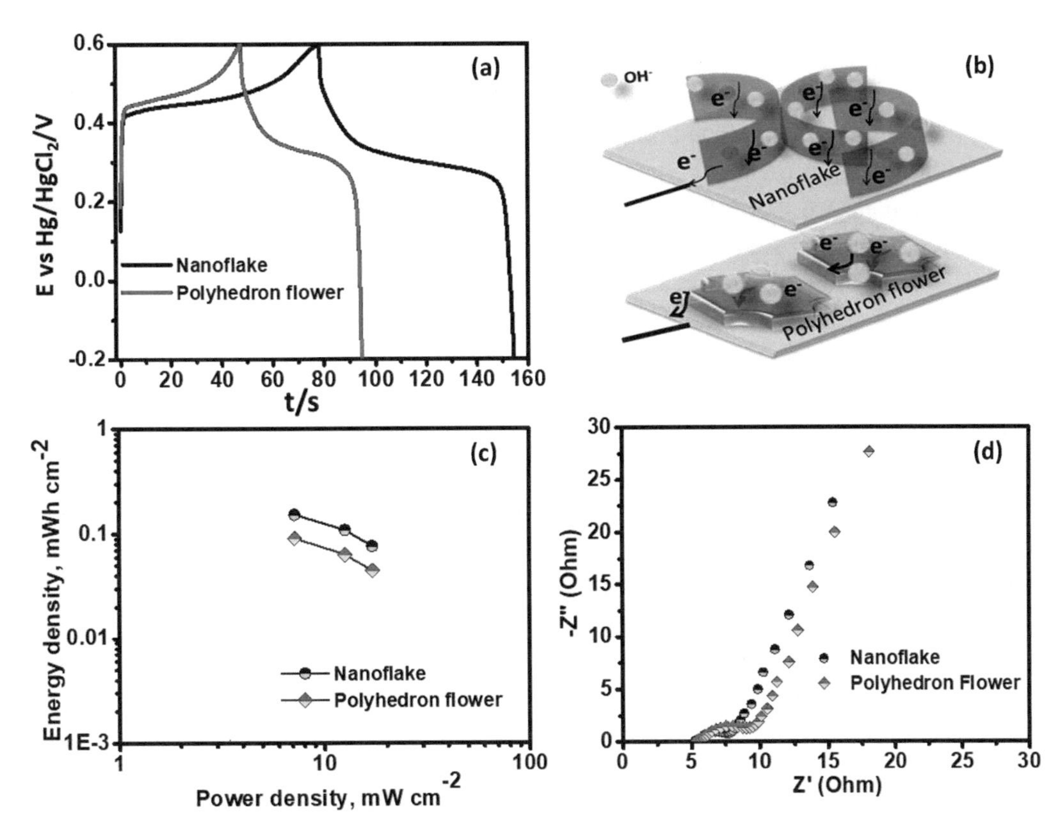

FIGURE 1.11 Comparative electrochemical performance of $CO_2(CO_3)(OH)_2$ (a) GCD curve at 20 mA cm^{-2}, (b) Illustration of charge storage mechanism, (c) Ragone plot, (d) Nyquist plot. (Reproduced with permission from Ref. [85], © 2017, Elsevier Ltd.)

density owing to larger inner surface charge storage with lower charge transfer resistance compared to polyhedron flower morphology.

Next, when manipulating the pore size distribution during electrode material synthesis, it is crucial to choose a suitable electrolyte because enlargement of pore size does not directly increase the charge storage capacity, instead the optimized charge storage capacity is achieved when the pore size is most similar to the electrolyte ion size [7]. A classic example is the routine of using 1-ethyl-3-methylimidazoliumbis(trifluoromethylsulfonyl)imdide (EMI-TFSI) ionic liquid electrolyte as the electrolyte for CDC-based active material and the pore diameter of both EMI^+ (0.76 nm) and $TFSI^-$ (0.79 nm) ions are in the typical range of CDCs' pore size (0.65–1.1 nm) [86]. An interesting perspective that is notable during the fabrication of hybrid composite or binary/ternary TM-based material is the choice of materials, as material selection is an important stage for the fabrication of energy storage devices, especially supercapacitors and supercapatteries. Therefore, during material selection, it is essential to know the strengths and weaknesses of the material so that a material that can overcome the previous material's weakness is selected to turn its weakness into strength. For instance, Jing et al. selected MnO_2 as the base metal oxide for ternary due to its excellent theoretical capacitance of 1,370 F g^{-1} and excellent cyclic stability, yet lower conductivity (1.00×10^{-8} S cm^{-1}) [36]. Thus, V_2O_5 was selected to incorporate MnO_2 because it has higher conductivity (1.17×10^{-3} S cm^{-1}) and also it can stabilize the MnO_2 configuration by formation of interpenetrating network for better ion diffusion process. Aside from that, this combination was also made to prevent the electrolyte ion corroding from V_2O_5. Lastly, the NiO was incorporated into the V_2O_5 to utilize its magnifying theoretical capacitance without compromising its thermal and cyclic stability.

Aside from electrode material, electrolyte selection is also critical as it determines the potential windows for the device to operate. For instance, aqueous electrolytes operate in the potential window maximum 1.0 V, while organic electrolytes, and ionic liquid electrolytes work in a potential window of 2.0–3.0 V, and 3.0–6.0 V respectively [7]. A different potential window affects the energy storage capacity of the device as energy density is proportional to the square of voltage (potential window). Although an aqueous electrolyte is commonly used in the fabrication of a supercapattery, the decomposition of water in the device beyond 1.0 V hinders its device performance [7]. The organic electrolyte, as the alternative, serves a wider working potential yet it suffers from degradation and self-discharging issues if impurities are not properly removed [7]. The organic electrolyte also tends to create safety issues when it is used commercially because of its toxicity, volatility, and flammability [87]. Thus, an ionic liquid electrolyte was introduced to counter the issues faced by both an aqueous or organic electrolyte, as it has great thermal and chemical stability with higher working potential. However, ionic liquids have diminutive ionic conductivity (<15 mS cm^{-1} at ambient temperature) with high viscosity that gives rise to elevated internal resistance which deteriorates the electrochemical performance of the supercapattery [86, 87]. Therefore, redox additives were introduced into the ionic electrolyte to enhance its conductivity and concurrently to contribute extra capacitance to the device. Recent research on usage of redox active electrolytes is increasing significantly in supercapattery applications due to their capability in the enhancement of charge and energy storage capacity. For instance, Meng et al. incorporated 0.1 M of iron-based redox couple ($Fe^{2+/3+}$) into aqueous electrolyte (H_2SO_4) to produce a redox active electrolyte and tested it with MoO_3//PANI nanofiber asymmetric supercapacitor [88]. From the study it was revealed the redox active electrolyte with PANI nanofiber exhibited fascinating charge capacity (specific capacitance of 3,330 F g^{-1} at 1 A g^{-1}) and when incorporated with MoO_3, the device exhibited splendid energy density up to 54 Wh kg^{-1}, even at high power density of 900 W kg^{-1} with maintained specific capacitance over 1,000 charge/discharge cycles [88]. In another research study, $BiBr_3$ dual redox mediators were added to the electrolyte of a traditional carbon-based electrode and revealed that the device exhibited battery-like behavior due to the dual redox coupled with outstanding electrochemical performance (specific capacitance 1,150 F g^{-1} with energy density 61.8 Wh kg^{-1} at the current density of 2 A g^{-1}) with zero performance degradation over 10,000 charge/discharge cycles [89].

1.6 CONCLUSION

The supercapattery is a new exploration of a hybrid device, which is closely related and inspired by hybrid capacitors, with the enhancement of physiochemical properties that bridge batteries with supercapacitors for a better fit with current and future demand on energy storage systems. Thus, the charge storage mechanism exhibited by the device is half battery-like (non-capacitive redox reaction) and half capacitive-like (EDL or capacitive redox reaction) owing to the electrode material used. Aside from the charge storage mechanism, electrode material also plays a crucial role in determining the overall electrochemical performance of the supercapattery because it is the major contributor of capacitance and determiner of the device's cyclic stability. Thus, advanced research should be done by researchers during their electrode material selection to select the appropriate type, and to further investigate newly explored materials like MXenes. Another concern about electrode materials is that only certain fabrication methods, like hydrothermal and electrodeposition methods, can be used to synthesize binder-free electrodes. However, most of the electrode materials are in powder form, so this is often coated on the current collector using binders like poly (vinylidene fluoride) (PVDF), polyacrylonitrile (LA_{133}), styrene-butadiene rubber and sodium carboxymethyl cellulose mixture (SBR/CMC), and nafion resulting in the drastic drop in the electrochemical properties due to the unwanted dead weight of the binder and increased internal resistance [90, 91]. The usage of binders also gives rise to safety concerns in the commercial stage because they use dangerous organic solvents like N-methylpyrrolidone (NMP). Thus, more alternative preparations of binder-free electrodes should be explored to minimize ecological damage and concurrently enhance the current energy storage device fabrication technology. Lastly, focusing on electrode material alone is not sufficient to fabricate a magnifying supercapattery but consideration of the electrolyte used is also critical as it determines the working potential of the device together and greatly affects its energy storage capacity. Utilization of redox-active electrolytes in supercapatteries should be focused on more in the near future to help the supercapattery receive an energy storage capacity close to commercially available batteries without compromising its power density and cyclic stability, due to electrolyte degradation.

REFERENCES

[1] C. Figueres, H.J. Schellnhuber, G. Whiteman, J. Rockström, A. Hobley, S. Rahmstorf, Three years to safeguard our climate, *Nature* 546 (2017) 593–595. https://doi.org/10.1038/546593A

[2] D. Maradin, Advantages and disadvantages of renewable energy sources utilization, *Int. J. Energy Econ. Policy* 11 (2021) 176–183. https://doi.org/10.32479/ijeep.11027

[3] L. Yu, G.Z. Chen, Supercapatteries as high-performance electrochemical energy storage devices, *Electrochem. Energy Rev.* 3 (2020) 271–285. https://doi.org/10.1007/S41918-020-00063-6/TABLES/1

[4] K. Subramani, N. Sudhan, R. Divya, M. Sathish, All-solid-state asymmetric supercapacitors based on cobalt hexacyanoferrate-derived CoS and activated carbon, *RSC Adv.* 7 (2017) 6648–6659. https://doi.org/10.1039/C6RA27331A

[5] M.Z. Iqbal, U. Aziz, Supercapattery: Merging of battery-supercapacitor electrodes for hybrid energy storage devices, *J. Energy Storage.* 46 (2022) 103823. https://doi.org/10.1016/J.EST.2021.103823

[6] L.H.J. Raijmakers, D.L. Danilov, R.A. Eichel, P.H.L. Notten, A review on various temperature-indication methods for Li-ion batteries, *Appl. Energy.* 240 (2019) 918–945. https://doi.org/10.1016/J.APENERGY.2019.02.078

[7] M. Pershaanaa, S. Bashir, S. Ramesh, K. Ramesh, Every bite of Supercap: A brief review on construction and enhancement of supercapacitor, *J. Energy Storage.* 50 (2022) 104599. https://doi.org/10.1016/J.EST.2022.104599

[8] A.R. Baird, E.J. Archibald, K.C. Marr, O.A. Ezekoye, Explosion hazards from lithium-ion battery vent gas, *J. Power Sources.* 446 (2020). https://doi.org/10.1016/J.JPOWSOUR.2019.227257

[9] A. Mishra, A. Mehta, S. Basu, S.J. Malode, N.P. Shetti, S.S. Shukla, M.N. Nadagouda, T.M. Aminabhavi, Electrode materials for lithium-ion batteries, *Mater. Sci. Energy Technol.* 1 (2018) 182–187. https://doi.org/10.1016/J.MSET.2018.08.001

[10] V. Ruiz, A. Pfrang, A. Kriston, N. Omar, P. Van den Bossche, L. Boon-Brett, A review of international abuse testing standards and regulations for lithium ion batteries in electric and hybrid electric vehicles, *Renew. Sustain. Energy Rev.* 81 (2018) 1427–1452. https://doi.org/10.1016/J.RSER.2017.05.195

[11] N. Muralidharan, R. Essehli, W. Yan, X. Jia, S. Yang, X. Chen, W. Yang, Y. Zhang, Advanced electrode materials for lithium-ion battery: Silicon-based anodes and Co-less-Ni-rich cathodes, *J. Phys. Conf. Ser.* 2133 (2021) 12003. https://doi.org/10.1088/1742-6596/2133/1/012003

[12] W. An, B. Gao, S. Mei, B. Xiang, J. Fu, L. Wang, Q. Zhang, P.K. Chu, K. Huo, Scalable synthesis of ant-nest-like bulk porous silicon for high-performance lithium-ion battery anodes, *Nat. Commun.* 10 (2019) 1–11. https://doi.org/10.1038/s41467-019-09510-5

[13] Z. Li, H. Zhao, P. Lv, Z. Zhang, Y. Zhang, Z. Du, Y. Teng, L. Zhao, Z. Zhu, Watermelon-like structured SiOx–TiO_2@C nanocomposite as a high-performance lithium-ion battery anode, *Adv. Funct. Mater.* 28 (2018) 1605711. https://doi.org/10.1002/ADFM.201605711

[14] L. Kouchachvili, W. Yaïci, E. Entchev, Hybrid battery/supercapacitor energy storage system for the electric vehicles, *J. Power Sources.* 374 (2018) 237–248. https://doi.org/10.1016/j.jpowsour.2017.11.040

[15] S.H. Nagarajarao, A. Nandagudi, R. Viswanatha, B.M. Basavaraja, M.S. Santosh, B.M. Praveen, A. Pandith, Recent developments in supercapacitor electrodes: A mini review, *ChemEngineering.* 6 (2022). https://doi.org/10.3390/chemengineering6010005

[16] Y. Wang, L. Zhang, H. Hou, W. Xu, G. Duan, S. He, K. Liu, S. Jiang, Recent progress in carbon-based materials for supercapacitor electrodes: A review, *J. Mater. Sci.* 56 (2020) 173–200. https://doi.org/10.1007/S10853-020-05157-6

[17] M.E. Şahin, F. Blaabjerg, A. Sangwongwanich, A Comprehensive Review on Supercapacitor Applications and Developments, *Energies.* 15 (2022) 674. https://doi.org/10.3390/EN15030674

[18] J. Chaparro-Garnica, D. Salinas-Torres, M.J. Mostazo-López, E. Morallón, D. Cazorla-Amorós, Biomass waste conversion into low-cost carbon-based materials for supercapacitors: A sustainable approach for the energy scenario, *J. Electroanal. Chem.* 880 (2021) 114899. https://doi.org/10.1016/J.JELECHEM.2020.114899
[19] W. Hu, M. Zheng, B. Xu, Y. Wei, W. Zhu, Q. Li, H. Pang, Design of hollow carbon-based materials derived from metal–organic frameworks for electrocatalysis and electrochemical energy storage, *J. Mater. Chem. A.* 9 (2021) 3880–3917. https://doi.org/10.1039/D0TA10666F
[20] R. Kumar, E. Joanni, S. Sahoo, J.J. Shim, W.K. Tan, A. Matsuda, R.K. Singh, An overview of recent progress in nanostructured carbon-based supercapacitor electrodes: From zero to bi-dimensional materials, *Carbon N. Y.* 193 (2022) 298–338. https://doi.org/10.1016/J.CARBON.2022.03.023
[21] M.E.F. Sahin, F. Blaabjerg, A. Sangwongwanich, A review on supercapacitor materials and developments, *Turkish J. Mater.* 5 (2020) 10–24.
[22] M. Vinayagam, R. Suresh Babu, A. Sivasamy, A.L. Ferreira de Barros, Biomass-derived porous activated carbon from Syzygium cumini fruit shells and Chrysopogon zizanioides roots for high-energy density symmetric supercapacitors, *Biomass and Bioenergy.* 143 (2020) 105838. https://doi.org/10.1016/J.BIOMBIOE.2020.105838
[23] K. Charoensook, C.L. Huang, H.C. Tai, V.V.K. Lanjapalli, L.M. Chiang, S. Hosseini, Y.T. Lin, Y.Y. Li, Preparation of porous nitrogen-doped activated carbon derived from rice straw for high-performance supercapacitor application, *J. Taiwan Inst. Chem. Eng.* 120 (2021) 246–256. https://doi.org/10.1016/J.JTICE.2021.02.021
[24] Poonam, K. Sharma, A. Arora, S.K. Tripathi, review of supercapacitors: Materials and devices, *J. Energy Storage.* 21 (2019) 801–825. https://doi.org/10.1016/J.EST.2019.01.010
[25] D. Mandal, P.L. Mahapatra, R. Kumari, P. Kumbhakar, A. Biswas, B. Lahiri, A. Chandra, C.S. Tiwary, Convert waste petroleum coke to multi-heteroatom self-doped graphene and its application as supercapacitors, *Emergent Mater.* 4 (2021) 531–544. https://doi.org/10.1007/S42247-020-00159-1
[26] M. Khandelwal, C. Van Tran, J. Lee, J. Bin In, Nitrogen and boron co-doped densified laser-induced graphene for supercapacitor applications, *Chem. Eng. J.* 428 (2022) 131119. https://doi.org/10.1016/J.CEJ.2021.131119
[27] B. De, S. Banerjee, K.D. Verma, T. Pal, P.K. Manna, K.K. Kar, Carbon nanotube as electrode materials for supercapacitors, Springer Ser. *Mater. Sci.* 302 (2020) 229–243. https://doi.org/10.1007/978-3-030-52359-6_9/COVER
[28] T. Kshetri, D.T. Tran, D.C. Nguyen, N.H. Kim, K. Tak Lau, J.H. Lee, Ternary graphene-carbon nanofibers-carbon nanotubes structure for hybrid supercapacitor, *Chem. Eng. J.* 380 (2020) 122543. https://doi.org/10.1016/J.CEJ.2019.122543
[29] C. Xiong, T. Li, Y. Zhu, T. Zhao, A. Dang, H. Li, X. Ji, Y. Shang, M. Khan, Two-step approach of fabrication of interconnected nanoporous 3D reduced graphene oxide-carbon nanotube-polyaniline hybrid as a binder-free supercapacitor electrode, *J. Alloys Compd.* 695 (2017) 1248–1259. https://doi.org/10.1016/J.JALLCOM.2016.10.253
[30] J. Zhang, Y. Cui, G. Shan, Metal oxide nanomaterials for pseudocapacitors, *Appl. Phys.* (2019). https://doi.org/10.48550/arxiv.1905.01766
[31] P. Bhojane, Recent advances and fundamentals of Pseudocapacitors: Materials, mechanism, and its understanding, *J. Energy Storage.* 45 (2022) 103654. https://doi.org/10.1016/J.EST.2021.103654
[32] L. Zhou, C. Li, X. Liu, Y. Zhu, Y. Wu, T. van Ree, Metal oxides in supercapacitors, *Met. Oxides Energy Technol.* (2018) 169–203. https://doi.org/10.1016/B978-0-12-811167-3.00007-9
[33] Y. Jiang, J. Liu, Definitions of pseudocapacitive materials: A brief review, *Energy Environ. Mater.* 2 (2019) 30–37. https://doi.org/10.1002/EEM2.12028

[34] Y. Liu, S.P. Jiang, Z. Shao, Intercalation pseudocapacitance in electrochemical energy storage: Recent advances in fundamental understanding and materials development, *Mater. Today Adv.* 7 (2020) 100072. https://doi.org/10.1016/J.MTADV.2020.100072
[35] M. Sarno, Nanotechnology in energy storage: The supercapacitors, *Stud. Surf. Sci. Catal.* 179 (2020) 431–458. https://doi.org/10.1016/B978-0-444-64337-7.00022-7
[36] J. Wang, F. Zheng, Y. Yu, P. Hu, M. Li, J. Wang, J. Fu, Q. Zhen, S. Bashir, J.L. Liu, Symmetric supercapacitors composed of ternary metal oxides (NiO/V2O5/MnO2) nanoribbon electrodes with high energy storage performance, *Chem. Eng. J.* 426 (2021) 131804. https://doi.org/10.1016/J.CEJ.2021.131804
[37] L.T. Gong, M. Xu, R.P. Ma, Y.P. Han, H.B. Xu, G. Shi, High-performance supercapacitor based on MOF derived porous NiCo2O4 nanoparticle, *Sci. China Technol. Sci.* 63 (2020) 1470–1477. https://doi.org/10.1007/S11431-020-1658-7
[38] L. Wan, P. Wang, Recent progress on self-supported two-dimensional transition metal hydroxides nanosheets for electrochemical energy storage and conversion, *Int. J. Hydrogen Energy.* 46 (2021) 8356–8376. https://doi.org/10.1016/J.IJHYDENE.2020.12.061
[39] X. Wang, H. Song, S. Ma, M. Li, G. He, M. Xie, X. Guo, Template ion-exchange synthesis of Co-Ni composite hydroxides nanosheets for supercapacitor with unprecedented rate capability, *Chem. Eng. J.* 432 (2022) 134319. https://doi.org/10.1016/J.CEJ.2021.134319
[40] M.S. Vidhya, G. Ravi, R. Yuvakkumar, D. Velauthapillai, M. Thambidurai, C. Dang, B. Saravanakumar, Nickel–cobalt hydroxide: A positive electrode for supercapacitor applications, *RSC Adv.* 10 (2020) 19410–19418. https://doi.org/10.1039/D0RA01890B
[41] J. Tang, T. Mathis, X. Zhong, X. Xiao, H. Wang, M. Anayee, F. Pan, B. Xu, Y. Gogotsi, Optimizing ion pathway in titanium carbide MXene for practical high-rate supercapacitor, *Adv. Energy Mater.* 11 (2021) 2003025. https://doi.org/10.1002/AENM.202003025
[42] C.W. Wu, B. Unnikrishnan, I.W.P. Chen, S.G. Harroun, H.T. Chang, C.C. Huang, Excellent oxidation resistive MXene aqueous ink for micro-supercapacitor application, *Energy Storage Mater.* 25 (2020) 563–571. https://doi.org/10.1016/J.ENSM.2019.09.026
[43] M. Hu, H. Zhang, T. Hu, B. Fan, X. Wang, Z. Li, Emerging 2D MXenes for supercapacitors: Status, challenges and prospects, *Chem. Soc. Rev.* 49 (2020) 6666–6693. https://doi.org/10.1039/D0CS00175A
[44] Y. Guan, S. Jiang, Y. Cong, J. Wang, Z. Dong, Q. Zhang, G. Yuan, Y. Li, X. Li, A hydrofluoric acid-free synthesis of 2D vanadium carbide (V2C) MXene for supercapacitor electrodes, 2D Mater. 7 (2020) 025010. https://doi.org/10.1088/2053-1583/AB6706
[45] H. He, Q. Xia, B. Wang, L. Wang, Q. Hu, A. Zhou, Two-dimensional vanadium carbide (V2CTx) MXene as supercapacitor electrode in seawater electrolyte, *Chinese Chem. Lett.* 31 (2020) 984–987. https://doi.org/10.1016/J.CCLET.2019.08.025
[46] R. Syamsai, A.N. Grace, Synthesis, properties and performance evaluation of vanadium carbide MXene as supercapacitor electrodes, *Ceram. Int.* 46 (2020) 5323–5330. https://doi.org/10.1016/J.CERAMINT.2019.10.283
[47] A. Djire, H. Zhang, J. Liu, E.M. Miller, N.R. Neale, Electrocatalytic and optoelectronic characteristics of the two-dimensional titanium nitride Ti 4 N 3 T x MXene, *ACS Appl. Mater. Interfaces.* 11 (2019) 11812–11823. https://doi.org/10.1021/ACSAMI.9B01150/SUPPL_FILE/AM9B01150_SI_001.PDF
[48] A. Djire, A. Bos, J. Liu, H. Zhang, E.M. Miller, N.R. Neale, Pseudocapacitive storage in nanolayered Ti2NTx MXene using Mg-ion electrolyte, *ACS Appl. Nano Mater.* 2 (2019) 2785–2795. https://doi.org/10.1021/ACSANM.9B00289/SUPPL_FILE/AN9B00289_SI_001.PDF
[49] S. Venkateshalu, J. Cherusseri, M. Karnan, K.S. Kumar, P. Kollu, M. Sathish, J. Thomas, S.K. Jeong, A.N. Grace, New method for the synthesis of 2D vanadium nitride (MXene) and its application as a supercapacitor electrode, *ACS Omega.* 5 (2020) 17983–17992. https://doi.org/10.1021/ACSOMEGA.0C01215/ASSET/IMAGES/MEDIUM/AO0C01215_M003.GIF

[50] J. Xiao, J. Wen, J. Zhao, X. Ma, H. Gao, X. Zhang, A safe etching route to synthesize highly crystalline Nb2CTx MXene for high performance asymmetric supercapacitor applications, *Electrochim. Acta.* 337 (2020) 135803. https://doi.org/10.1016/J.ELECTACTA.2020.135803

[51] X. Wang, D. Zhang, H. Zhang, L. Gong, Y. Yang, W. Zhao, S. Yu, Y. Yin, D. Sun, In situ polymerized polyaniline/MXene (V2C) as building blocks of supercapacitor and ammonia sensor self-powered by electromagnetic-triboelectric hybrid generator, *Nano Energy.* 88 (2021) 106242. https://doi.org/10.1016/J.NANOEN.2021.106242

[52] Y. Chae, S.J. Kim, S.Y. Cho, J. Choi, K. Maleski, B.J. Lee, H.T. Jung, Y. Gogotsi, Y. Lee, C.W. Ahn, An investigation into the factors governing the oxidation of two-dimensional Ti 3 C 2 MXene, *Nanoscale.* 11 (2019) 8387–8393. https://doi.org/10.1039/C9NR00084D

[53] X. Zhao, A. Vashisth, E. Prehn, W. Sun, S.A. Shah, T. Habib, Y. Chen, Z. Tan, J.L. Lutkenhaus, M. Radovic, M.J. Green, Antioxidants Unlock Shelf-Stable Ti3C2Tx (MXene) Nanosheet Dispersions, *Matter.* 1 (2019) 513–526. https://doi.org/10.1016/J.MATT.2019.05.020

[54] M. Morshed, J. Wang, M. Gao, C. Cong, Z. Wang, Polyaniline and rare earth metal oxide composition: A distinctive design approach for supercapacitor, *Electrochim. Acta.* 370 (2021) 137714. https://doi.org/10.1016/J.ELECTACTA.2021.137714

[55] G. Sowmiya, G. Velraj, Designing a ternary composite of PPy-PT/ TiO2 using TiO2, and multipart-conducting polymers for supercapacitor application, *J. Mater. Sci. Mater. Electron.* 31 (2020) 14287–14294. https://doi.org/10.1007/S10854-020-03985-5

[56] A. Moyseowicz, G. Gryglewicz, High-performance hybrid capacitor based on a porous polypyrrole/reduced graphene oxide composite and a redox-active electrolyte, *Electrochim. Acta.* 354 (2020) 136661. https://doi.org/10.1016/J.ELECTACTA.2020.136661

[57] S. Li, Y. Ruan, Q. Xie, Morphological modulation of NiCo2Se4 nanotubes through hydrothermal selenization for asymmetric supercapacitor, *Electrochim. Acta.* 356 (2020) 136837. https://doi.org/10.1016/J.ELECTACTA.2020.136837

[58] Y. Wei, M. Zheng, W. Luo, B. Dai, J. Ren, M. Ma, T. Li, Y. Ma, All pseudocapacitive MXene-MnO2 flexible asymmetric supercapacitor, *J. Energy Storage.* 45 (2022) 103715. https://doi.org/10.1016/J.EST.2021.103715

[59] M. Mandal, R. Nagaraj, K. Chattopadhyay, M. Chakraborty, S. Chatterjee, D. Ghosh, S.K. Bhattacharya, A high-performance pseudocapacitive electrode based on CuO–MnO2 composite in redox-mediated electrolyte, *J. Mater. Sci.* 56 (2020) 3325–3335. https://doi.org/10.1007/S10853-020-05415-7

[60] Y. Shang, S. Ma, Y. Wei, H. Yang, Z. Xu, Flower-like ternary metal of Ni-Co-Mn hydroxide combined with carbon nanotube for supercapacitor, *Ionics.* 26 (2020) 3609–3619. https://doi.org/10.1007/S11581-020-03496-7

[61] K. Li, X. Wang, X. Wang, M. Liang, V. Nicolosi, Y. Xu, Y. Gogotsi, All-pseudocapacitive asymmetric MXene-carbon-conducting polymer supercapacitors, *Nano Energy.* 75 (2020) 104971. https://doi.org/10.1016/J.NANOEN.2020.104971

[62] S.J. Dillon, K. Sun, Microstructural design considerations for Li-ion battery systems, *Curr. Opin. Solid State Mater. Sci.* 16 (2012) 153–162. https://doi.org/10.1016/J.COSSMS.2012.03.002

[63] F.L.E. Usseglio-Viretta, D.P. Finegan, A. Colclasure, T.M.M. Heenan, D. Abraham, P. Shearing, K. Smith, Quantitative relationships between pore tortuosity, pore topology, and solid particle morphology using a novel discrete particle size algorithm, *J. Electrochem. Soc.* 167 (2020) 100513. https://doi.org/10.1149/1945-7111/AB913B

[64] E. Suvaci, E. Özel, Hydrothermal Synthesis, *Encycl. Mater. Tech. Ceram. Glas.* 1–3 (2021) 59–68. https://doi.org/10.1016/B978-0-12-803581-8.12096-X

[65] Z. Rák, D.W. Brenner, Negative surface energies of nickel ferrite nanoparticles under hydrothermal conditions, *J. Nanomater.* 2019 (2019). https://doi.org/10.1155/2019/5268415

[66] L. Ma, Z. Wei, X. Zhu, J. Liang, X. Zhang, Synthesis and photocatalytic properties of co-doped Zn1-xCoxMn2O hollow nanospheres, *J. Nanomater.* 2019 (2019). https://doi.org/10.1155/2019/4257270

[67] T. Wang, Y. Su, M. Xiao, M. Zhao, T. Zhao, J. Shen, One-step hydrothermal synthesis of a CoTe@rGO electrode material for supercapacitors, *Trans. Tianjin Univ.* 28 (2022) 112–122. https://doi.org/10.1007/S12209-021-00306-7/FIGURES/10
[68] S. Saha, S. Das, Nanomaterials in thin-film form for new-generation energy storage device applications, *Chem. Solut. Synth. Mater. Des. Thin Film Device Appl.* (2021) 561–583. https://doi.org/10.1016/B978-0-12-819718-9.00017-0
[69] Z. Stević, I. Radovanović, M. Rajčić-Vujasinović, S. Bugarinović, V. Grekulović, Synthesis and characterization of specific electrode materials for solar cells and supercapacitors, *J. Renew. Sustain. Energy.* 5 (2013) 041816. https://doi.org/10.1063/1.4817716
[70] S.S. Ray, R. Gusain, N. Kumar, Water purification using various technologies and their advantages and disadvantages, in *Carbon Nanomaterial-Based Adsorbents for Water Purification Fundamentals and Applications.* (2020) 37–66. https://doi.org/10.1016/B978-0-12-821959-1.00003-9
[71] M.H. Tahmasebi, A. Vicenzo, M. Hashempour, M. Bestetti, M.A. Golozar, K. Raeissi, Nanosized Mn-Ni oxide thin films via anodic electrodeposition: A study of the correlations between morphology, structure and capacitive behaviour, *Electrochim. Acta.* 206 (2016) 143–154. https://doi.org/10.1016/J.ELECTACTA.2016.04.087
[72] Y. Liu, L. Wang, K. Jiang, S. Yang, Traditional electrodeposition preparation of nonstoichiometric tin-based anodes with superior lithium-ion storage, *ACS Omega.* 4 (2019) 2410–2417. https://doi.org/10.1021/ACSOMEGA.8B03535/ASSET/IMAGES/LARGE/AO-2018-03535D_0006.JPEG
[73] Q. Hu, M. Tang, M. He, N. Jiang, C. Xu, D. Lin, Q. Zheng, Core-shell MnO2@CoS nanosheets with oxygen vacancies for high-performance supercapattery, *J. Power Sources.* 446 (2020) 227335. https://doi.org/10.1016/J.JPOWSOUR.2019.227335
[74] J.A. Switzer, G. Hodes, Electrodeposition and chemical bath deposition of functional nanomaterials, *MRS Bull.* 35 (2010) 743–750. https://doi.org/10.1557/S0883769400051253
[75] Y. Anil Kumar, S. Singh, D.K. Kulurumotlakatla, H.J. Kim, A MoNiO4 flower-like electrode material for enhanced electrochemical properties via a facile chemical bath deposition method for supercapacitor applications, *New J. Chem.* 44 (2019) 522–529. https://doi.org/10.1039/C9NJ05529K
[76] F. Zhao, W. Huang, D. Zhou, Chemical bath deposition synthesis of nickel cobalt oxides/sulfides for high-performance supercapacitors electrode materials, *J. Alloys Compd.* 755 (2018) 15–23. https://doi.org/10.1016/J.JALLCOM.2018.04.304
[77] P.E. Lokhande, U.S. Chavan, A. Pandey, Materials and fabrication methods for electrochemical supercapacitors: Overview, *Electrochem. Energy Rev.* 3 (2020) 155–186. https://doi.org/10.1007/S41918-019-00057-Z
[78] E.S. Agudosi, E.C. Abdullah, A. Numan, N.M. Mubarak, S.R. Aid, R. Benages-Vilau, P. Gómez-Romero, M. Khalid, N. Omar, Fabrication of 3D binder-free graphene NiO electrode for highly stable supercapattery, *Sci. Reports* 10 (2020) 1–13. https://doi.org/10.1038/s41598-020-68067-2
[79] J. Iqbal, A. Numan, R. Jafer, S. Bashir, A. Jilani, S. Mohammad, M. Khalid, K. Ramesh, S. Ramesh, Ternary nanocomposite of cobalt oxide nanograins and silver nanoparticles grown on reduced graphene oxide conducting platform for high-performance supercapattery electrode material, *J. Alloys Compd.* 821 (2020) 153452. https://doi.org/10.1016/J.JALLCOM.2019.153452
[80] S. Nayak, A.A. Kittur, S. Nayak, Nickel phosphide-polyaniline binary composite as electrode material using chitosan biopolymer electrode binder for supercapattery applications, *Mater. Today Proc.* 54 (2022) 912–922. https://doi.org/10.1016/J.MATPR.2021.11.221
[81] F. Yu, C. Zhang, F. Wang, Y. Gu, P. Zhang, E.R. Waclawik, A. Du, K. Ostrikov, H. Wang, A zinc bromine "supercapattery" system combining triple functions of capacitive, pseudocapacitive and battery-type charge storage, *Mater. Horizons.* 7 (2020) 495–503. https://doi.org/10.1039/C9MH01353A

[82] B. Zhuang, Z. Wu, W. Chu, Y. Gao, Z. Cao, T. Bold, N. Yang, High-performance lithium-ion supercapatteries constructed using Li3V2(PO4)3/C mesoporous nanosheets, *ChemistrySelect.* 4 (2019) 9822–9828. https://doi.org/10.1002/SLCT.201902966

[83] M.M. Ndipingwi, C.O. Ikpo, A.C. Nwanya, K.C. Januarie, M.E. Ramoroka, O. V. Uhuo, K. Nwambaekwe, S.T. Yussuf, E.I. Iwuoha, Engineering the chemical environment of lithium manganese silicate by Mn ion substitution to boost the charge storage capacity for application in high efficiency supercapattery, *Electrochim. Acta.* 414 (2022) 140180. https://doi.org/10.1016/J.ELECTACTA.2022.140180

[84] S. Balasubramaniam, A. Mohanty, S.K. Balasingam, S.J. Kim, A. Ramadoss, Comprehensive insight into the mechanism, material selection and performance evaluation of supercapatteries, *Nano-Micro Lett.* 2020 121. 12 (2020) 1–46. https://doi.org/10.1007/S40820-020-0413-7

[85] K.V. Sankar, Y. Seo, S.C. Lee, S. Liu, A. Kundu, C. Ray, S.C. Jun, Cobalt carbonate hydroxides as advanced battery-type materials for supercapatteries: Influence of morphology on performance, *Electrochim. Acta.* 259 (2018) 1037–1044. https://doi.org/10.1016/J.ELECTACTA.2017.11.009

[86] L. Yu, G.Z. Chen, Ionic liquid-based electrolytes for supercapacitor and supercapattery, *Front. Chem.* 7 (2019) 272. https://doi.org/10.3389/FCHEM.2019.00272/BIBTEX

[87] L. Guan, G.Z. Chen, A.K. Croft, D.M. Grant, Perspective—Redox ionic liquid electrolytes for supercapattery, *J. Electrochem. Soc.* 169 (2022) 030529. https://doi.org/10.1149/1945-7111/AC5BA8

[88] W. Meng, Y. Xia, C. Ma, X. Du, Electrodeposited polyaniline nanofibers and MoO3 Nanobelts for High-Performance Asymmetric Supercapacitor with Redox Active Electrolyte, *Polym.* 12 (2020) 2303. https://doi.org/10.3390/POLYM12102303

[89] Y. Wang, Z. Chang, M. Qian, T. Lin, F. Huang, A bridge between battery and supercapacitor for power/energy gap by using dual redox-active ions electrolyte, *Chem. Eng. J.* 375 (2019) 122054. https://doi.org/10.1016/J.CEJ.2019.122054

[90] M.M. Faisal, S.R. Ali, S. Pushpan, K.K. Singh, N. Pineda-Aguilar, E.M. Sánchez, J.A. Hernández-Magallane, K.C. Sanal, Effect of binder on the performance of zinc-tin-sulfide nano flakes for the high-performance supercapattery devices, *Int. J. Energy Res.* 46 (2022) 12787–12803. https://doi.org/10.1002/ER.8045

[91] R. Wang, L. Feng, W. Yang, Y. Zhang, Y. Zhang, W. Bai, B. Liu, W. Zhang, Y. Chuan, Z. Zheng, H. Guan, Effect of different binders on the electrochemical performance of metal oxide anode for lithium-ion batteries, *Nanoscale Res. Lett.* 12 (2017) 1–11. https://doi.org/10.1186/S11671-017-2348-6/TABLES/2

CHAPTER 2

Rare-Earth Doped Cathode Materials for Solid Oxide Fuel Cells

Amol Nande

Guru Nanak College of Science, Ballarpur, India

Vijay Chaudhari

S. N. Mor College of Arts & Commerce & Smt. G. D Sarda Science College, Tumsar, India

J. D. Punde

S S Girls' College, Gondia, India

Sanjay J. Dhoble

R. T. M. Nagpur University, Nagpur, India

2.1 INTRODUCTION

The global demand for energy is ever increasing because of continuous industrial development and population growth. The energy required worldwide is mostly met in the traditional ways, that is, the combustion of fossil fuels. As a result of this, many environmental problems like climate change, air pollution, oil spills, and acid rain, among others, arise [1]. Hence, it is now extremely necessary to go for alternative ways of producing clean energy and to switch from fossil fuel-based technology [2]. Solar, wind, and hydro-power are some of the technologies which are rapidly growing and gaining popularity all over the world. But the widespread development of these renewable sources is limited because they often rely on the weather. Also, they are less efficient and non-portable. Fuel cells are promising technologies that are envisaged as fulfilling most of these requirements.

A fuel cell is a device that produces electrical energy through an electrochemical redox reaction, that is, it converts chemical energy directly into electrical energy with negligible emission of pollutants. Solid oxide fuel cells (SOFCs) have attracted much interest from researchers over the past few years since they show high efficiency, and high energy as well

DOI: 10.1201/9781003315261-3

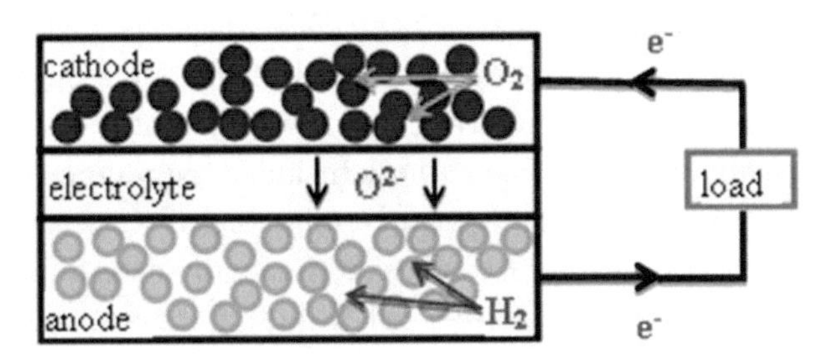

FIGURE 2.1 Schematic of the operating principle of a fuel cell.

as power density, and fuel flexibility. High efficiency and fuel adaptability are not the only major advantages of fuel cells, but they are also attractive as energy sources because they are clean, reliable, and almost entirely non-polluting. Furthermore, since there are no moving parts, they are vibration-free. The noise pollution associated with power generation can also be eliminated.

A fuel cell in its simplest design involves two electrodes (cathode and anode) on either side of an electrolyte. Hydrogen and oxygen pass over each of the electrodes and electricity, heat and water are produced using a chemical reaction.

The operating principle of an SOFC, with O_2 as oxidant and H_2 as fuel, is illustrated in Figure 2.1. The oxygen supplied at the cathode is reduced, forming oxygen ions that then migrate to the anode through the oxygen-ion conducting electrolyte. The oxygen ions react with hydrogen at the anode to form water and release electrons.

Theoretically, a fuel cell can produce electricity as long as fuel is constantly supplied, so, the fuel cell does not require recharging, unlike a battery. The electrical conversion efficiency of fuel cells is greater than any conventional energy conversion technology and is independent of size. Practically, SOFCs can produce output up to 40–60% when unassisted, and up to 70% in pressurized hybrid systems compared to 30–40% from internal combustion engines and modern thermal power plants [3]. Hence, fuel cells are useful in a wide range of applications, from small-scale power sources and small auxiliary power units to large power systems.

2.2 HISTORICAL BACKGROUND

Fuel cells have been known to humankind for more than 180 years. First of all, William Robert Grove, a Welsh scientist, developed an improved wet-cell battery in the year 1838 from which time the fuel cell came into existence. He used the concept of electrolysis to produce electricity by using oxygen and hydrogen as fuels [4]. Then Grove developed a device that would combine hydrogen and oxygen to produce electricity. He used a platinum electrode immersed in nitric acid and a zinc electrode in zinc sulfate to generate about 12 A of current at about 1.8 V. Practically, two platinum electrodes with one end of each immersed in a container of sulfuric acid and the other ends separately sealed in containers of oxygen and hydrogen, would provide a constant current flowing between the electrodes. Eventually, Grove developed the first gas battery which was then known as the fuel cell. In 1889, Mond and Carl Langer described their experiments with a fuel cell using coal-derived

Mond-gas. They attained 6 A per square foot at 0.73 V. They reported difficulties in using liquid electrolytes and succeeded only by using an electrolyte in a quasi-solid form. Emil Baur of Switzerland conducted extensive research into different types of fuel cells during the first half of the twentieth century. Baur's work included high-temperature devices using molten silver as an electrode and a unit that used a solid electrolyte of clay and metal oxides. In the year 1893, Friedrich Wilhelm Ostwald experimentally determined the relationship between the different components of the fuel cell, including the electrodes, electrolyte, oxidizing and reducing agent, anions, and cations. Ostwald's work opened doors in the area of fuel cell research by supplying information to future fuel cell researchers. In 1905, Haber filed the first patent on fuel cells, using glass and porcelain as the electrolyte materials, and, depending on the temperature of operation, platinum or gold as the electrode materials have been employed [5]. Following these, several workers, including Baur and Schottky in the 1930s, diverted their attention towards the development of a fundamental understanding of solid electrolytes and proposed practical solid electrolyte fuel cells. However, initial developments of fuel cells were concentrated on low-temperature devices such as alkaline, phosphoric acid, and polymer membrane cells. It is only since the 1960s that solid oxide fuel cells have been considered as a possible future energy technology [6]. After 1960 various factors resulted in renewed interest in fuel cell technology, while advances in the preparation and production of ceramic materials led to a resurgence of interest in SOFCs. One of the problems with these cells during the 1960s has been their poor efficiency, mainly due to thick electrolyte layers. Advances in preparation and production methods that led to the development of considerably thinner electrolytes facilitating a significant improvement in performance have only occurred since the 1970s. In the last two decades, numerous designs of SOFCs have been investigated, which include various tubular and planar designs.

However, even with more than a century of research and developmental work, the widespread commercialization of fuel cells has remained elusive. The roadblocks to this realization are primarily the cost and reliability of these devices. To attain commercial viability, several efforts have been made in the selection of materials and better design of fuel cell components.

2.3 TYPES OF FUEL CELLS

Fuel cells are classified primarily based on the electrolyte they use. This classification determines the kind of chemical reactions that take place in the cell, the kind of catalysts required, the temperature range in which the cell operates, the fuel required, and many other factors. These characteristics, in turn, affect the applications for which these cells are most suitable. From Table 2.1, it is evident that the choice of electrolyte determines the operating temperature of the fuel cell. Furthermore, the physicochemical and thermomechanical properties of materials used as the cell components (that is, electrodes, electrolyte, interconnect, etc.) decide the operating temperature and life of a fuel cell. Aqueous electrolytes are limited to operating at low temperatures of about 200°C or even lower than this because of their high vapor pressure and rapid degradation at higher temperatures.

Low-temperature fuel cells (< 200°C) need pure hydrogen and so all the fuels must be converted to hydrogen before entering the fuel cell. In addition, Co poisoning of the anode

TABLE 2.1 Summary of Major Differences of the Fuel Cell Types [7]

	Proton-Exchange Membrane Cell (PEMFC)	Alkaline Fuel Cell (AFC)	Phosphoric Acid Fuel Cell (PAFC)	Molten Carbonate Fuel Cell (MCFC)	Solid Oxide Fuel Cells (SOFC)
Electrolyte	Hydrated Polymeric Ion Exchange Membranes	Mobilized or Immobilized Potassium Hydroxide	Immobilized Liquid Phosphoric Acid in SiC	Immobilized Liquid Molten Carbonate in $LiAlO_2$	Perovskite (Ceramics)
Electrodes	Carbon	Transition metals	Carbon	Nickel and Nickel Oxide	Perovskite and perovskite/ metal cermets
Catalyst	Platinum	Platinum	Platinum	Electrode material	Electrode material
Interconnect	Carbon or metal	Metal	Graphite	Stainless steel or Nickel	Nickel, ceramic, or steel
Operating Temperature	40–80°C	65–220°C	200°C	650°C	600–1000°C
Charge Carrier	H^+	OH^-	H^+	CO_3^{2-}	O^{2-}
Efficiency	50%	60%	40%	45–50%	60%

catalyst (mostly platinum) in these fuel cells is the greatest drawback. Both these factors lead to a very low efficiency (40–50%, when operated on methanol and hydrocarbon fuels and 50%, when operated on pure hydrogen fuel). In contrast, high-temperature fuel cells (> 600°C), CO, and even CH_4 can be internally converted to hydrogen or even directly oxidized electrochemically, and hence, they are extremely low emissive. They exhibit high efficiency (45–60%, for common fuels such as natural gas and 90%, for heat recovery) [8]. Moreover, SOFCs are simple, reliable, environmentally friendly, and highly efficient (up to 85% energy efficiency when combined with gas turbines) compared to engines and modern thermal power plants (30%) [3].

2.4 OPERATING PRINCIPLE OF SOFC

The working principle of an SOFC is schematically shown in Figure 2.1. As shown in Figure 2.1, cathode and anode electrodes are separated by a solid electrolyte that conducts oxygen ion (O^{2-}) through it. Initially, reduction of oxygen gases (O_2) takes place at the cathode and O^{2-} ions thus formed are transferred to the anode through the electrolyte as given by the reaction in Equation (2.1):

$$\frac{1}{2}O_2 + 2e' + V_o^{\bullet\bullet} \Leftrightarrow O_o^x \tag{2.1}$$

Simultaneously the oxidation process occurs at the surface of the anode, where either hydrogen (H_2) or hydrocarbons, is oxidized to H_2O or CO_2, and electrons (e^-) as shown in Equation (2.2):

$$H_2 + O_o^x \rightarrow H_2O + V_o^{\bullet\bullet} + 2e' \tag{2.2}$$

The overall reaction of the fuel cell yields heat, water, and carbon dioxide (when hydrocarbon fuels are used), and releases electrons to the external circuit, as in Equation (2.3) [9]

$$H_2 + \frac{1}{2}O_2 \rightarrow H_2O \tag{2.3}$$

By the use of a SOFC, Gibb's free energy change ΔG of this reaction can be directly transformed into electric energy as in Equation (2.4):

$$\Delta G = 2F \cdot E \tag{2.4}$$

where F is the Faraday constant (96485 C), and E is the cell voltage. The cell voltage $E°$ calculated using the above equation is 1.1 V, which is the highest voltage of an SOFC when operated at its full rates. If higher voltages than this are to be generated from fuel cells, then several cells have to be connected in series (in a stack). The open circuit voltage (E_0) of a cell can also be calculated using the difference in oxygen partial pressures at the cathode (P_{O_2}) and at the anode (P_{O_2}) with the help of the Nernst equation as given in Equation (2.5) [10]:

$$E_0 = -\frac{\Delta G}{nF} = \frac{RT}{nF}\ln\frac{(P_{O_2})\text{cathode}}{(P_{O_2})\text{anode}} \tag{2.5}$$

where ΔG is the free energy change, R is the gas constant, T is the absolute temperature, n is the electron equivalent of oxygen (n = 4), F is the Faraday constant, P_{O_2}(cathode) the partial pressure of the oxygen at cathode, and P_{O_2}(anode) at the anode.

2.5 INTERMEDIATE TEMPERATURE SOLID OXIDE FUEL CELL (IT-SOFC)

The operating temperature of SOFCs is in the range of 850–1000°C, and therefore, the components of the stack need to be ceramic and a tubular or box section design, which results in low volumetric power density [11]. Therefore, there is a need to reduce the cell operating temperature to an intermediate temperature range of 500–700°C [12].

Lowering the operating temperature of the fuel cell is useful in avoiding many drawbacks as listed below [13]:

- Allows use of a wider range of materials leading to cheaper fabrication.
- Affords more rapid start-up and shutdown.
- Reduces corrosion rate of metallic components.
- Improves durability (sintering and component inter-diffusion is accelerated at high temperatures).
- Facilitates robust construction through the use of compressive seals and metallic interconnects.

2.6 COMPONENTS OF SOFC

A fuel cell consists of mainly anode and cathode separated by electrolyte material that acts as an oxygen ion conductor. Along with this, there are other cell components contributing to the individual cell performance as discussed below [14].

2.6.1 Anode

The anode acts as an oxidizing site for hydrogen molecules. Its electronic conductivity should be sufficiently high to facilitate efficient electron transport to the current collector. It must have high catalytic activity for the electrochemical oxidation of H_2/CO and other fuel at the anode. It should be chemically stable and compatible with electrolytes and other cell components. It must have a low cost [13]. The anode performs oxidation of fuel electrochemically by catalyzing the reaction. Anodes are mostly porous composites of ceramic and metals (cermets), this allows the conduction of electrons through the structure by increasing active three-phase boundaries (TPBs). A TPB is the common interface at which the electronic and ionic conducting phases co-exist with the open pore containing fuel and where the reaction takes place in most cermets. The use of cermets helps in improving thermal expansion coefficients (TECs) matching between the electrolyte and collector. The most commonly used cermets material for most commercial applications is Ni-YSZ.

On the other hand, the use of mixed ionic and electronic conducting (MIEC) ceramic materials is also being studied at present to extend the TPB from a one-dimensional interface to a two-dimensional area.

2.6.2 Cathode

The main function of the cathode is to provide a site for the electrochemical reduction of oxygen, although it performs several other functions like transport of charged species to the electrolyte and distribution of the electrical current associated with the oxygen reduction reaction. Like the anode, the cathode should also have specific properties as mentioned below [15]:

- High electronic conductivity (more than 100 S cm^{-1} under oxidizing atmosphere).
- Matched TEC and chemical compatibility with the electrolyte and interconnect materials.
- Adequate porosity allowing gaseous oxygen to readily diffuse through the cathode to the cathode/electrolyte interface.
- Stability under an oxidizing atmosphere during fabrication and operation.
- High catalytic activity for the oxygen reduction reaction (ORR) and
- Low cost.

The electrochemical reactions kinetics of cathode at high operating temperature is relatively fast, which offers minimal polarization losses during cell operations. At this temperature, electronic conducting cathodes are preferably used as they operate mostly on a triple-phase boundary mechanism. Typically, $(La,Sr)MnO_{3-\delta}$ (LSM) is used as cathode material operating at high temperatures with yttria-stabilized zirconia (YSZ) as electrolyte material. Reducing the operating temperature to an intermediate value causes an increase in the polarization losses because at an intermediate temperature, the chemical kinetics become much slower for LSM. Hence, MIECs, which show high oxygen ion transport even at intermediate temperatures, are preferred over LSM. Other key features of operating at intermediate temperatures include increasing the choice of materials, and improving reliability and durability, as well as reducing the production cost[16]. Hence, in an effort to realize fuel cell technology at intermediate temperatures, the development of variable cathode compositions which work effectively at lower temperatures is very important [17].

2.6.3 Electrolyte

The electrolyte provides a pathway for oxygen ions to be transported between the two electrodes. The electrolyte properties mostly have a direct impact on fuel cell performance as they contribute to the ohmic internal resistance. Ideally, characteristic properties of an SOFC electrolyte must include high oxide ion conductivity (typically $>1 \times 10^{-3}$ S cm^{-1}), low electronic conductivity, good thermal and chemical stability, closely matched TEC with the electrodes and other cell components, a fully dense structure to maximize conductivity and minimize reactant cross-over, and low cost. Ceria- and zirconia-based electrolytes are most suitable for this and hence have been extensively studied as electrolytes for SOFCs [18]. Gadolinium-doped ceria $Ce_{0.9}Gd_{0.1}O_{1.95}$ (CGO) is the most extensively studied ceria-based electrolyte in the intermediate temperature range [19].

2.6.4 Interconnects

Interconnects offer a major contribution to fuel cell performance by separating fuel from air. They provide ohmic contact between each layer of the fuel cell stack. Some of their important role in a fuel cell involves the distribution of reactant gas evenly across the face of each electrode, and mechanical support to the overall cell structure [20]. At high temperature, the high-Cr or -Ni alloy or ceramic interconnects are used but both types make the interconnects more expensive. On the other hand, metallic interconnects, being stronger, can resist rapid temperature changes associated with fast start-up; they also make it easier to incorporate internal structures into the stack such as internal manifolds, flow-field geometries, or sealing wells. Some the examples are chromium-based alloy Cr-5Fe-1Y_2O_3 and the nickel-based alloy Inconel 600. Operation in intermediate temperature regions allows relatively low-cost ferritic stainless-steel alloys to be used.

2.6.5 Sealants

Sealants are necessary to protect against fuel and air leaking within a fuel cell and also to avoid leakage from the edges of cells. The sealant should be an electrical insulator that is chemically compatible with the gaseous environment and materials with which it is in contact. The sealant comes in two categories: (1) bonded seals, and (2) compressive seals. Both types of sealants are electrically neutral, that is, chemically compatible with their adjacent materials and other environment. They have a thermal expansion coefficient that is compatible with the adjacent components. This is required to attain fast start-up and shut-down times. Glass or glass-ceramic materials (e.g. silicates, borosilicates, boro-aluminosilicates) are called bonded sealants. Thermal expansion coefficients of these materials need to be carefully matched with the surrounding cell components to avoid cracking as they are extremely fragile and prone to cracking during cell operation. Comparatively, compressive sealant provides liberty for variations in TEC.

2.7 DEVELOPMENT OF CATHODES FOR SOFC

The cathode acts as a key component in SOFC performance. The required properties for selecting materials as cathode candidates have already been mentioned above. In general, fuel cells do not have any ideal operating temperature. A particular fuel cell technology is a function of the requisite performance, lifetime, cost, fuel, size, weight, efficiency, start-up time, waste heat quality, and so on. By lowering the operating temperature, the choice of materials widens, allowing cheaper fabrication, particularly concerning interconnects and balance-of-plant (BoP) components. Lower temperature operation also facilitates more rapid start-up and shut-down, reduces corrosion rate of metallic components, and improves durability [13].

At high operating temperatures, the kinetics associated with cathode reactions are sufficiently rapid, hence, cell losses are relatively minor. However, decreasing operating temperature increases the polarization losses of the cell as the kinetics associated with the oxygen reduction reaction and charge transport at the cathode surface are much slower. The development of alternative cathode compositions that function effectively at lower temperatures is, therefore, an important step in the realization of technologically feasible intermediate-temperature solid oxide fuel cells (IT-SOFCs) [17].

A large number of oxides that have exhibited improved electrochemical performance at relatively lower temperatures and high stability during operation are proposed as potential cathode materials for SOFC applications. Amongst them, perovskite and related oxides are extensively explored for a range of applications including IT-SOFC electrodes and high-temperature superconductors [21, 22].

2.7.1 Perovskite Cathode Materials

Perovskites are ABO3-type compounds, where A and B are cations of different sizes. A-site cations have a relatively larger ionic radius, with a dodecahedral oxygen framework with coordination number (CN) 12. The bond between the A-site cations and the O anions is strongly ionic. The B-site cations are smaller in size, and are in an octahedral framework of oxygen (CN = 6), with a strong covalent bond (Figure 2.2 (a)). The ideal perovskite structure is cubic. If the A-site cation is considered to be the corner of a unit cell, then the B-site cation would be in the body center position, and the oxygen ions are arranged in an Face-Centred Cubic (FCC) configuration (Figure 2.2 (b)).

Perovskites used in SOFC applications comprise A-site cation, often as a rare-earth metal from the lanthanide family (La, Sm, Pr) with a typical oxidation state +3. The A-site may be particularly replaced with an alkaline earth metal (Sr, Ca), having +2 oxidation

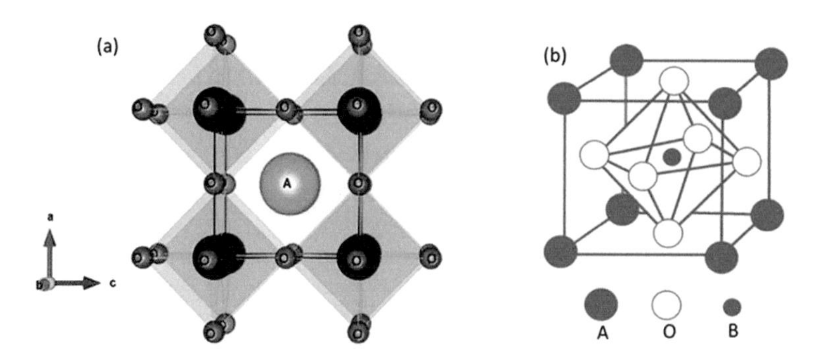

FIGURE 2.2 Unit cell of the ABO_3 perovskite structure showing (a) showing corner shared BO_6 octahedra and A cation at the interstitial position, and (b) showing B cation and BO_6 octahedra at the center of the unit cell. (Reproduced with permission from Ref. [23], © The Royal Society of Chemistry 2021 under the Creative Commons Licence.)

state [24]. The B-site cation is typically a transition metal, (Co, Fe, Ni, Mn) which can easily change oxidation states. Oxygen electro-reduction on the surface of perovskite oxides, such as LSM in solid oxide fuel cells, depends on the mixed valence states of the B-site cation. Partial substitution of divalent dopants at lanthanum lattice sites under oxidizing conditions results in the creation of commensurate Mn^{4+} species to achieve charge neutrality. The Mn^{3+}/Mn^{4+} couples give rise to a high electronic (p-type) conductivity via $Mn^{3+} \leftrightarrow Mn^{4+}$ electron transfer. Furthermore, if oxygen vacancies are retained in the perovskite LSM then the oxygen ion conductivity can be increased. The low oxygen ion conductivity in LSM at relatively lower temperatures is restricting its use as cathode material in IT-SOFCs.

As mentioned earlier, the conventional SOFC cathode, strontium-doped lanthanum manganite (LSM), is unsuitable for IT-SOFCs operating below 800°C. Jiang has reviewed and updated the development, understanding, and achievements of the LSM-based cathodes for SOFCs [25]. Particularly, their structure, non-stoichiometry, defect model, and the relation between the microstructure and their properties (electrical, thermal, mechanical, chemical, and interfacial) to understand the electrochemical performance and long-term stability are critically reviewed. The poor electrochemical activity of an LSM cathode at intermediate temperatures is mainly attributed to its negligible ionic conductivity. Consequently, the electrochemical reaction is strictly limited to the TPB [26, 27].

Currently, several oxides useful for SOFC cathode applications with the materials design rules are coming up with improved performance at relatively lower temperatures and with resistance to degradation during operation [28]. Hammouche et al. [29] investigated the $La_{1-x}Sr_xMnO_3$ system while x varying from 0 to 0.5 from the structural and thermal characteristics viewpoints. The Sr-doped lanthanum cobaltites ($La_{1-x}CoO_{3-\delta}$) are studied intensively from their high electrical and ionic conductivity point of view [30]. The $La_{1-x}Sr_xCoO_{3-\delta}$ solid solutions show a large thermal expansion coefficient (TEC) but react readily with the YSZ electrolyte [31]. Accordingly, other Sr-doped lanthanide cobaltites $Ln_{1-x}Sr_xCoO_{3-\delta}$ are examined as intermediate-temperature SOFC cathodes [32–34]. The perovskite-related Ruddlesden-Popper (RP) series of layered oxides ($A_{n+1}B_nO_{3n+1}$) are proposed as an important class of candidate materials for the next-generation SOFC cathodes. They exhibit good electrochemical catalytic and electrical properties as well as thermal and mechanical stabilities. The $LnBaCo_2O_{5+\delta}$ (Ln = Pr, Nd, Sm, Eu, Gd, Tb, Dy, Ho) exhibit structural anisotropy, which allows the generation of new original magneto-transport properties as a consequence of the ordering of the oxygen vacancies [35].

As discussed above, until now, most of the studies related to cathode materials for SOFC are devoted to perovskite-type oxides. The above-mentioned perovskite-type materials are becoming less popular, despite high electrochemical performance, because of high chemical expansion coefficients, susceptibility to Cr poisoning, and long-term stability [36]. In the recent past, therefore, a new family of oxides with general formulation $A_2BO_{4+\delta}$ are attracting attention. These materials, in terms of magnates, ferrates, cuprates, and nickelates, exhibit higher/better thermo-chemical stability than those of the corresponding perovskite-type oxides [37]. The structure of K_2NiF_4-type exhibits some oxygen non-stoichiometry. Preliminary results are promising in terms of oxygen diffusion and surface exchange coefficients. A literature survey pertinent to the cathode materials having K_2NiF_4-type structure, first members of the RP series (for $n = 1$, A_2BO_4), and most relevant to the materials under the present study, is given in the subsequent section.

2.7.2 K_2NiF_4-Type Cathode Materials

K_2NiF_4-type materials are emerging as promising alternatives to the traditional perovskites for use as cathodes in IT-SOFCs. These compounds have the general formula A_2BO_4 (A = rare earth/alkaline earth; B = transition metal). It consists of an alternate stack of perovskite (ABO_3) layers and rock salt (AO) layers as shown in Figure 2.3.

The K_2NiF_4-type materials show diverse defect chemistry allowing both hypo- and hyper-stoichiometry in terms of their oxygen content, which makes them very attractive for the fine-tuning of their electrical properties. In particular, oxygen hyper-stoichiometry can be achieved through the incorporation of excess oxide ions into interstitial sites within the rock-salt layers [38]. Conversely, oxygen vacancies can be created in the perovskite layer. The charge neutrality condition due to excess oxygen at interstitial sites is maintained by the formation of an electron-hole ($B_B^{\bullet}$). Such oxides are of interest as oxygen electrode material for applications in various electrochemical devices. The high electronic conductivity of A_2BO_4 (B = Co, Ni, Cu) compounds and their solid solutions is due to the mixed valence of the B-site metal cation. In addition, a high concentration of oxygen interstitials offers the possibility of rapid oxygen transport through the ceramic material, and thus provides a new type of MIEC cathode materials [15].

The partial substitution of Sr at the A-site affects the oxygen stoichiometry rather than the A-site vacancies. In other words, the constancy of charge in A_2BO_4 type oxides is achieved through the change in the oxidation state of B (transition metal) cation instead of cation vacancies upon partial replacement of trivalent A cations by divalent alkaline earth metal cations. In the recent past, extensive efforts have been directed towards studying nickelates and cuprates with K_2NiF_4-type structures [39, 40]. The rare-earth nickelates, with general formula $Ln_2NiO_{4+\delta}$ (Ln = La, Pr, Sm, Nd) are projected as a promising cathode in conjunction with known solid oxide-ion conductors such as YSZ, gadolinium-doped

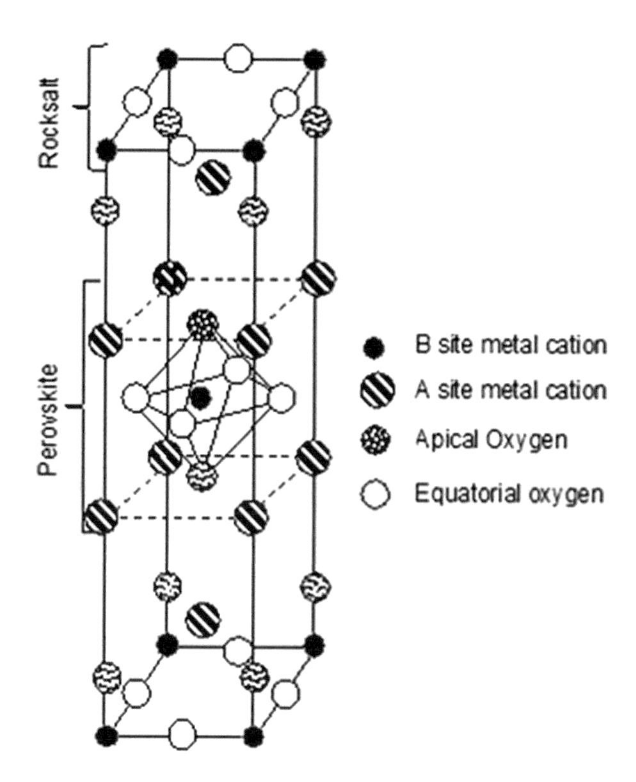

FIGURE 2.3 K_2NiF4-type structure showing alternating AO and ABO_3 layers.

ceria (GDC), and $La_{0.8}Sr_{0.2}Ga_{0.8}Mg_{0.2}O_{2.55}$ (LSGM) owing to good TEC compatibility. Several studies have been reported in the literature to support the above-mentioned facts.

Some of the important performance-determining factors of K_2NiF_4-type cathode materials are: chemical and thermal stability, compatibility with traditional electrolytes, electrical conductivity, oxygen transport behavior, and electrochemical properties. In terms of chemical reactivity of the K_2NiF_4-type electrode materials with those traditional electrolytes, such as YSZ, CGO ($Gd_{0.1}Ce_{0.9}O_{1.95}$), LSGM ($La_{0.9}Sr_{0.1}Ga_{0.8}Mg_{0.2}O_3$), among others, no reactivity was found, at least up to 1100°C for several hours of heating. The reactivity of these materials may depend on their morphology and, therefore, it might be related to the route of their preparation. Consequently, a comparison of reactivity with electrolyte materials, between nickelates prepared by different methods, could give a clue about the applicability of these compounds as electrodes for SOFCs. From all the above-mentioned studies, it can be said that the chemical stability of K_2NiF_4-type materials is much higher than that of the corresponding perovskite-type oxides.

In addition to a high electrochemical activity for the oxygen reduction reaction (ORR), the cathode material should have a compatible TEC with the established electrolytes. The perovskite materials, such as (Sr/Ba, La) (Co, Fe)O_3, suffer thermodynamical instability and/or dimensionally unstable issues under large oxygen chemical potential gradients, limiting their applicability in SOFC [41]. On the contrary, the K_2NiF_4-type materials exhibit much more compatible TEC values to the well-known oxygen ion conducting electrolytes. The TEC values of most of the K_2NiF_4-type materials, varying in the range

(75–79) × 10^{-6} K^{-1} at 300–1100°C are significantly lower than those of perovskite materials. Especially, the compatible TEC of K_2NiF_4-type oxides is an advantage from the viewpoint of possible use in high-temperature electrochemical devices. The changes in oxygen stoichiometry contribute to the apparent thermal expansion. Concurrently, chemical expansion affects the material's compatibility and dimensional stability in high-temperature electrochemical devices. Based on several studies, it can be said that the TEC's behavior depends not only on the lattice oxygen non-stoichiometry, but also on the doping regime, the structure evolution, and so on.

Oxides with K_2NiF_4 structure exhibit anisotropic electrical conductivity owing to their specific crystallography characteristics. As discussed earlier, the structure of K_2NiF_4-type oxides, for example, La_2NiO_4, can be described as an ordered intergrowth of perovskite $LaNiO_3$ and rock-salt LaO layers stacked along the crystallographic *c*-axis. The NiO_6 octahedra share corners in the *ab* plane forming a two-dimensional network. These two factors are responsible for the anisotropic electrical transport phenomena. Further, Ln_2MO_4 (Ln = La, Pr, Nd; M = Ni, Cu, Co) are characterized by reversible absorption and desorption of oxygen in the interstitial position (1/4,1/4,1/4) leading to oxygen non-stoichiometry [42]. The combination of in-plane metallic and out-of-plane non-metallic transport behavior observed is heavily dependent on the oxygen non-stoichiometry. The low-temperature electrical conductivity (below room temperature) of these layered quasi-two-dimensional perovskite-like materials is extensively studied from a superconductivity application viewpoint. The high-temperature charge transport properties of these materials have been, however, moderately investigated.

A comparison of the effective conductivity of different $Ln_{2-x}Sr_xMO_4$ (M = Ni, Co, Fe, Mn) compounds revealed that the nickelates invariably exhibit the highest conductivity and the lowest activation energy, whereas the cobaltites give the lowest conductivity, and the highest activation energy at intermediate temperature [43]. The cobaltites exhibit comparable values with the analogues nickelates above 600°C. On the other hand, the ferrites and manganites give relatively lower conductivity, less than 20 Scm^{-1} at 600°C in air. As discussed earlier, the partial substitution of Sr^{2+} at the A-site of A_2BO_4-type oxides can significantly affect the oxygen stoichiometry to achieve charge neutrality. Such created extrinsic defects significantly affect the conductivity.

The kinetics of the ORR on the surface of the cathode can be evaluated through polarization resistance (R_p) studies on the electrodes. The magnitude of the polarization resistance is determined, generally, by the intrinsic physicochemical properties of the cathode material. The reduction in R_p is most of the time reported in the materials with improved chemical stability, high electrical conductivity, oxygen diffusion, and surface exchange coefficient, and so on.

The cathode polarization arises due to: (1) diffusion of O_2 through the electrode layer, (2) adsorption followed by the electrochemical reduction of O_2 to O^{2-}, (3) transport of O^{2-} across the electrode layer into the electrolyte, and (4) charge transfer across the current collector/electrode and electrode/electrolyte interfaces. The first criterion can be achieved by fabricating a porous cathode layer. The second and fourth are possible by extending the electrochemically active three-phase boundary (TPB) (gas/electrolyte/electrode). Whereas the third criterion is related to the oxygen-ion conductivity of the cathode layer. A detailed literature survey reveals that the effective TPB length is the key factor to manipulate R_p [44]. An extension of the electrochemically active TPB is, therefore, one of the approaches to reduce electrode polarization. To achieve the above criterion (2)–(4) attempts are made to develop electrode material with high ionic as well as electronic conductors. In this regard, two approaches that have been opted for are: (1) mixing electron conducting cathode as composite cathode, and (2) developing or searching mixed-ion and electron conductors.

From all of the above studies discussed, one can reach the point that rare-earth doped materials should be considered the best options as cathode materials for solid oxide fuel cell applications. Nonetheless, more intensive research works need to be done to realize this technology in practice, and for commercial as well as domestic use in the future.

2.8 FUTURE CHALLENGES AND WORK

To improve further the electricity generation ability of fuel cells from their current efficiency, researchers are focusing their attention mainly on:

- Reducing operating temperature, which ultimately helps in avoiding cell degradation and also in increasing the choice of material selection for the cathode.
- Finding novel and inexpensive rare-earth doped and other cathode materials capable of having elevated conducting and catalytic properties, yet cell operating at reduced temperature and achieving higher efficiency.

For future study, the electrode material should be well chosen to obtain excellent balance for all its properties (electronic conductivity, high catalytic, and TEC properties). In the past, single-phase oxides have been intensively studied as a cathode of SOFCs. However, achieving single-phase oxides is itself a huge challenge. Alternatively, multiple-phase composites can be considered as they are more promising. Particularly, nano-composites can bring some unusual effects on the physicochemical properties in the cathode material, and hence, electrode performance could be highly improved.

2.9 CONCLUSION

Assuredly, IT-SOFCs are highly efficient and durable alternative renewable sources of energy production. Active materials including electrodes and electrolytes play a crucial role in the overall performance of IT-SOFCs. Availability of appropriate cathode materials is the major constraint and, thus, the main attention of research is on reducing R_p of a thermo-chemically stable cathode, to improve the electrochemical performance and ultimately increase the power output of IT-SOFCs. Being ionic as well as electronic conductors, MIEC cathodes are considered as most promising and, hence, acceptable in IT-SOFCs. Further, MIEC cathodes belonging to the K_2NiF_4-type structure (rare-earth doped) are preferred because apart from a compatible thermal expansion coefficient with electrolyte, they are most thermo-chemically stable in both the oxidizing and the moderately reducing atmospheres. The use of rare-earth elements as dopants in various oxides shows very promising and improved results, hence, they are intentionally preferred over others by most researchers.

REFERENCES

[1] Y. Zhang, R. Knibbe, J. Sunarso, Y. Zhong, W. Zhou, Z. Shao, Z. Zhu, Recent progress on advanced materials for solid-oxide fuel cells operating below 500 °C, *Adv. Mater.* 29 (2017) 1700132. https://doi.org/10.1002/adma.201700132

[2] X. Zhang, S. H. Chan, G. Li, H. K. Ho, J. Li, Z. Feng, A review of integration strategies for solid oxide fuel cells, *J. Power Sources.* 195 (2010) 685–702. https://doi.org/10.1016/j.jpowsour.2009.07.045

[3] S. C. Singhal, Science and technology of solid-oxide fuel cells, *MRS Bull.* 25 (2000) 16–21. https://doi.org/10.1557/mrs2000.13

[4] R. M. Ormerod, Solid oxide fuel cells, *Chem. Soc. Rev.* 32 (2003) 17–28. https://doi.org/10.1039/b105764m

[5] M. Cimenti, J. Hill, Direct utilization of liquid fuels in SOFC for portable applications: Challenges for the selection of alternative anodes, *Energies.* 2 (2009) 377–410. https://doi.org/10.3390/en20200377

[6] A. Lashtabeg, S. J. Skinner, Solid oxide fuel cells—A challenge for materials chemists?, *J. Mater. Chem.* 16 (2006) 3161–3170. https://doi.org/10.1039/B603620A

[7] A. M. Abdalla, S. Hossain, P. M. Petra, M. Ghasemi, A. K. Azad, Achievements and trends of solid oxide fuel cells in clean energy field: A perspective review, *Front. Energy.* 14 (2020) 359–382. https://doi.org/10.1007/s11708-018-0546-2

[8] A. B. Stambouli, E. Traversa, Solid oxide fuel cells (SOFCs): A review of an environmentally clean and efficient source of energy, *Renew. Sustain. Energy Rev.* 6 (2002) 433–455. https://doi.org/10.1016/S1364-0321(02)00014-X

[9] A. Kumar Yadav, C. R. Gautam, A review on crystallization behaviour of perovskite glass ceramics, *Adv. Appl. Ceram.* 113 (2014) 193–207. https://doi.org/10.1179/1743676113Y.0000000134

[10] N. S. Mohd Sabri, S. Izman, D. Kurniawan, Perovskite materials for intermediate temperature solid oxide fuel cells cathodes: A review, in *AIP Conf. Proc.*, 2020, p. 030013. https://doi.org/10.1063/5.0015824

[11] S. C. Singhal, K. Kendall, High-temperature solid oxide fuel cells: Fundamentals, *Des. Appl.* (2003). https://doi.org/10.1016/B978-1-85617-387-2.X5016-8

[12] B. C. H. Steele, Material science and engineering: The enabling technology for the commercialization of fuel cell systems, *J. Mater. Sci.* 36 (2001) 1053–1068. https://doi.org/10.1023/A:1004853019349

[13] S. P. Jiang, S. H. Chan, A review of anode materials development in solid oxide fuel cells, *J. Mater. Sci.* 39 (2004) 4405–4439. https://doi.org/10.1023/B:JMSC.0000034135.52164.6b

[14] I. Sreedhar, B. Agarwal, P. Goyal, S.A. Singh, Recent advances in material and performance aspects of solid oxide fuel cells, *J. Electroanal. Chem.* 848 (2019) 113315. https://doi.org/10.1016/j.jelechem.2019.113315

[15] C. Sun, R. Hui, J. Roller, Cathode materials for solid oxide fuel cells: A review, *J. Solid State Electrochem.* 14 (2010) 1125–1144. https://doi.org/10.1007/s10008-009-0932-0

[16] B. Wei, Z. Lü, T. Wei, D. Jia, X. Huang, Y. Zhang, J. Miao, W. Su, Nanosized $Ce_{0.8}Sm_{0.2}O_{1.9}$ infiltrated $GdBaCo_2O_{5+\delta}$ cathodes for intermediate-temperature solid oxide fuel cells, *Int. J. Hydrogen Energy.* 36 (2011) 6151–6159. https://doi.org/10.1016/j.ijhydene.2011.02.061

[17] J. M. Ralph, A.C. Schoeler, M. Krumpelt, Materials for lower temperature solid oxide fuel cells, *J. Mater. Sci.* 36 (2001) 1161–1172. https://doi.org/10.1023/A:1004881825710

[18] W. Vielstich, A. Lamm, H. A. Gasteiger, H. Yokokawa, eds., *Handbook of Fuel Cells*, Wiley, 2010. https://doi.org/10.1002/9780470974001

[19] V. V. Kharton, F. M. Figueiredo, L. Navarro, E. N. Naumovich, A. V. Kovalevsky, A. A. Yaremchenko, A. P. Viskup, A. Carneiro, F. M. B. Marques, J. R. Frade, Ceria-based materials for solid oxide fuel cells, *J. Mater. Sci.* 36 (2001) 1105–1117. https://doi.org/10.1023/A:1004817506146
[20] D. J. L. Brett, A. Atkinson, N. P. Brandon, S. J. Skinner, Intermediate temperature solid oxide fuel cells, *Chem. Soc. Rev.* 37 (2008) 1568. https://doi.org/10.1039/b612060c
[21] R. V. Vovk, M. A. Obolenskii, A. A. Zavgorodniy, A. V. Bondarenko, I. L. Goulatis, A.V. Samoilov, A. Chroneos, Effect of high pressure on the fluctuation conductivity and the charge transfer of $YBa_2Cu_3O7-_{\delta}$ single crystals, *J. Alloys Compd.* 453 (2008) 69–74. https://doi.org/10.1016/j.jallcom.2006.11.169
[22] K. Larbaoui, A. Tadjer, B. Abbar, H. Aourag, B. Khelifa, C. Mathieu, State of the art simulations in electronic structure and total energy for the high temperature superconductor $YBa_2Cu_3O_7$, *J. Alloys Compd.* 403 (2005) 1–14. https://doi.org/10.1016/j.jallcom.2005.04.203
[23] M. Yoshimura, K. Sardar, Revisiting the valence stability and preparation of perovskite structure type oxides ABO_3 with the use of Madelung electrostatic potential energy and lattice site potential, *RSC Adv.* 11 (2021) 20737–20745. https://doi.org/10.1039/D1RA01979A
[24] N. Q. Minh, Ceramic fuel cells, *J. Am. Ceram. Soc.* 76 (1993) 563–588. https://doi.org/10.1111/j.1151-2916.1993.tb03645.x
[25] S. P. Jiang, Development of lanthanum strontium manganite perovskite cathode materials of solid oxide fuel cells: A review, *J. Mater. Sci.* 43 (2008) 6799–6833. https://doi.org/10.1007/s10853-008-2966-6
[26] E. Perry Murray, $(La,Sr)MnO_3$–$(Ce,Gd)O_{2-x}$ composite cathodes for solid oxide fuel cells, *Solid State Ionics.* 143 (2001) 265–273. https://doi.org/10.1016/S0167-2738(01)00871-2
[27] E. P. Murray, T. Tsai, S. A. Barnett, Oxygen transfer processes in $(La,Sr)MnO_3/Y_2O_3$-stabilized ZrO_2 cathodes: An impedance spectroscopy study, *Solid State Ionics.* 110 (1998) 235–243. https://doi.org/10.1016/S0167-2738(98)00142-8
[28] A. Chroneos, R. V. Vovk, I. L. Goulatis, L. I. Goulatis, Oxygen transport in perovskite and related oxides: A brief review, *J. Alloys Compd.* 494 (2010) 190–195. https://doi.org/10.1016/j.jallcom.2010.01.071
[29] A. Hammouche, E. Siebert, A. Hammou, Crystallographic, thermal and electrochemical properties of the system $La_{1-x}Sr_xMnO_3$ for high temperature solid electrolyte fuel cells, *Mater. Res. Bull.* 24 (1989) 367–380. https://doi.org/10.1016/0025-5408(89)90223-7
[30] T. Kawada, J. Suzuki, M. Sase, A. Kaimai, K. Yashiro, Y. Nigara, J. Mizusaki, K. Kawamura, H. Yugami, Determination of oxygen vacancy concentration in a thin film of $La_{0.6}Sr_{0.4}CoO_{3-\delta}$ by an electrochemical method, *J. Electrochem. Soc.* 149 (2002) E252. https://doi.org/10.1149/1.1479728
[31] L. Tai, Structure and electrical properties of $La_{1-x}Sr_xCo_{1-y}Fe_yO_3$. Part 1. The system $La_{0.8}Sr_{0.2}Co_{1-y}Fe_yO_3$, *Solid State Ionics.* 76 (1995) 259–271. https://doi.org/10.1016/0167-2738(94)00244-M
[32] E. Lust, P. Möller, I. Kivi, G. Nurk, S. Kallip, Electrochemical characteristics of $La_{0.6}Sr_{0.4CoO3-\delta}$, $Pr_{0.6}Sr_{0.4CoO3-\delta}$ and $Gd_{0.6}Sr_{0.4CoO3-\delta}$ on $Ce_{0.85}Sm_{0.15}O_{1.925}$ electrolyte, *J. Solid State Electrochem.* 9 (2005) 882–889. https://doi.org/10.1007/s10008-005-0040-8
[33] K. T. Lee, A. Manthiram, Characterization of $Nd_{1-x}Sr_xCoO_{3-\delta}$ ($0 \leq x \leq 0.5$) cathode materials for intermediate temperature SOFCs, *J. Electrochem. Soc.* 152 (2005) A197. https://doi.org/10.1149/1.1828243
[34] C. Xia, $Sm_{0.5}Sr_{0.5}CoO_3$ cathodes for low-temperature SOFCs, *Solid State Ionics.* 149 (2002) 11–19. https://doi.org/10.1016/S0167-2738(02)00131-5
[35] A. Maignan, C. Martin, D. Pelloquin, N. Nguyen, B. Raveau, Structural and magnetic studies of ordered oxygen-deficient perovskites $lnbaco_2o_{5+\delta}$, closely related to the "112" structure, *J. Solid State Chem.* 142 (1999) 247–260. https://doi.org/10.1006/jssc.1998.7934
[36] M. C. Tucker, H. Kurokawa, C. P. Jacobson, L. C. De Jonghe, S. J. Visco, A fundamental study of chromium deposition on solid oxide fuel cell cathode materials, *J. Power Sources.* 160 (2006) 130–138. https://doi.org/10.1016/j.jpowsour.2006.02.017

[37] M. Al Daroukh, Oxides of the AMO_3 and A_2MO_4-type: Structural stability, electrical conductivity and thermal expansion, *Solid State Ionics.* 158 (2003) 141–150. https://doi.org/10.1016/S0167-2738(02)00773-7

[38] A. Orera, P. R. Slater, New chemical systems for solid oxide fuel cells, *Chem. Mater.* 22 (2010) 675–690. https://doi.org/10.1021/cm902687z

[39] F. Mauvy, C. Lalanne, J-M. Bassat, J-C. Grenier, H. Zhao, L. Huo, P. Stevens, Electrode properties of $Ln_2NiO_{4+\delta}$ (Ln=La, Nd, Pr), *J. Electrochem. Soc.* 153 (2006) A1547. https://doi.org/10.1149/1.2207059

[40] A. V. Kovalevsky, V. V. Kharton, A. A. Yaremchenko, Y. V. Pivak, E. V. Tsipis, S. O. Yakovlev, A. A. Markov, E. N. Naumovich, J. R. Frade, Oxygen permeability, stability and electrochemical behavior of $Pr_2NiO_{4+\delta}$-based materials, *J. Electroceramics.* 18 (2007) 205–218. https://doi.org/10.1007/s10832-007-9024-7

[41] Y. H. Lim, J. Lee, J. S. Yoon, C. E. Kim, H. J. Hwang, Electrochemical performance of $Ba_{0.5}Sr_{0.5}Co_xFe_{1-x}O_{3-\delta}$ (x=0.2–0.8) cathode on a ScSZ electrolyte for intermediate temperature SOFCs, *J. Power Sources.* 171 (2007) 79–85. https://doi.org/10.1016/j.jpowsour.2007.05.050

[42] J. D. Jorgensen, B. Dabrowski, S. Pei, D. R. Richards, D. G. Hinks, Structure of the interstitial oxygen defect in $La_2NiO_{4+\delta}$, *Phys. Rev. B.* 40 (1989) 2187–2199. https://doi.org/10.1103/PhysRevB.40.2187

[43] V. Vashook, E. Girdauskaite, J. Zosel, T. Wen, H. Ullmann, U. Guth, Oxygen non-stoichiometry and electrical conductivity of $Pr_{2-x}Sr_xNiO_{4\pm\delta}$ with x=0–0.5, *Solid State Ionics.* 177 (2006) 1163–1171. https://doi.org/10.1016/j.ssi.2006.05.018

[44] Y. Leng, S. H. Chan, Q. Liu, Development of LSCF-GDC composite cathodes for low-temperature solid oxide fuel cells with thin film GDC electrolyte, *Int. J. Hydrogen Energy.* 33 (2008) 3808–3817. https://doi.org/10.1016/j.ijhydene.2008.04.034

CHAPTER 3

Future Materials for Thermoelectric and Hydrogen Energy

Ashwani Kumar

Institute Instrumentation Centre, IIT Roorkee, Roorkee, India

Durvesh Gautam

Ch. Charan Singh University, Meerut, India

Ramesh Chandra

Institute Instrumentation Centre, IIT Roorkee, Roorkee, India

Yogendra K. Gautam

Ch. Charan Singh University, Meerut, India

3.1 INTRODUCTION TO RENEWABLE ENERGY SOURCES

The world is facing an increasing demand for energy, and there is a growing recognition of the importance of transitioning towards sustainable energy systems. Renewable energy sources are gaining popularity as a solution to this challenge due to the numerous benefits they offer over traditional fossil-fuel-based sources. Renewable energy sources, such as solar, wind, hydro, and geothermal energy are replenished naturally and continuously. These technologies are advancing rapidly and becoming more cost-effective, making them a feasible option for meeting the world's rising energy needs. In 2020, global renewable energy capacity reached 2,799 GW, indicating a 45% increase over the last five years [1]. Renewable energy sources have several advantages over traditional fossil fuels, including reducing greenhouse gas emissions, improving energy security, and reducing dependence on imported energy sources. Furthermore, renewable energy sources can provide economic benefits such as job creation and reduced energy costs. They also have the potential to bring access to energy in rural and remote communities that currently lack electricity [2]. However, despite the benefits of renewable energy sources, there are still challenges to

DOI: 10.1201/9781003315261-4

their widespread adoption, including infrastructure development, technological innovation, and policy and regulatory support. Governments worldwide are enacting policies and incentives to promote the growth of renewable energy sources, including feed-in tariffs, tax credits, and renewable portfolio standards [3]. As renewable energy technologies continue to evolve, they hold the potential to revolutionize the global energy system and help mitigate the impact of climate change. This chapter aims to provide an overview of materials for thermoelectric and hydrogen energy and their current status, along with the challenges and opportunities for their development.

3.2 THERMOELECTRIC ENERGY

Thermoelectric energy is a process that converts temperature differences into electrical energy. The discovery of the thermoelectric effect dates back to the nineteenth century, and since then, researchers have extensively explored the potential applications of thermoelectric materials and devices in waste heat recovery, power generation, and refrigeration [4]. Thermoelectric device operation is based on the Seebeck effect, which generates an electrical potential difference when a temperature gradient is applied to a material with a Seebeck coefficient. The coefficient measures the amount of voltage generated per unit temperature difference, resulting in a flow of electrical current [4]. Thermoelectric materials usually consist of semiconductors with a high electrical conductivity and a low thermal conductivity, allowing for efficient conversion of thermal energy into electrical energy. However, current thermoelectric materials have low conversion efficiencies and are limited by their operating temperature range. Recent advances in materials science and nanotechnology have led to the development of new thermoelectric materials with improved conversion efficiencies and higher operating temperatures, including complex oxides, skutterudites, and half-Heusler alloys [5]. Thermoelectric energy has potential applications in waste heat recovery, where it can convert waste heat into electricity, thereby reducing energy waste and greenhouse gas emissions. It also has potential applications in power generation, especially in remote or off-grid locations where traditional power sources are unavailable [4].

Despite the challenges of widespread adoption, such as the need for further research and development of new materials, the high cost of materials and manufacturing, and the limited efficiency of current thermoelectric devices, ongoing research offers promise for thermoelectric energy to become a crucial component in the transition towards a more sustainable energy system.

The Seebeck, Peltier, and Thomson effects are three significant and quite well-known factors that contribute to the thermoelectric phenomena. In 1821, Thomas Johann Seebeck found whenever a conductor is exposed to a temperature gradient, a voltage is produced. This phenomenon is known as the Seebeck effect and is defined as Equation (3.1),

$$V = \alpha \Delta T \tag{3.1}$$

Where V denotes thermoelectric voltage, T denotes a temperature gradient, and α denotes the so-called Seebeck coefficient, as seen in Figure 3.1 [6].

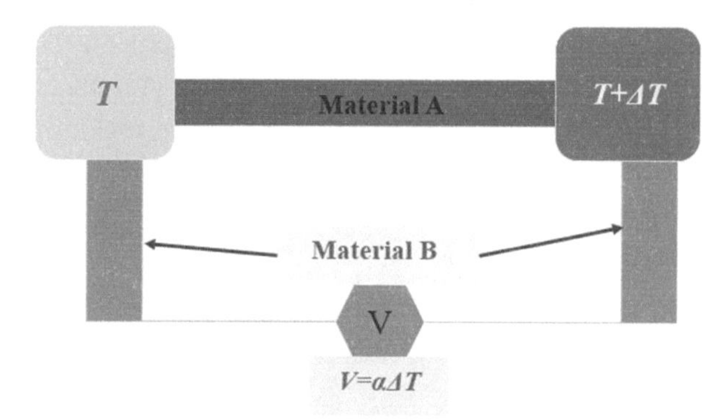

FIGURE 3.1 A diagram showing the Seebeck effect. The temperature difference between the two junctions is ΔT.

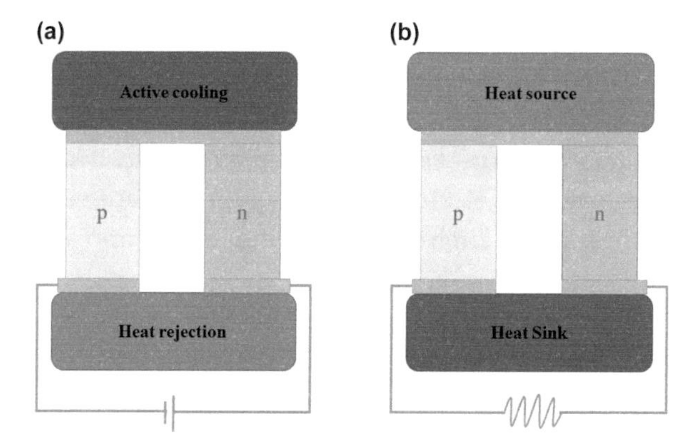

FIGURE 3.2 Illustration of thermoelectric modules: (a) cooling module, (b) power generation module.

The Peltier effect is the opposite of the Seebeck effect and describes the change in temperature brought on by a voltage gradient. The temperature gradient and electric field in a homogeneous conductor are connected via the Thomson effect [6, 7].

Thermoelectric modules for cooling (see Figure 3.2(a)) or power generation systems (see Figure 3.2(b)) can be constructed based on the thermoelectric effects discussed above. The figure of merit of the thermoelectric material describes the effectiveness of thermoelectric devices [6, 7]. This is determined by several transport coefficients, given in Equation (3.2):

$$\mathrm{ZT} = \frac{\sigma S^2 T}{k_e + k_l} \tag{3.2}$$

where,

σ is the electrical conductivity,
S is the Seebeck coefficient,
T is the mean operating temperature and
κ is the thermal conductivity.

The subscripts of e and l in κ refer to electronic and lattice contributions, respectively.

The efficiency of the thermoelectric cooler or power generator improves with greater figures of merit. For many industrial and energy applications, it is therefore of great importance to increase the figure of merit for thermoelectric materials. In actuality, the development of rising ZT may be used to describe the history of thermoelectric materials.

It is evident from Equation (3.2) that increasing electrical conductivity, the Seebeck coefficient, and lower thermal conductivity are the general directions that should be taken to enhance the figure of merit (ZT). It is difficult to increase ZT because, σ, S, κ, and are all related to one another and all depend heavily on the crystal structure, electrical structure, and carrier concentration of the material. Additionally, for a suitable thermoelectric material to be employed in practical thermoelectric generators, excellent mechanical, metallurgical, and thermal qualities are also necessary in addition to a high figure of merit throughout a broad working temperature range [6].

3.2.1 Thermoelectric Materials

Thermoelectric materials possess the ability to convert heat energy directly into electrical energy, or vice versa, and offer potential applications in waste heat recovery, power generation, and refrigeration [8]. Typically, these materials are composed of semiconductors with high electrical conductivity and low thermal conductivity, allowing for efficient conversion of thermal energy into electrical energy. However, their conversion efficiencies are limited by their operating temperature range [9]. Recent advancements in materials science and nanotechnology have resulted in the development of new thermoelectric materials with improved conversion efficiencies and higher operating temperatures, such as complex oxides, skutterudites, and half-Heusler alloys [5]. For instance, complex oxides, including $Ca_3Co_4O_9$ and $SrTiO_3$, exhibit high thermoelectric figures of merit and hold promise for high-temperature applications [10], whereas skutterudites, like $CoSb_3$ and $FeSb_3$, offer high thermal stability, and are promising for mid-temperature applications [11, 12]. Additionally, half-Heusler alloys, such as ZrNiSn and HfNiSn, are highly efficient in low-temperature applications [13]. Nanoscale engineering of thermoelectric materials has also been explored to enhance their performance. Two approaches are reducing the size of the material to nanoscale dimensions, which increases the number of energy-carrying carriers and reduces thermal conductivity [14], and introducing nanostructured interfaces or barriers that can increase the Seebeck coefficient or reduce thermal conductivity, thereby enhancing the thermoelectric figure of merit [8]. Despite the potential of thermoelectric materials, there are still challenges to their widespread adoption, including the need for further research and development of new materials, the high cost of materials and manufacturing, and the limited efficiency of current thermoelectric devices. Nevertheless, with continuous research, thermoelectric materials could become a significant component in transitioning to a more sustainable energy system.

3.2.2 Transition Metal Oxide-Based Film Systems for Thermoelectric Energy

Transition metal oxides (TMOs) have recently attracted attention as potential candidates for thermoelectric energy conversion applications due to their unique electronic and thermal properties. TMOs are typically composed of transition metal cations in a crystal lattice of oxygen anions, and they exhibit a variety of interesting physical properties such as high electrical conductivity, high Seebeck coefficient, and low thermal conductivity [15].

In particular, TMO-based thin film systems have been the subject of intense research in recent years. These systems consist of TMO films deposited on a substrate and can be engineered to have specific electronic and thermal properties through the choice of TMO composition, deposition method, and post-deposition processing. One promising TMO-based film system for thermoelectric energy conversion is the perovskite oxide-based system. Perovskite oxides are a family of TMOs with a specific crystal structure, and they exhibit a variety of interesting physical properties such as high electrical conductivity and high Seebeck coefficient [16]. Perovskite oxide-based thin films have been shown to exhibit high thermoelectric power factors, making them promising materials for thermoelectric applications [17]. Another TMO-based film system that has received significant attention is the spinel oxide-based system. Spinel oxides are a family of TMOs with a specific crystal structure, and they exhibit a variety of interesting physical properties such as high electrical conductivity and low thermal conductivity [18]. Spinels have been shown to exhibit high thermoelectric power factors, making them promising materials for thermoelectric application [18].

In addition to perovskite and spinel oxide-based systems, other TMO-based film systems such as layered oxide-based systems and complex oxide-based systems have also been investigated for their thermoelectric properties [16]. Although TMO-based film systems show promise for thermoelectric energy conversion, there are still significant challenges to their widespread adoption, such as the need for further research and development of new materials, the high cost of materials and manufacturing, and the limited efficiency of current thermoelectric devices. Nevertheless, ongoing research suggests that TMO-based film systems have the potential to play a significant role in the transition towards a more sustainable energy system.

3.2.3 Future Challenges and Future Needs for Thermoelectric Energy

Thermoelectric energy has the potential to play a significant role in the transition to a more sustainable energy system, but there are still several challenges that need to be addressed to make it a more viable option. Some of the future challenges and needs for thermoelectric energy are discussed below.

3.2.3.1 Improvement in the Efficiency of Thermoelectric Devices

One of the main challenges facing the widespread adoption of thermoelectric energy is the limited efficiency of current thermoelectric devices. The efficiency of thermoelectric devices is determined by the thermoelectric figure of merit (ZT), which is a measure of the ability of a material to convert heat into electricity. Current state-of-the-art materials have ZT values of around 2 to 3, but higher ZT values are needed for practical applications [15].

3.2.3.2 Development of New Materials

Another challenge is the need for new materials with higher ZT values. Many of the materials currently used in thermoelectric devices are expensive and/or toxic, which limits their practical use. New materials that are abundant, inexpensive, and environmentally friendly are needed to make thermoelectric energy a more viable option [19].

3.2.3.3 Scale-Up and Commercialization

Thermoelectric energy is currently limited by the high cost of materials and manufacturing, which makes it difficult to scale up and commercialize. Cost-effective manufacturing methods and large-scale production techniques are needed to make thermoelectric energy more economically viable.

3.2.3.4 Integration with Other Energy Systems

Thermoelectric energy needs to be integrated with other energy systems to maximize its potential. For example, thermoelectric devices can be used to recover waste heat from industrial processes or to generate electricity from geothermal energy sources. Integration with other energy systems can help to increase the efficiency and viability of thermoelectric energy.

3.2.3.5 Improving the Durability and Reliability of Thermoelectric Devices

Thermoelectric devices need to be durable and reliable in order to be practical for real-world applications. Improvements in materials and manufacturing processes can help to increase the durability and reliability of thermoelectric devices.

In summary, advances in materials science, manufacturing techniques, and integration with other energy systems will be crucial in the development of thermoelectric energy sources for a sustainable energy future.

3.3 HYDROGEN ENERGY

Hydrogen has long been recognized as a promising alternative to fossil fuels for clean and sustainable energy [20]. As a highly abundant and versatile element, hydrogen can be produced from a variety of sources, including renewable sources such as solar and wind energy. When used in fuel cells, hydrogen can generate electricity without producing harmful emissions, with only water and heat as byproducts. This makes hydrogen an attractive option for a range of applications, from transportation to industrial processes and power generation [21]. In recent years, hydrogen energy has gained increasing attention as a key component of the transition towards a low-carbon economy. The potential benefits of hydrogen are clear: it can help to reduce greenhouse gas emissions, enhance energy security, and diversify energy supply. Governments and industries around the world are investing in research and development to unlock the full potential of hydrogen energy, with the goal of making it a viable and competitive alternative to fossil fuels [22–24]. However, there are also significant challenges that need to be addressed in order to realize the full potential of hydrogen energy. These include the cost of production, storage, and transport, the need for supportive infrastructure and regulations, and the development of new technologies for hydrogen generation and use. In addition, there are still many unknowns around the safety, reliability, and scalability of hydrogen energy systems, which require further research and testing. Despite these challenges, the potential of hydrogen energy is immense, and it will play an increasingly important role in the transition towards a sustainable energy future [24]. This chapter explores the different aspects of hydrogen energy, from its production and storage to its applications in transportation, industry, and power generation. It also discusses the challenges and barriers that need to be overcome to accelerate the adoption of hydrogen energy and highlight the potential for future research and development in this exciting field.

The key important factor in hydrogen production is the hydrogen economy, which refers to the overall infrastructure and systems required for the widespread production, distribution, and utilization of hydrogen as an energy carrier. Hydrogen production technologies play a crucial role in realizing the potential of the hydrogen economy.

There are several commercially available hydrogen production technologies, while others are still in the development phase. These technologies utilize both fossil and non-fossil fuels to produce hydrogen. Some of the commonly used methods include:

- **Steam reforming**: This is the most widely used method for hydrogen production and involves the reaction of steam with hydrocarbons, such as natural gas or methane. Steam reforming has high efficiency and large-scale industrial applications.
- **Partial oxidation**: In this process, partial combustion of hydrocarbons occurs in the presence of a limited oxygen supply, resulting in hydrogen production. It is commonly used for producing hydrogen-rich gases.
- **Autothermal reforming**: This method combines partial oxidation and steam reforming in a single reactor. It offers advantages in terms of efficiency and process flexibility.
- **Pyrolysis**: Pyrolysis involves the thermal decomposition of hydrocarbons or other organic materials to produce hydrogen gas. It is a developing technology with potential for decentralized hydrogen production.
- **Plasma technology**: Plasma arc and plasma reforming techniques use electrical energy to produce high-temperature plasma, which can dissociate hydrocarbons and generate hydrogen. These methods are still under development but show promise for efficient hydrogen production.

Apart from these fossil-fuel-based methods, water electrolysis is another important technology for hydrogen production. Water electrolysis involves splitting water molecules into hydrogen and oxygen using an electrical current. This process can be powered by renewable energy sources, such as solar or wind, making it an environmentally friendly option for hydrogen production [25].

Currently, the highest hydrogen fuel production levels are achieved through steam reforming, gasification, and partial oxidation technologies that utilize fossil fuels. However, these technologies face challenges related to total energy consumption and carbon emissions. The combustion or reaction of fossil fuels in these processes releases carbon dioxide, a greenhouse gas contributing to climate change. Addressing these challenges is crucial to ensure sustainable and low-carbon hydrogen production. As the hydrogen economy continues to evolve, research and development efforts are focused on improving existing technologies, developing novel methods, and integrating renewable energy sources with hydrogen production processes. These advancements aim to enhance efficiency, reduce costs, and minimize environmental impacts, ultimately driving the transition towards a cleaner and more sustainable hydrogen economy.

3.3.1 Materials for Hydrogen Production

Hydrogen can be produced from a variety of sources, including fossil fuels, biomass, and water. However, the most promising sources of hydrogen for sustainable energy production are renewable sources, such as solar and wind power, which can be used to produce hydrogen through water splitting. Water is precious and an abundantly available natural resource. The current renewable energy situation places a significant deal of emphasis on the creation of hydrogen through the splitting of water [26–29]. In photoelectrochemical (PEC) water splitting, electricity (applied externally), and solar energy are combined to make hydrogen with very little to no greenhouse gas emission. Direct water splitting takes place when certain photocatalysts are used in photochemical water splitting, which is carried out in the presence of sunlight. More photocatalysts demonstrate improved photochemical water-splitting efficiency. During the photochemical water-splitting process, semiconductor-based photocatalysts primarily perform four tasks:

1) Light absorption,
2) Charge carrier generation,
3) Charge carrier separation, and
4) Charge carrier transport to the semiconductor-liquid junction (SCLJ) for the surface reaction.

Semiconductors with the proper band gap can absorb light. Light is absorbed, causing the electron-hole pairs to be excited and subsequently separated. The hole is then left behind as the electron moves towards the photocathode. In PEC water splitting, the recombination of electron-hole pairs poses a significant barrier. For the reduction or oxidation of water molecules to occur, it is critical that the semiconductor's conduction band (CB) and valence band (VB) are in the correct places. For the hydrogen evolution reaction (HER), the semiconductor's CB must be more negative than the H^+ ion's reduction potential ($E_{CB} < E^0{}_{red}$), whereas the water oxidation potential must be lower than the VB ($E_{VB} > E^0{}_{ox}$). A potential window of 1.23 V is necessary for solar-driven PEC reactions at photo electrodes to split water into H_2 and O_2. The water-splitting reaction includes oxidizing water by using four electrons for the oxygen evolution reaction (OER) and reducing the protons for the HER with two electrons. The Equations (3.3)–(3.5) illustrate the half-cell reactions:

$$\text{Anode: } 2H_2O(l) = O_2(g) + 4H^+ + 4e^- \tag{3.3}$$

$$\text{Cathode: } 4H^{+} + 4e^{-} = 2H_2\left(g\right) \tag{3.4}$$

$$\text{Overall: } 2H_2O\left(l\right) = O_2\left(g\right) + H_2\left(g\right) \tag{3.5}$$

The potential for H_2 evolution is 0 V_{NHE}, while for O_2 evolution, it is 1.23 V, giving ΔE^0 = 1.23 V. The variation in water's oxidation and reduction potential is also influenced by the pH of the electrolyte. According to the Nernst equation, the slope of the pH versus potential graph has a value of 59 mV/pH, which means that a unit change in pH reduces the water's oxidation/reduction potential by 59 mV. The water-splitting reaction in the acidic electrolyte is shown in Equations (3.6) and (3.7):

$$\text{Anode: } 2H_2O + 4h^{+} = O_2 + 4H^{+}\ \Delta E^{0}{}_{ox} = 1.23\ V_{RHE} \tag{3.6}$$

$$\text{Cathode: } 4H^{+} + 4e^{-} = 2H_2\ \Delta E^{0}{}_{red} = 0\ V_{RHE} \tag{3.7}$$

The water-splitting reaction in the basic electrolyte is presented in Equations (3.8) and (3.9):

$$\text{Anode: } 2OH^{-} + 4h^{+} = O_2 + 2H_2O\ \Delta E^{0}{}_{ox} = -0.404\ V_{RHE} \tag{3.8}$$

$$\text{Cathode: } 4H^{+} + 4e^{-} = 2H_2\ \Delta E^{0}{}_{red} = -0.826\ V_{RHE} \tag{3.9}$$

A critical component of PEC water splitting is the required characteristics of the semiconductor photoactive materials or semiconductor thin films. By using interactions between light and matter to photoexcite semiconductor sheets, the required redox potential gap of 1.23 V is attained. Typically, in this situation, the charge carriers need to have enough overpotential to carry out the reactions at the photoanode and photocathode. Under experimental settings, the semiconductor can produce a photovoltage of around 400 mV below its bandgap [30]. For a semiconductor to successfully push the OER and HER over the potential barrier and absorb light from solar radiation, its optimal bandgap range is between 1.6 and 2.6 eV. [31, 32]. Most n-type semiconductors, such as TiO_2, CdS, $SrTiO_3$, g-C_3N_4, and ZrO_2, have a very suitable band-edge position for the OER. However, with their wide bandgap, the PEC efficiency is very low compared to narrow bandgap semiconductors [28]. Narrow bandgap p-type semiconductors such as Cu_2O, Cu_2S, Sb_2Se_3, p-Si, and $CuInS_2$ have CB edges suitable for photocathodes, and their bandgaps are optimal for light absorption [33]. On the other hand, n-type semiconductors, such as $BiVO_4$, WO_3, and Fe_2O_3, have suitable visible light absorption and feasible VB edges for the oxygen evolution reaction [34].

One of the potential methods to create hydrogen is the use of nanomaterials (NMs) in photo-induced (PC and PEC) water splitting. The use of NMs in water splitting enables the production of hydrogen in a simpler, more affordable, and environmentally friendly manner. The total effectiveness of hydrogen generation can be increased by modifying a photocatalyst with the right nanomaterials. By adjusting the size and shape of nanomaterials, photocatalyst bandwidths may be tailored. Because the primary issue of photo-induced water-splitting processes is the charge separation and transportation of the carrier (electron-hole pair), NMs with very high surface-to-volume ratios help to separate the charge while also suppressing the recombination of the electron-to-hole pair. Photoelectrodes made of nanomaterials provide several benefits for PEC water splitting, namely, tunable optical band gap, larger surface area, smaller light reflection loss, shorter carrier diffusion length, and electronic structure. Processes of water splitting occur at the photoelectrode/electrolyte interface, hence NMs with advantageous surface characteristics become crucial. Numerous surface engineering techniques can be used to fine-tune the surface characteristics of NMs, which in turn can increase the photocatalytic efficiency of photocatalysts (in terms of photostability and photoelectrochemical stability, the kinetics of surface redox reactions, and charge separation/recombination efficiency). Consequently, the use of nanomaterials in photo-induced water splitting to create ecologically friendly and clean sources of energy appears promising. Finding new, affordable materials and techniques with great efficiency and stability is essential for the objective of producing hydrogen on a big scale. Different nanomaterials including metal-based (Co, Ni, Fe, Mo, Cu, W, etc.) and metal-free (B, C, N, O, S, Se, etc.) catalysts are discussed by Zou et al. [35]. Joy et al. [36] reported the application of various classes of NMs, specifically, titanium oxide (TiO_2), zinc oxide (ZnO), tungsten trioxide (WO_3), hematite (α-Fe_2O_3), quantum dots, graphene-based metal-free NMs, and bismuth vanadate ($BiVO_4$), to generate hydrogen via photocatalytic water splitting. Among various TiO_2-based nanomaterials, TiO_2 nanotube has superior efficiency. Black TiO_2 nanoparticles with small band gaps reduce the electron-hole recombination process and lead the high quantum efficiency [36–40].

3.3.2 Hydrogen Storage

One of the main challenges of using hydrogen as a fuel is its low energy density, which makes it difficult to store and transport. To be practical, hydrogen needs to be stored in a compact, lightweight, and safe manner, while also being easily retrievable when needed. Materials research is critical for developing hydrogen storage technologies that meet these requirements. There are several ways to store hydrogen, including compressed gas, liquid hydrogen, and solid-state storage [41]. Compressed gas is the simplest and most common method of hydrogen storage, but it requires high-pressure tanks that can be heavy and bulky [42]. Liquid hydrogen offers higher energy density than compressed gas, but it requires cryogenic temperatures and specialized storage tanks that can be expensive and complex [43]. Solid-state storage, on the other hand, offers the potential for higher energy density and easier handling, but it requires materials that can store hydrogen reversibly and safely [44]. Metal hydrides and porous materials, such as carbon-based materials and metal-organic frameworks (MOFs), are two promising classes of materials for solid-state hydrogen storage [45–47]. Metal hydrides, such as magnesium hydride and lithium borohydride, can store hydrogen through reversible chemical reactions, but they suffer from slow kinetics and high operating temperatures [48]. Porous materials, such as activated carbon and MOFs, can store hydrogen through physical adsorption, but they suffer from low hydrogen uptake and poor reversibility [49, 50] To overcome these limitations, researchers are exploring new materials and approaches for solid-state hydrogen storage.

For example, they are investigating new metal hydrides that offer faster kinetics and lower operating temperatures, as well as new porous materials that offer higher hydrogen uptake and better reversibility. They are also exploring hybrid approaches that combine the benefits of metal hydrides and porous materials, such as metal-organic polyhedra (MOPs) that can store hydrogen through both chemical and physical adsorption. Overall, hydrogen storage remains a key challenge for the widespread adoption of hydrogen energy. Materials research is critical for developing new storage technologies that offer higher energy density, easier handling, and lower cost. By exploring new materials and approaches, researchers can help to overcome the limitations of current hydrogen storage technologies and enable the transition to a more sustainable energy future.

3.3.2.1 Metal Hydride

Metal hydrides are a class of materials that have the ability to reversibly store hydrogen through chemical reactions. The most common metal hydrides for hydrogen storage are based on light metals such as magnesium, lithium, and sodium [48]. These metal hydrides have high hydrogen storage capacity and can operate at relatively low temperatures, making them attractive for practical hydrogen storage applications [51–54]. Metal hydrides can store hydrogen through two types of reactions: reversible and irreversible. Reversible reactions involve the formation of hydride compounds during hydrogen absorption and the release of hydrogen during desorption. Irreversible reactions involve the formation of non-reversible compounds during hydrogen absorption, which require additional steps to release the stored hydrogen. The reversible reactions in metal hydrides can be classified into two types: complex metal hydrides and intermetallic hydrides. Complex metal hydrides are formed by the reaction of metal hydrides with other compounds, such as ammonia borane, which can increase the hydrogen storage capacity and improve the kinetics of hydrogen release [55]. Intermetallic hydrides are formed by the reaction of two or more metals to form a solid solution with the ability to store hydrogen. One of the main advantages of metal hydrides for hydrogen storage is their high hydrogen storage capacity. For example, magnesium hydride (MgH_2) has a theoretical hydrogen storage capacity of 7.6 wt%, which is higher than that of compressed hydrogen gas or liquid hydrogen. In addition, metal hydrides have low flammability and can be stored and transported safely.

However, metal hydrides also have some limitations for hydrogen storage. One of the main challenges is their slow kinetics, which can limit the rate of hydrogen absorption and release. In addition, metal hydrides can be sensitive to air and moisture, which can degrade their performance and reduce their hydrogen storage capacity over time.

To overcome these limitations, researchers are exploring new metal hydrides and hybrid materials that offer faster kinetics and better stability. For example, they are investigating new complex metal hydrides that can improve the hydrogen storage capacity and kinetics of metal hydrides, as well as hybrid materials that combine the benefits of metal hydrides and porous materials, such as metal-organic frameworks (MOFs) and carbon-based materials. These new materials and approaches offer the potential for higher energy density and better performance and could help to overcome the limitations of current metal hydride-based hydrogen storage technologies.

3.3.2.2 Chemical Hydrogen Storage Materials

Chemical hydrogen storage materials are a class of materials that store hydrogen through chemical reactions, rather than physical compression or adsorption. These materials have the potential to offer high hydrogen storage capacity and fast kinetics, making them attractive for hydrogen storage applications [49]. Chemical hydrogen storage materials can be classified into two main categories: reversible and irreversible. Reversible materials store hydrogen through reversible chemical reactions, where the hydrogen can be released and reabsorbed without significant degradation of the material. Irreversible materials, on the other hand, store hydrogen through irreversible chemical reactions, which require additional steps to release the stored hydrogen. One of the most promising reversible chemical hydrogen storage materials is ammonia borane (AB), which has a high hydrogen storage capacity of 19.6 wt% and releases hydrogen through a reversible dehydrogenation reaction [56, 57]. However, the reaction kinetics are slow, and the material requires high temperatures and pressures for complete hydrogen release. Other reversible chemical hydrogen storage materials include metal amides and imides, which can store hydrogen through reversible hydrogenation and dehydrogenation reactions. These materials have high hydrogen storage capacities, but their reaction kinetics can be slow, and they often require high temperatures and pressures for hydrogen release. Irreversible chemical hydrogen storage materials include metal hydrazides and metal borohydrides. These materials store hydrogen through irreversible decomposition reactions, which require additional steps to regenerate the material and release the stored hydrogen. While these materials have high hydrogen storage capacities, their irreversible reactions can limit their practical application. To overcome the limitations of current chemical hydrogen storage materials, researchers are exploring new materials and approaches that offer higher energy density, faster kinetics, and better stability. One promising approach is the use of catalysts to enhance the reaction kinetics of chemical hydrogen storage materials. Catalysts can increase the rates of hydrogen release and uptake, as well as improve the thermodynamic properties of the materials. Overall, chemical hydrogen storage materials offer the potential for high hydrogen storage capacity and fast kinetics, but their practical application is still limited by their reaction kinetics, stability, and regeneration processes. Ongoing research is focused on developing new materials and approaches that can overcome these limitations and enable the widespread use of chemical hydrogen storage materials for hydrogen energy applications.

3.3.2.3 Sorbent Materials

Sorbent materials are a class of materials that can be used to store hydrogen through adsorption or absorption. These materials have the potential to offer high hydrogen storage capacities and fast kinetics, making them attractive for hydrogen storage applications [58]. Adsorbent materials rely on physical adsorption to store hydrogen, where hydrogen molecules are attracted to and held on the surface of the material through weak Van der Waals forces. Absorbent materials, on the other hand, rely on chemical absorption to store hydrogen, where hydrogen molecules are incorporated into the bulk of the material through chemical reactions. One of the most promising adsorbent materials for hydrogen storage is activated carbon. Activated carbon has a high surface area and porosity, which allows it to adsorb large amounts of hydrogen at room temperature and pressure. However, the hydrogen uptake of activated carbon is limited by its low binding energy, which can lead to low hydrogen storage densities. Other adsorbent materials being investigated for hydrogen storage include metal-organic frameworks (MOFs), covalent organic frameworks (COFs), and zeolites. MOFs and COFs are porous materials composed of metal ions or organic molecules linked by strong covalent bonds. These materials offer high surface areas and tunable pore sizes, which can be optimized for hydrogen storage. Zeolites, on the other hand, are crystalline aluminosilicates with regular pore structures that can adsorb hydrogen through their micropores. Absorbent materials for hydrogen storage include metal hydrides and complex metal hydrides, which can absorb large amounts of hydrogen through reversible chemical reactions. Metal hydrides are composed of metal ions and hydrogen atoms, which can form chemical bonds to store hydrogen [58]. Complex metal hydrides, on the other hand, are composed of metal ions and complex anions, which can also form chemical bonds to store hydrogen. To overcome the limitations of current sorbent materials for hydrogen storage, researchers are exploring new materials and approaches that offer higher hydrogen storage capacities, faster kinetics, and better stability. One promising approach is the use of hybrid materials, which combine the advantages of adsorbent and absorbent materials to improve hydrogen storage performance. For example, metal-organic frameworks with incorporated metal hydrides have been shown to offer high hydrogen storage capacities and fast kinetics. Overall, sorbent materials offer the potential for high hydrogen storage capacity and fast kinetics, but their practical application is still limited by their binding energies, stability, and regeneration processes. Ongoing research is focused on developing new materials and approaches that can overcome these limitations and enable the widespread use of sorbent materials for hydrogen energy applications.

3.3.3 Materials for Hydrogen Detection

Hydrogen is a colorless, odorless, and highly flammable gas that requires careful monitoring to ensure safe handling and use [59]. Various materials have been developed to detect the presence of hydrogen in the air, including metal oxide sensors, palladium-based sensors, and carbon nanotube sensors [60–68]. Metal oxide sensors are one of the most commonly used materials for hydrogen detection. These sensors work by measuring changes in the electrical conductivity of metal oxide materials in the presence of hydrogen. Metal oxide materials such as tin oxide, zinc oxide, and tungsten oxide have been used for hydrogen sensing applications due to their high sensitivity and low cost [60, 61]. Palladium-based sensors are another commonly used material for hydrogen detection [65]. These sensors work by measuring changes in the electrical resistance of thin palladium films in the presence of hydrogen. Palladium is highly sensitive to hydrogen, and its resistance changes in response to the adsorption and desorption of hydrogen atoms on its surface. Carbon nanotube sensors are a relatively new material for hydrogen detection. These sensors work by measuring changes in the electrical resistance of carbon nanotubes in the presence of hydrogen [62]. Carbon nanotubes have a high surface area and sensitivity to hydrogen, which enables them to detect even low concentrations of hydrogen in the air. In addition to these materials, other materials such as conducting polymers, metal nanoparticles, and graphene have also been investigated for hydrogen detection applications [66–69]. Conducting polymers can be easily synthesized and offer high sensitivity and selectivity for hydrogen detection. Metal nanoparticles, such as gold and platinum nanoparticles, can also be used for hydrogen detection due to their unique optical and electronic properties [66]. Graphene is a two-dimensional material that offers high sensitivity and selectivity for hydrogen detection and can be integrated into various sensor platform [69]. Overall, the development of materials for hydrogen detection is important for ensuring the safe and efficient handling and use of hydrogen. Ongoing research is focused on developing new materials and approaches that offer higher sensitivity, selectivity, and stability for hydrogen detection, which will enable the widespread use of hydrogen as a clean and sustainable energy source.

3.3.4 Future Challenges and Future Needs for Hydrogen Energy

Despite the potential of hydrogen as a clean and sustainable energy source, there are still several challenges that must be addressed to enable its widespread use. Some of the key challenges and needs for hydrogen energy are summarized in the following subsections.

3.3.4.1 Cost Reduction

One of the biggest challenges for hydrogen energy is reducing its cost to make it more competitive with other energy sources. This includes reducing the cost of hydrogen production, storage, and transportation.

3.3.4.2 Infrastructure Development

Another challenge for hydrogen energy is the need for infrastructure development. This includes the development of hydrogen production, storage, and transportation facilities, as well as the integration of hydrogen into existing energy systems.

3.3.4.3 Safety

Hydrogen is a highly flammable gas that requires careful handling and storage. Ensuring the safety of hydrogen production, storage, and transportation systems is critical for enabling its widespread use.

3.3.4.4 Durability and Stability

Many hydrogens storage and conversion materials, such as metal hydrides and catalysts, have limited durability and stability, which can affect their performance and increase their cost over time. Developing more durable and stable materials is important for reducing cost and improving the efficiency of hydrogen energy systems.

3.3.4.5 Scalability

As hydrogen energy systems become more widespread, there is a need for materials and technologies that can be scaled up to meet growing demand. This includes developing materials and technologies that can be mass-produced at low cost.

3.3.4.6 Integration with Renewable

Hydrogen can be produced from renewable energy sources such as wind and solar power, which can help to reduce greenhouse gas emissions and improve energy security. However, integrating renewable energy sources with hydrogen production and storage systems is a complex challenge that requires further research and development.

Addressing these challenges and needs will require ongoing research and development efforts, as well as collaboration between industry, academia, and government. However, the potential benefits of hydrogen as a clean and sustainable energy source make it a promising area for future innovation and investment.

3.3.5 Conclusion

The chapter explores the advancements in materials that have the capacity to enhance the efficiency of thermoelectric devices and reduce the cost of hydrogen production, storage, and transportation. By leveraging the capabilities of new materials, we have the opportunity to achieve significant reductions in greenhouse gas emissions, enhance energy security, and improve overall efficiency in energy systems. When looking towards the future, continued research and development efforts, along with collaborative partnerships between academia, industry, and government, will be essential for realizing the full potential of these materials in clean energy applications. It is evident that the development of future materials holds immense promise for advancing thermoelectric power generation and hydrogen energy. The potential of these materials to contribute to the reduction of greenhouse gas emissions, enhance energy security, and improve overall efficiency underscores their critical importance. The chapter has provided a comprehensive overview of the current state of research, highlighting the merits and challenges of different materials, synthesis techniques, process parameters, and composition in the context of thermoelectric and hydrogen energy technologies.

In summary, the exploration and utilization of new materials offer a transformative path towards a cleaner and more sustainable energy landscape. By fostering collaboration and innovation, we can surmount existing challenges and fully unlock the potential of these materials, propelling us towards a future characterized by clean, efficient, and abundant energy for all.

REFERENCES

[1] Renewable capacity statistics, 2021. www.irena.org.
[2] Renewable power generation costs in 2019, International Renewable Energy Agency (IRENA), Abu Dhabi, 2020. www.irena.org.
[3] REN21, 2021. https://www.ren21.net/reports/ren21-reports/
[4] D. Michael Rowe, *Thermoelectrics handbook: Macro to Nano*, CRC/Taylor & Francis, 2006.
[5] X. Zianni, D. Narducci, Synergy between defects, charge neutrality and energy filtering in hyper-doped nanocrystalline materials for high thermoelectric efficiency, *Nanoscale.* 11 (2019) 7667–7673. https://doi.org/10.1039/c8nr09263j
[6] I.M. Abdel-Motaleb, S.M. Qadri, Thermoelectric devices: Principles and future trends, n.d.
[7] J.C. Zheng, Recent advances on thermoelectric materials, *Front Phys China.* 3 (2008) 269–279. https://doi.org/10.1007/s11467-008-0028-9

[8] M.S. Dresselhaus, G. Chen, M.Y. Tang, R. Yang, H. Lee, D. Wang, Z. Ren, J.P. Fleurial, P. Gogna, New directions for low-dimensional thermoelectric materials, *Adv. Mater.* 19 (2007) 1043–1053. https://doi.org/10.1002/adma.200600527

[9] R. Venkatasubramanian, E. Siivola, T. Colpitts, B. O'Quinn, Thin-®lm thermoelectric devices with high room-temperature ®gures of merit, 2001. www.nature.com

[10] G. Ren, J. Lan, C. Zeng, Y. Liu, B. Zhan, S. Butt, Y.H. Lin, C.W. Nan, High performance oxides-based thermoelectric materials, *JOM.* 67 (2015) 211–221. https://doi.org/10.1007/s11837-014-1218-2

[11] M.V. Daniel, L. Hammerschmidt, C. Schmidt, F. Timmermann, J. Franke, N. Jöhrmann, M. Hietschold, D.C. Johnson, B. Paulus, M. Albrecht, Structural and thermoelectric properties of $FeSb_3$ skutterudite thin films, *Phys Rev B Condens Matter Mater Phys.* 91 (2015). https://doi.org/10.1103/PhysRevB.91.085410

[12] Z.Y. Liu, J.L. Zhu, X. Tong, S. Niu, W.Y. Zhao, A review of $CoSb_3$-based skutterudite thermoelectric materials, *J. Adv. Ceramics.* 9 (2020) 647–673. https://doi.org/10.1007/s40145-020-0407-4

[13] B. Sahni, A. Alam, *Double Half-Heusler Alloys X_2Ni_2InSb (X= Zr/Hf) with promising thermoelectric performance: Role of varying structural phases*, (2023). https://doi.org/10.48550/arXiv.2301.00598

[14] T.C. Harman, P.J. Taylor, M.P. Walsh, B.E. LaForge, Quantum dot superlattice thermoelectric materials and devices, *Science* 297 (2002) 2229–2232. https://doi.org/10.1126/science.1072886

[15] G.J. Snyder, E.S. Toberer, Complex thermoelectric materials, *Nat. Mater.* 7 (2008) 105–114. https://doi.org/10.1038/nmat2090

[16] R. Tian, T. Zhang, D. Chu, R. Donelson, L. Tao, S. Li, Enhancement of high temperature thermoelectric performance in Bi, Fe co-doped layered oxide-based material Ca3Co4O 9+δ, *J. Alloys Compd.* 615 (2014) 311–315. https://doi.org/10.1016/j.jallcom.2014.06.190

[17] T. Wu, P. Gao, Development of perovskite-type materials for thermoelectric application, *Materials.* 11 (2018). https://doi.org/10.3390/ma11060999

[18] M.H.N. Assadi, J.J. Gutiérrez Moreno, M. Fronzi, High-performance thermoelectric oxides based on spinel structure, *ACS Appl. Energy Mater.* 3 (2020) 5666–5674. https://doi.org/10.1021/acsaem.0c00640

[19] B. Jiang, X. Liu, Q. Wang, J. Cui, B. Jia, Y. Zhu, J. Feng, Y. Qiu, M. Gu, Z. Ge, J. He, Realizing high-efficiency power generation in low-cost PbS-based thermoelectric materials, *Energy Environ. Sci.* 13 (2020) 579–591. https://doi.org/10.1039/c9ee03410b

[20] P.P. Edwards, V.L. Kuznetsov, W.I.F. David, Hydrogen energy, *Philosophical Transactions of the Royal Society A: Mathematical, Physical and Engineering Sciences.* 365 (2007) 1043–1056. https://doi.org/10.1098/rsta.2006.1965

[21] M. Momirlan, T.N. Veziroglu, Current status of hydrogen energy, *Renew. Sustain. Energy Rev.* 6 (2002) 141–179. https://doi.org/10.1016/S1364-0321(02)00004-7

[22] J.O. Abe, A.P.I. Popoola, E. Ajenifuja, O.M. Popoola, Hydrogen energy, economy and storage: Review and recommendation, *Int. J. Hydrog. Energy.* 44 (2019) 15072–15086. https://doi.org/10.1016/j.ijhydene.2019.04.068

[23] Y. Manoharan, S.E. Hosseini, B. Butler, H. Alzhahrani, B.T.F. Senior, T. Ashuri, J. Krohn, Hydrogen fuel cell vehicles; current status and future prospect, *Appl. Sci.* 9 (2019) 2296. https://doi.org/10.3390/app9112296

[24] J. Mergel, M. Carmo, D. Fritz, Status on technologies for hydrogen production by water electrolysis, in: *Transition to Renewable Energy Systems.*, Wiley-VCH Verlag GmbH & Co. KGaA, Weinheim, Germany, 2013: pp. 423–450. https://doi.org/10.1002/9783527673872.ch22

[25] A. Ursua, L.M. Gandia, P. Sanchis, Hydrogen production from water electrolysis: Current status and future trends, *Proc. IEEE.* 100 (2012) 410–426. https://doi.org/10.1109/JPROC.2011.2156750

[26] S.T. Kochuveedu, Photocatalytic and photoelectrochemical water splitting on TiO 2 via photosensitization, *J. Nanomater.* 2016 (2016) 1–12. https://doi.org/10.1155/2016/4073142

[27] A. Hossain, K. Sakthipandi, A.K.M. Atique Ullah, S. Roy, Recent progress and approaches on carbon-free energy from water splitting, *Nano-Micro Lett.* 11 (2019) 103. https://doi.org/10.1007/s40820-019-0335-4

[28] B. Rohland, J. Nitsch, H. Wendt, Hydrogen and fuel cells — The clean energy system, *J. Power Sources.* 37 (1992) 271–277. https://doi.org/10.1016/0378-7753(92)80084-O

[29] Y. Naimi, A. Antar, Hydrogen generation by water electrolysis, *Adv. Hydrog. Gener. Technol., InTech* (2018). https://doi.org/10.5772/intechopen.76814

[30] J. Barber, Photosynthetic energy conversion: Natural and artificial, *Chem. Soc. Rev.* 38 (2009) 185–196. https://doi.org/10.1039/b802262n

[31] T. Hardo Panintingjati Brotosudarmo, L. Limantara, R. Dwi Chandra, Heriyanto, Chloroplast pigments: Structure, function, assembly and characterization, in: *Plant Growth and Regulation – Alterations to Sustain Unfavorable Conditions*, IntechOpen, 2018. https://doi.org/10.5772/intechopen.75672

[32] J. Barber, Photosystem II: An enzyme of global significance, *Biochem. Soc. Trans.* 34 (2006) 619–631. https://doi.org/10.1042/BST0340619

[33] J. Barber, Photosystem II: The water-splitting enzyme of photosynthesis, *Cold Spring Harb. Symp. Quant. Biol.* 77 (2012) 295–307. https://doi.org/10.1101/sqb.2012.77.014472

[34] J. Barber, Photosystem II: The engine of life, *Q. Rev. Biophys.* 36 (2003) 71–89. https://doi.org/10.1017/S0033583502003839

[35] X. Zou, Y. Zhang, Noble metal-free hydrogen evolution catalysts for water splitting, *Chem. Soc. Rev.* 44 (2015) 5148–5180. https://doi.org/10.1039/c4cs00448e

[36] J. Joy, J. Mathew, S.C. George, Nanomaterials for photoelectrochemical water splitting – Review, *Int. J. Hydrog. Energy.* 43 (2018) 4804–4817. https://doi.org/10.1016/j.ijhydene.2018.01.099

[37] T.T. Isimjan, S. Rohani, A.K. Ray, Photoelectrochemical water splitting for hydrogen generation on highly ordered TiO_2 nanotubes fabricated by using Ti as cathode, *Int. J. Hydrog. Energy.* 37 (2012) 103–108. https://doi.org/10.1016/j.ijhydene.2011.04.167

[38] T. Takashima, N. Moriyama, Y. Fujishiro, J. Osaki, S. Takeuchi, B. Ohtani, H. Irie, Visible-light-induced water splitting on a hierarchically constructed Z-scheme photocatalyst composed of zinc rhodium oxide and bismuth vanadate, *J. Mater. Chem. A Mater.* 7 (2019) 10372–10378. https://doi.org/10.1039/c8ta12316k

[39] F. Meng, J. Li, S.K. Cushing, J. Bright, M. Zhi, J.D. Rowley, Z. Hong, A. Manivannan, A.D. Bristow, N. Wu, Photocatalytic water oxidation by hematite/reduced graphene oxide composites, *ACS Catal.* 3 (2013) 746–751. https://doi.org/10.1021/cs300740e

[40] A. Jelinska, K. Bienkowski, M. Jadwiszczak, M. Pisarek, M. Strawski, D. Kurzydlowski, R. Solarska, J. Augustynski, Enhanced photocatalytic water splitting on very thin WO3 films activated by high-temperature annealing, *ACS Catal.* 8 (2018) 10573–10580. https://doi.org/10.1021/acscatal.8b03497

[41] M. Felderhoff, C. Weidenthaler, R. Von Helmolt, U. Eberle, Hydrogen storage: The remaining scientific and technological challenges, *Phys. Chem. Chem. Phys.* 9 (2007) 2643–2653. https://doi.org/10.1039/b701563c

[42] T. Hua, R. Ahluwalia, J-K. Peng, M. Kromer, S. Lasher, K. McKenney, K. Law, J. Sinha, Technical assessment of compressed hydrogen storage tank systems for automotive applications nuclear engineering division, Argonne National Laboratory, 2010 https://www.energy.gov/eere/fuelcells/articles/technical-assessment-compressed-hydrogen-storage-tank-systems-automotive

[43] G. Valenti, Hydrogen liquefaction and liquid hydrogen storage, in: *Compendium of Hydrogen Energy*, Elsevier, 2016: pp. 27–51. https://doi.org/10.1016/b978-1-78242-362-1.00002-x

[44] C. Pistidda, Solid-state hydrogen storage for a decarbonized society, *Hydrogen.* 2 (2021) 428–443. https://doi.org/10.3390/hydrogen2040024

[45] P. Jena, Materials for hydrogen storage: Past, present, and future, *J. Phys. Chem. Lett.* 2 (2011) 206–211. https://doi.org/10.1021/jz1015372

[46] J. Ren, N.M. Musyoka, H.W. Langmi, M. Mathe, S. Liao, Current research trends and perspectives on materials-based hydrogen storage solutions: A critical review, *Int. J. Hydrog. Energy.* 42 (2017) 289–311. https://doi.org/10.1016/j.ijhydene.2016.11.195

[47] L.F. Chanchetti, D.R. Leiva, L.I. Lopes de Faria, T.T. Ishikawa, A scientometric review of research in hydrogen storage materials, *Int. J. Hydrog. Energy.* 45 (2020) 5356–5366. https://doi.org/10.1016/j.ijhydene.2019.06.093

[48] P. Muthukumar, M.P. Maiya, S.S. Murthy, Experiments on a metal hydride-based hydrogen storage device, *Int. J. Hydrog. Energy.* 30 (2005) 1569–1581. https://doi.org/10.1016/j.ijhydene.2004.12.007

[49] T.A. Semelsberger, Chemical hydrogen storage materials, 2015. https://www.energy.gov/sites/default/files/2015/02/f19/fcto_h2_storage_summit_semelsberger.pdf

[50] K. Jastrzębski, P. Kula, Emerging technology for a green, sustainable energy-promising materials for hydrogen storage, from nanotubes to graphene—A review, *Materials (Basel).* 14 (2021) 2499. https://doi.org/10.3390/ma14102499

[51] L. Jinzhe, A. M. Lider, V. N. Kudiiarov, An overview of progress in Mg-based hydrogen storage films, *Chinese Phys. B.* 28 (2019) 098801. https://doi.org/10.1088/1674-1056/ab33f0

[52] C. Zhou, J. Zhang, R.C. Bowman, Z.Z. Fang, Roles of Ti-based catalysts on magnesium hydride and its hydrogen storage properties, *Inorganics.* 9 (2021) 36. https://doi.org/10.3390/inorganics9050036

[53] B. Li, J. Li, H. Zhao, X. Yu, H. Shao, Mg-based metastable nano alloys for hydrogen storage, *Int. J. Hydrog. Energy.* 44 (2019) 6007–6018. https://doi.org/10.1016/j.ijhydene.2019.01.127

[54] Y. Shang, C. Pistidda, G. Gizer, T. Klassen, M. Dornheim, Mg-based materials for hydrogen storage, *J. Magnes. Alloy.* 9 (2021) 1837–1860. https://doi.org/10.1016/j.jma.2021.06.007

[55] Y. Lu, H. Kim, K. Jimura, S. Hayashi, K. Sakaki, K. Asano, Strategy of thermodynamic and kinetic improvements for Mg hydride nanostructured by immiscible transition metals, *J. Power Sources.* 494 (2021) 229742. https://doi.org/10.1016/j.jpowsour.2021.229742

[56] H.L. Jiang, Q. Xu, Catalytic hydrolysis of ammonia borane for chemical hydrogen storage, *Catal Today.* 170 (2011) 56–63. https://doi.org/10.1016/j.cattod.2010.09.019

[57] S. Akbayrak, S. Özkar, Ammonia borane as hydrogen storage materials, *Int. J. Hydrog. Energy.* 43 (2018) 18592–18606. https://doi.org/10.1016/j.ijhydene.2018.02.190

[58] L. Wang, R.T. Yang, New sorbents for hydrogen storage by hydrogen spillover – A review, *Energy Environ Sci.* 1 (2008) 268–279. https://doi.org/10.1039/b807957a

[59] T. Hübert, L. Boon-Brett, G. Black, U. Banach, Hydrogen sensors - A review, *Sens. Actuat. B Chem.* 157 (2011) 329–352. https://doi.org/10.1016/j.snb.2011.04.070

[60] S. Phanichphant, Semiconductor metal oxides as hydrogen gas sensors, in: *Procedia Eng*, Elsevier Ltd, 2014: pp. 795–802. https://doi.org/10.1016/j.proeng.2014.11.677

[61] N. Kumar, S. Haviar, P. Baroch, Nanostructured metal-oxide based hydrogen gas sensor prepared by magnetron sputtering, in: *Studentská Vědecká Konf.*, 2018: pp. 21–22.

[62] C. McConnell, S.N. Kanakaraj, J. Dugre, R. Malik, G. Zhang, M.R. Haase, Y.Y. Hsieh, Y. Fang, D. Mast, V. Shanov, Hydrogen sensors based on flexible carbon nanotube-palladium composite sheets integrated with ripstop fabric, *ACS Omega.* 5 (2020) 487–497. https://doi.org/10.1021/acsomega.9b03023

[63] V. Schroeder, S. Savagatrup, M. He, S. Lin, T.M. Swager, Carbon nanotube chemical sensors, *Chem Rev.* 119 (2019) 599–663. https://doi.org/10.1021/acs.chemrev.8b00340

[64] A. Ilnicka, J.P. Lukaszewicz, Graphene-based hydrogen gas sensors: A review, *Processes.* 8 (2020) 633. https://doi.org/10.3390/pr8050633

[65] I. Darmadi, F.A.A. Nugroho, C. Langhammer, High-performance nanostructured palladium-based hydrogen sensors – Current limitations and strategies for their mitigation, *ACS Sens.* 5 (2020) 3306–3327. https://doi.org/10.1021/acssensors.0c02019

[66] N. L. Torad, M. M. Ayad, Gas sensors based on conducting polymers, in: *Gas sensors*, IntechOpen, 2020. https://doi.org/10.5772/intechopen.89888

[67] T.H. Eom, T. Kim, H.W. Jang, Hydrogen sensing of graphene-based chemoresistive gas sensor enabled by surface decoration, *J. Sens. Sci. Technol.* 29 (2020) 382–387. https://doi.org/10.46670/JSST.2020.29.6.382

[68] L. Zhou, F. Kato, N. Nakamura, Y. Oshikane, A. Nagakubo, H. Ogi, MEMS hydrogen gas sensor with wireless quartz crystal resonator, Sensors Actuators, *B Chem.* 334 (2021). https://doi.org/10.1016/j.snb.2021.129651

[69] T. Wang, D. Huang, Z. Yang, S. Xu, G. He, X. Li, N. Hu, G. Yin, D. He, L. Zhang, A review on graphene-based gas/vapor sensors with unique properties and potential applications, *Nano-Micro Lett.* 8 (2016) 95–119. https://doi.org/10.1007/s40820-015-0073-1

II

Phosphors and Luminescent Materials

CHAPTER 4

Quantum Cutting in Photoluminescence Downconversion Phosphors

Abhijeet R. Kadam
R. T. M. Nagpur University, Nagpur, India

Nirupama S. Dhoble
Sevadal Mahila Mahavidhyalaya, Nagpur, India

Sanjay J. Dhoble
R. T. M. Nagpur University, Nagpur, India

4.1 INTRODUCTION

A visible photon has half as much energy as a vacuum-ultraviolet (VUV) photon. A single VUV photon can theoretically result in the generation of two visible photons, a process known as quantum cutting (QC), quantum splitting (QS), or photon cascade emission (PCE). The idea behind the efficiency improvement in QC materials is that a QC phosphor can produce two visible photons for every absorbed VUV photon [1–3]. Quantum cutting has been proven in a variety of ways. The energy-level diagrams for two (imaginary) types of rare-earth (RE) ions (I and II) are shown in Figure 4.1, which also demonstrates the idea of downconversion (DC). According to Figure 4.1(a), efficient visible QC using a single RE ion and two-photon emission from a high energy level is theoretically feasible. Nevertheless, rival visible QC on a single RE ion can also happen and preclude effective infrared (IR) and ultraviolet (UV) emissions (the narrow lines of Figure 4.1(a)). Simplified energy-level diagrams for three DC mechanisms that involve energy transfer (ET) among two distinct RE ions, I and II, are shown in Figure 4.1(b)–(d). Ions of type I emit from levels that are far below the surface. An activator ion of type II is where ET occurs. Two-photon emission from ion pairs is depicted in Figure 4.1(b) by cross-relaxation (CR) from ions I to II (symbolized by (1)) and ET from ions I to II (symbolized by (2)), both of which are coupled with emission from ion II. A CR mechanism is accompanied by photon emission

DOI: 10.1201/9781003315261-6

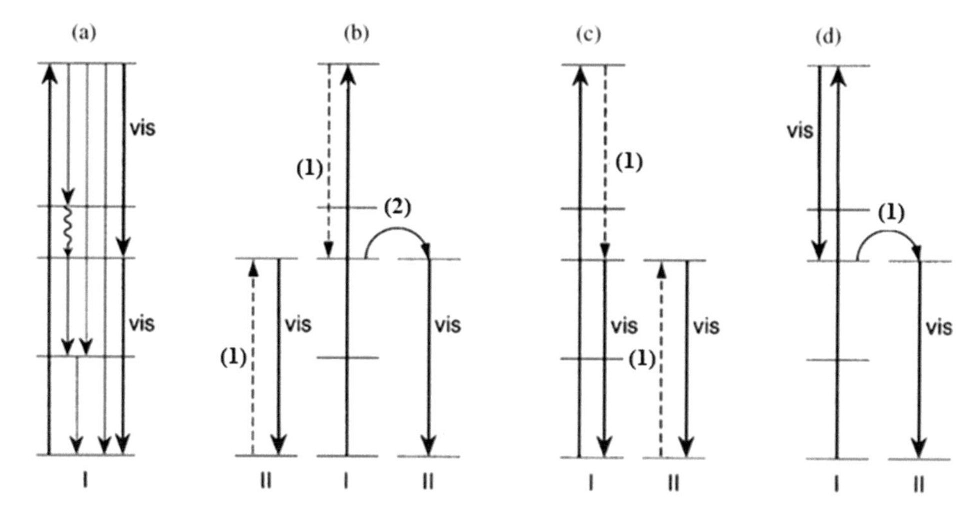

FIGURE 4.1 Energy level diagrams for two (hypothetical) types of RE ions (I and II) showing the concept of DC. Type I is an ion for which emission from a high energy level can occur. Type II is an ion to which ET takes place. (a) QC on a single ion I by the sequential emission of two visible photons. (b) The possibility of QC by a two-step ET. In the first step (indicated by (1)), a part of the excitation energy is transferred from ion I to ion II by CR. Ion II returns to the ground state by emitting one photon of visible light. Ion I is still in an excited stated and can transfer the remaining energy to a second ion of type II (indicated by (2)), which also emits a photon in the visible spectral region, giving a QE of 200%. (c) and (d) The remaining two possibilities involve only one ET step from ion I to ion II. This is sufficient to obtain visible QC if one of two visible photons can be emitted by ion I. (Reproduced with permission from Ref. [6], © 2009 Elsevier B.V.)

from both ions I and II in Figure 4.1(c) and (d.) A theoretical visible QE of 200% can be attained in all three scenarios, if the two-step ET process is effective, since the prior IR and UV losses present in a single ion could be eliminated [4, 5].

Researchers have been working on efficient and affordable photovoltaic (PV) solar cell technologies for the past few decades in an effort to convert solar energy into electricity [7–10]. QC phosphors allow for the downshifting and quantum cutting of luminous materials to convert high energy solar photons into lower-energy but above bandgap photons, so minimizing the energy loss caused by thermalization of hot charge carriers, notably in low band gap semiconductors like c-Si [11, 12].

Because of their high energy level structure and excellent quantum efficiency (QE) of light emission, lanthanide (Ln) ions have recently been used to perform near-infrared (NIR) QC [13–16]. The characteristics of quantum-cutting luminous materials are affected by a variety of parameters, including the host matrix composition and structure, the kind and concentration of dopant ions, and so on [17, 18]. An important part of the luminescence qualities is played specifically by the host material. The phosphor material's capacity for quantum-cutting can be improved through effective regulation of the aforementioned factors. The most popular component in PV solar cells is, by far, crystalline silicon. To increase the effectiveness of silicon-based (low bandgap) solar cells, it is therefore required to discover suitable doped luminous layers. Among the Ln ions, the Tb^{3+}/Yb^{3+} pair offers a significant sensitizer/acceptor combination for DC due to Tb^{3+} absorption at approximately 480 nm (close to the solar spectrum's maximum intensity) and its ET to two acceptor Yb^{3+} ions, which emit in the 970–1100 nm region, just slightly over the Si bandgap. Because of this, the combination of one Tb^{3+} and two Yb^{3+} ions is a potential system to research second-order ET processes. In this system, the former center, which is tuned for strong absorption, transfers its energy to the subsequent two, which are optimized for efficient emission, using a second-order cooperative, simultaneous QC process rather than a first order or sequential one [19, 20]. It has been investigated in several host matrices how the Tb^{3+}/Yb^{3+} combination emits QC energy [19, 20]. $Y_{0.74}Yb_{0.25}Tb_{0.01}PO_4$ powder, prepared by solid state reaction [21], lithium borate glass [22], lanthanum borogermanate glasses [23], Tb^{3+}/Yb^{3+} co-doped transparent oxyfluoride glass ceramics containing CaF_2 nanocrystals, with molar composition $60SiO_2$–$20Al_2O_3$–$20CaF_2$–0.3 Tb^{3+}x Yb^{3+} (QE of 155% for 0.3 mol% Tb and 26 mol% Yb) and 50 SiO_2–20 Al_2O_3–27 CaF_2–1 Tb^{3+}–x Yb_3 glass (QE of 120% with 5% Yb^{3+} doping), synthesized by conventional melting-quenching method [24]. These methods, however, need high temperatures. Co-precipitation and the sol-gel technique, as well as combustion synthesis, have the benefit of lower processing temperatures. For instance, $Y_{3-x-y}Tb_xYb_yGa_5O_{12}$ was made using a solution combustion technique, with a QE of 128% for 0.01 mol% Tb^{3+} and 0.015 mol% Yb^{3+} under 379 nm excitation [25], and Y_2O_3 powder samples were made using a co-precipitation process followed by an annealing process, with effective quantum efficiencies of 137% and 181% for 0.01 mol% [26]. These latter two materials showed improved properties in the luminescence measurements, however, the excitation was considerably below the solar spectrum's maximum intensity (about 500 nm). When stimulated at 474 nm with 1 mol% Tb^{3+} and 4 mol% Yb^{3+}, glass ceramic films of molar content 70 SiO_2 – 30 HfO_2 also appear to be a potential host. These films were made using the sol-gel dip-coating technique [27].

4.2 SYNTHESIS TECHNIQUES FOR QUANTUM-CUTTING PHOSPHORS

It is widely recognized that how materials are handled has a significant impact on how effectively they perform. The microstructural, luminescence, and QE characteristics of phosphors have all been significantly influenced by the synthesis procedure. Exploring phosphor preparation techniques continues to have important technological and scientific implications. Some methods are given below for the material preparation that is useful for the quantum-cutting phosphors.

4.2.1 Solid-State Reaction Method

Due to its simplicity, the solid-solid interaction among powders in the solid state has been exploited to create several inorganic compounds. Because of its relative simplicity and suitability for mass production, this process has so far been the most popular and often used methodology for the synthesis of QC phosphors, including fluorides and oxides. The fluorides of each element included in the synthesized fluoride are typically mixed, heated above their melting temperatures, and cooled in an inert gas atmosphere or vacuum to produce fluoride-based QC phosphors. Furthermore, fluorination using SF_6, HF, or F_2 is done to remove oxygen contamination. Traditional solid-solid reaction production of oxide-based QC phosphors demands temperatures greater than 1000°C due to the refractory character of alumina and RE oxides. Since diffusion control reaction is the mechanism for solid-state reactions, repetitive grinding and heating are necessary. Moreover, the regulated environment is required to control the valency of the activator and the host lattice's stoichiometry.

The typical solid-state approach for synthesizing QC phosphors has several drawbacks. First off, this method frequently produces materials with poor homogeneity and calls for high calcination temperatures. Second, the phosphor powders made using this process have grains that are several tens of micrometers in size. The bigger phosphor particles must be ground to produce the smaller phosphor particles. Additionally, using the traditional solid-state approach to prepare single phase compounds is challenging. Further contaminants and flaws can be easily introduced during this procedure, significantly decreasing luminescence efficiency.

4.2.2 Wet Chemical Method

Nitrates are typically used as materials for wet chemical methods since they are readily soluble in water and are chemically based. The wet chemical approach is another way to produce new materials, although it includes a few additional techniques as well, such as the sol-gel method, co-precipitation method, and combustion method. As the combustion technique has several submethods, we will discuss it separately at the end of this section. Hence, we'll discuss the main wet chemical process and the sol gel method individually here.

4.2.2.1 Sol-Gel Synthesis

The invention of the sol-gel techniques dates back to 1846, when Ebelenen found how to hydrolyze $Si(OEt)_4$ to produce SiO_2 gel. This technique did not start being further developed until the 1930s. This method's fundamental processing phases can be summed up as follows: condensation, hydrolysis, and gelation of the sol. Variations of precursors, solvents, ligands, various chemical addition sequences, further processing, and other alterations in sol-gel procedures have been described and widely used in the synthesis of, for example, glasses, ceramics, inorganic fillers, and coatings based on this synthetic approach. In recent years, there has been a lot of research going on, with a particular focus on the production of luminous materials using the sol-gel process. Nonetheless, the majority of inorganic materials based on the sol-gel method to date are nearly exclusively oxide networks. Nevertheless, there may be some issues with the synthesis process. Fluorides are interesting host materials to demonstrate QC effect. To manufacture fluorides using the practical sol-gel technique is, therefore, imperative.

4.2.2.2 Main Wet Chemical Method

In contrast to the sol-gel technique, this process does not require any kind of fuel for synthesis. For this method, the starting components needed must dissolve in a small amount of water. Each of the initial ingredients should be taken in a stochiometric ratio and separately dissolved in a beaker with the least amount of water possible, while being continuously stirred on a hotplate at around 80°C. Add them one at a time, swirling continuously as you add each one, and maintain the beaker on the hotplate once it has fully dissolved. Following this, a white precipitate is discovered and it is kept there to dry overnight at 90°C in a hot air oven. The dried sample is then collected, and it is annealed for a few hours to remove any impurities. The final product is then collected by cooling it to room temperature.

4.2.2.3 Co-Precipitation Method

As a similar method of wet chemical synthesis, this one also require salts that can be dissolved in water. In this procedure, all of the salts are soluble in a solution of double-distilled water, and RE elements are also dissolved in the water solution while ethanol is present. In order to prevent agglomeration, regulate particle size, and maintain homogeneity in particle size, ethanol is typically utilized as a surfactant [28]. With ethanol present, all components are combined. The precipitate is combined, and after that, it is collected by centrifuging. When have the desired precipitate is obtained, it is washed with distilled water multiple times to get rid of any impurities. After being washed, the precipitate is dried in an oven, followed by calcination.

4.2.3 Combustion Synthesis

One of the primary categories of wet chemical-based synthesis is combustion, which is further broken down into two types: simple combustion synthesis method and solution combustion synthesis method. The class of synthesis methodology used in the combustion synthesis method varies depending on the fuel that is being used. Both solution and simple combustion processes are utilized to prepare various fuels, such as urea, citric acid, glycine, ethanol, and DFH [29–33]. In the remainder of this chapter, both sample preparation procedures are described.

4.2.3.1 Simple Combustion Method

In the basic combustion process, all the initial components are weighed according to a stochiometric ratio, and the fuel is then employed in a specific ratio. All of the samples are pulverized and then, the solid solution of precursors is transferred to a pre-fired open furnace at 500–550°C. While the sample is burning, a huge flame is observed in the open furnace. Five minutes after this sample is burned entirely, the flame will subside. Remove this sample from the oven after that, and allow it to cool to room temperature. This sample can be crushed for five minutes after cooling and kept in a furnace for annealing for 24 hours to remove the impurities. After 24 hours, the sample is cooled down to normal temperature and used for additional investigation.

4.2.3.2 Solution Combustion Method

Another form of fuel-based combustion technology is the solution combustion. The same kinds of fuels that were employed in the simple combustion approach are also used in this procedure. Using a magnetic stirrer, a beaker is continuously stirred at a steady temperature to dissolve all of the initial components in distilled water. Repeat the process described in the basic combustion method after the starting materials and fuel have been mixed in the solution, which should now be in a china dish. It should be observed that the sample in both processes turns black when citric acid is employed as the fuel in the open furnace. It becomes a white powder after annealing and is then used for investigations.

All the above described synthesis techniques are very useful in preparation of quantum-cutting phosphors. In a later section of this chapter, phosphor materials synthesized using these procedures and their outcomes will be discussed.

4.3 SINGLE ION ACTIVATED PHOSPHORS

4.3.1 Er^{3+} Activated Phosphors

Although the most powerful emissions were identified in the UV area of the spectrum, the Er^{3+} ion also exhibits an energy level structure that could be deemed suitable for QC in the visible range [34]. The Judd-Ofelt theory led to the determination that the Er^{3+}-doped phosphors' maximum visible QE was 112% [35]; nevertheless, as a sizable portion of the emission intensity is positioned in the UV region, it is insufficient for a Xe-discharge lamp phosphor. However, it is feasible that efficient QC phosphors co-activated with Er^{3+} and RE^{3+} will be attainable, if a co-dopant can effectively transform this UV emission to visible light.

4.3.2 Tm^{3+} Activated Phosphors

Since Tm^{3+} ion has a diverse energy level structure, it was selected as the dopant for luminous materials. Tm^{3+} emission's spectrum spans a wide range of wavelengths, from blue to NIR light. By stimulating at various wavelengths, the abrupt atomic-like transitions of Tm^{3+} help to finish efficient QC operations. No luminescent materials that are Tm^{3+} activated have been documented thus far. As the Tm^{3+} was excited into the $^1G_4$ level, the QC techniques for Tm^{3+} in phosphate glasses and oxyfluoride glass ceramics were introduced; By exciting Tm^{3+} into $^1G_4$ level, the near-infrared QC of Tm^{3+} in $NaYF_4$ and YVO_4 hosts was investigated [36, 37]. The QC features that involve pumping electrons to higher Tm^{3+} levels, like $^1D_2$, were not present in these materials. Even though the study for the $YNbO_4$:Tm^{3+} phosphor described by Chen [38] comprised QC by exciting Tm^{3+} to various levels, the impact of Tm^{3+} concentration on QC characteristics and ET processes for a number of samples was lacking.

4.3.3 Gd^{3+} Activated Phosphors

Gd^{3+} ion QC has also been seen in a variety of phosphors. $ScPO_4$:1% Gd^{3+} exhibits the expected avalanche emissions from the transitions of $^6G_J \rightarrow {}^6P_J$ (600 nm) and $^6P_J \rightarrow {}^8S_{7/2}$ (313 nm). Figure 4.2 shows the QC mechanism graphically. A self-trapped exciton (STE) is produced following the formation of the electron-hole pair as a result of the VUV photon's absorption in an above band gap transition. As seen in Figure 4.2, the introduction of Gd^{3+} causes an ET from the STE to the excited states of Gd^{3+}. A STE Gd^{3+} ET mechanism comprising the ^6G_J state is strongly suggested by the substitution of the STE emission with the Gd^{3+} ^6G_J emission termed as host sensitization of Gd^{3+} [39, 40]. The ET from the STE to Gd^{3+} has been shown to be thermally triggered, most likely because of exciton mobility.

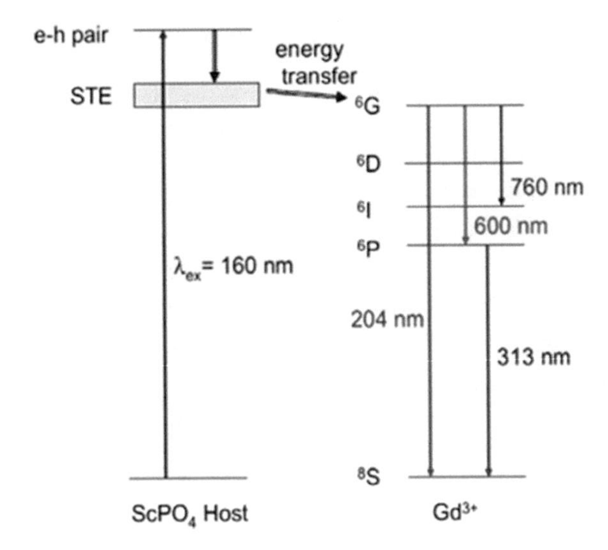

FIGURE 4.2 Schematic showing the creation of the initial electron–hole (e–h) pair, the relaxation to form the STE, and the subsequent energy flow to the excited states of Gd^{3+}. The relevant energy levels of Gd^{3+}, along with the Gd^{3+} transitions observed are identified. (Reproduced with permission from Ref. [6], © 2009 Elsevier B.V.)

When temperatures are low, transfer from the STE to killer centers effectively dominates ET, whereas at room temperature, the opposite is true. A model that posits two STE states separated in two by 280 cm^{-1}, a bottom triplet and an upper singlet, with radiative rates that differ by roughly two orders of magnitude, accurately describes the experimental observations. A 970 cm^{-1} thermal activation energy is obtained from the comparing with the model. In order to exploit STE sensitization of the ^6G_J state of Gd^{3+} for QC, it will be required to find materials for which STE emission arises at even wavelengths shorter than in the instance of $ScPO_4$ in order for a greater proportion of the ET to occur to the ^6G_J state or perhaps even larger states of Gd^{3+}. The effectiveness of ET to the ^6G_J state, meanwhile, is only approximately 30%, leading to a low QE of roughly 92%, which is much higher than the targeted 200%. In light of this, it is possible that materials based only on Gd^{3+} emission will not be discovered to be effective as visible quantum cutters. Gd^{3+}-doped materials might be more effective QC phosphors if the UV emission resulting from the $^6P_J \rightarrow {}^8S_{7/2}$ transition (313 nm) can be effectively transferred to some other ion with visible emission.

4.3.4 Pr^{3+}Activated Phosphors

For a number of uses, including laser applications, scintillator applications, field emission display devices, fiber optical communications, and UV lasers, the luminescence of the Pr^{3+} has been carefully studied. The PCE or QC phenomena from the Pr^{3+} $^1S_0$ state have received significant attention. Only when the 4f 5d states with the lowest energies are located above the $^1S_0$ state can the PCE mechanism of Pr^{3+} be seen. Potentially desirable hosts for the substitution include those with weak crystal fields, low phonon energies, high band gap energies, wide cation-anion distances, and high coordination numbers [6]. Unfortunately, it is currently not possible to develop an accurate quantum cutter based on the Pr^{3+} single ion.

4.4 DUAL ION CO-ACTIVATED PHOSPHORS

4.4.1 Gd^{3+}-Eu^{3+} Co-Activated Phosphors

At any Eu^{3+} concentration, ET from Gd^{3+} to Eu^{3+} is possible in QC. Nonetheless, it is evident that the ET process behaves strongly in relation to Eu^{3+} concentration. The combination of Gd^{3+} and Eu^{3+} was explained by Jaiswal et al. [41] with Gd^{3+} absorbing a VUV photon equivalent to $^8S_{7/2} \rightarrow {}^6G_J$. The high energy photon that was incident is split into two visible photons that are released by Eu^{3+} ions. The entire procedure is referred to as 'quantum cutting'. However, the absorption peaks of a VUV photon equivalent to $^8S_{7/2} \rightarrow {}^6G_J$ and $^8S_{7/2} \rightarrow {}^6D_J$ are not visible in the $KCaF_3$:Gd^{3+}, Eu^{3+} sample at the excitation wavelengths of 593 nm and 612 nm.

As there is no cross-relaxation after excitation at the ^6I_J level with 273 nm, there is never any QC, and the $^5D_J \rightarrow {}^7F_J$ transition emission of Eu^{3+} has a typical splitting proportion among $^5D_0$ and $^5D_{1,\,2}$, and $_3$. There were no excitation peaks seen between 140 nm and 200 nm wavelengths, despite the fact that the ^6G_J level can undergo two-step ET following 147 nm and 172 nm excitation. This indicates that high intensity VUV photons are not absorbed by the synthetic phosphor material. The intensity peak at 273 nm excitation is significantly higher than it is for 147 nm and 172 nm excitation and 593 nm emission, respectively.

4.4.2 Tb^{3+}-Yb^{3+} Co-Activated Phosphors

Figure 4.3(a) and Figure 4.3(b), which represent the diagrammatic representation of NIR QC luminescence in KYF_4:Tb^{3+}, Yb^{3+} phosphors under the excitation of 485 nm and 374 nm, correspondingly, are used to quantitatively examine the dynamics of NIR QC luminescence [42]. In a nutshell, Tb^{3+} ions are pushed from the $^7F_6$ level to the $^5D_4$ level under the excitation of 485 nm. An excited Tb^{3+} ion then cooperatively transfers energy to two nearby Yb^{3+} ions, causing two Yb^{3+} ions to emit at 975 nm due to $^2F_{5/2} \rightarrow {}^2F_{7/2}$ transitions. Similar to this, Tb^{3+} ions are first driven to the $^5D_3$ level when stimulated at 374 nm. Following a

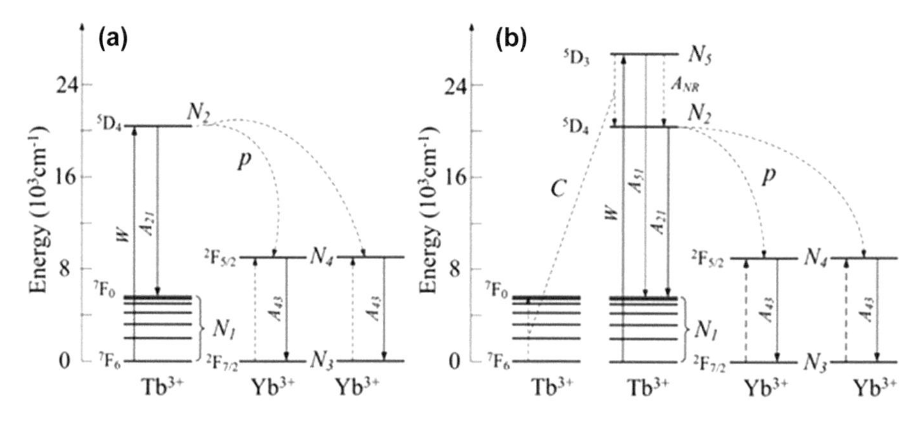

FIGURE 4.3 Schematic energy level diagrams of NIR QC luminescence of KYF_4: Tb^{3+},Yb^{3+} phosphors excited at (a) λ_{ex} = 485 nm and (b) λ_{ex} = 374 nm. (Reproduced with permission from Ref. [42], © 2021 Elsevier B.V.)

non-radiative transition ($^5D_3$ to $^5D_4$) and cross-relaxation ($^5D_3$ + $^7F_6$ to $^5D_4$ + $^7F_0$), Tb^{3+} ions at the $^5D_3$ level then relax to the $^5D_4$ level, where they cooperatively transfer energy to Yb^{3+} ions, which emit at 975 nm. Furthermore, the dynamic study of the NIR QC luminescence model ignores the reversible transfer from Yb^{3+} to Tb^{3+}. The population densities of varying levels of Tb^{3+} ions and Yb^{3+} ions are represented by the symbol N_i where (i = 1~5). The excitation rate, designated by the symbol W, is 2.7433 $*10^{-5}$ s^{-1}. The primary significant excitation transitions, radiative transitions, non-radiative transitions, cross-relaxation, and cooperative energy transfer processes are all described by rate equation models of QC luminescence under the weak excitation condition [42].

4.4.3 Pr^{3+}-Er^{3+} Co-Activated Phosphors

The intense green emission of Er^{3+} and several of its absorption bands that coincide with the Pr^{3+} $^1S_0$ emissions make it a desirable co-dopant. From 12 K to 290 K, temperature- and time-dependent ET mechanisms in Pr^{3+} and Er^{3+} co-activated $CaAl_{12}O_{19}$ crystal have been studied. Figure 4.4 displays the emission spectra of $CaAl_{12}O_{19}$:Pr^{3+}, Er^{3+} and the excitation spectrum of the $^4S_{3/2} \rightarrow {}^4I_{15/2}$ emission of Er^{3+} under 205 nm excitation. The majority of emission lines were discovered to originate from transitions of Pr^{3+}, with the exception of the 510–560 nm region, where Pr^{3+} and Er^{3+} emissions overlapped. Evidently, emissions positioned at 253 nm, 273 nm, 400 nm, 486 nm and at the 520–550 nm regions, allocated to the transitions of $^1S_0 \rightarrow {}^3F_4$, $^1S_0 \rightarrow {}^1G_4$, $^1S_0 \rightarrow {}^1I_6$, $^3P_0 \rightarrow {}^3H_4$, and $^3P_1/^3P_0 \rightarrow {}^3H_5$ of Pr^{3+}, overlapped with the Er^{3+} excitations $^4I_{15/2} \rightarrow {}^4D_{7/2}$, $^4I_{15/2} \rightarrow {}^2H_{11/2}$, $^4I_{15/2} \rightarrow {}^2H_{9/2}$, $^4I_{15/2} \rightarrow {}^4F_{7/2}$, and $^4I_{15/2} \rightarrow {}^2H_{11/2}/^4S_{3/2}$, correspondingly [43]. The huge donor-acceptor difference, on the other hand, produces a weak interaction, which results in a lower ET rate because both Pr^{3+} and Er^{3+} ions substituted Ca^{2+} ions in the lattice.

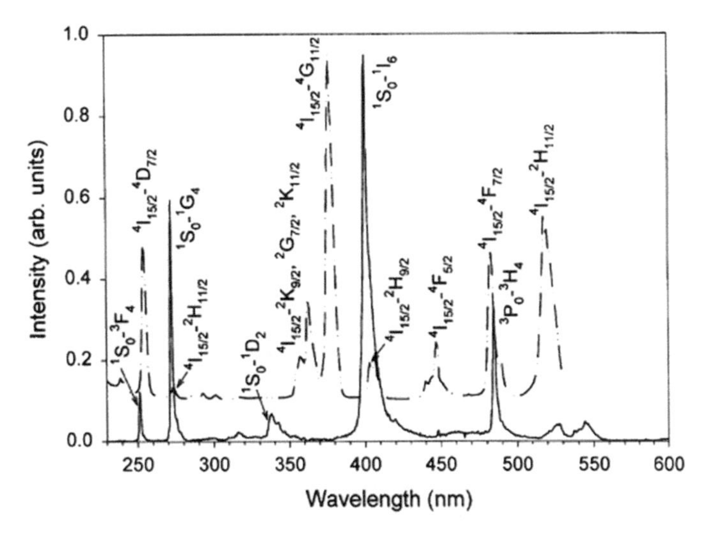

FIGURE 4.4 The excitation spectrum of $^4S_{3/2} \rightarrow {}^4I_{15/2}$ emission of Er^{3+} in $CaAl_{12}O_{19}$:Pr^{3+}, Er^{3+} and the emission spectrum of $CaAl_{12}O_{19}$:Pr^{3+}, Er^{3+} under 205 nm excitation. (Reproduced with the permission from Ref. [43], © 2001 Elsevier B.V.)

4.5 NEAR-INFRARED QUANTUM-CUTTING PHOSPHORS

Owing to their rich energy level structure and outstanding quantum efficiency of light emission, lanthanide (Ln) ions like Tb^{3+}-Yb^{3+}, Nd^{3+}-Yb^{3+}, Pr^{3+}-Yb^{3+}, and Er^{3+}/Ho^{3+}-Yb^{3+} have recently been used to achieve near-infrared (NIR) QC. Many different parameters, including such hosts with weak crystal fields, low phonon energy, high band gaps, big cation-anion distances, and large coordination numbers, affect the behavior of quantum-cutting luminous materials. Due to the possibility of visible QC by ET, whereby excited electrons of sensitizer ions transfer partial excitation energy to two or more excited electrons of activator ions, each sensitized electron produces a visible photon, special attention has been placed on the combination of two or three RE ions. QE is thereby increased through the quality control process, and theoretically, QE may increase to a maximum of 200% with optimal efficiency. Until now, a number of NIR QC phosphors has been reported so far. Some of them are discussed in the following subsections.

4.5.1 YBO_3: Ce^{3+} Yb^{3+}

Using samples with various amounts of Yb^{3+} doping, J. Chen et al. [44] published the photoluminescence (PL) and photoluminescence excitation (PLE) spectra. A prominent and broad band around 326 and 377 nm in the PLE spectra. Emission was seen around 414 nm in the Yb^{3+} $^2F_{5/2} \rightarrow {}^2F_{7/2}$ transition and ascribed to the 4f to 5d transition of Ce^{3+} (Figure 4.5(a)). A cooperative energy transfer (CET) procedure from Ce^{3+} to Yb^{3+} is theoretically feasible, as evidenced by the detection of the Ce^{3+} 4f to 5d band in the Yb^{3+} excitation spectrum.

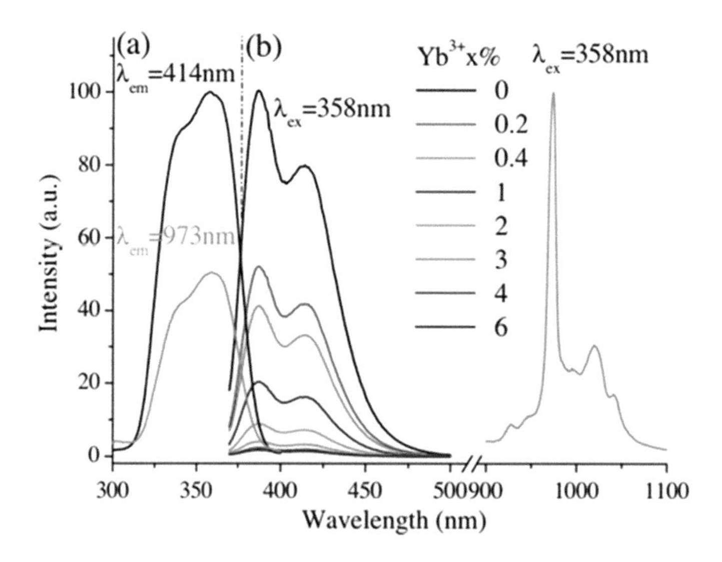

FIGURE 4.5 (a) Photoluminescence excitation spectra of YBO_3:Ce^{3+} (1%) sample (λ_{em} = 414 nm, Ce^{3+}: 5d → 4f transition) and YBO_3:Ce^{3+} (1%), Yb^{3+} (3%) sample (λ_{em} = 973 nm, Yb^{3+}: $^2F_{5/2} \rightarrow {}^2F_{7/2}$ transition); (b) Visible photoluminescence spectra of YBO_3:Ce^{3+} (1%), Yb^{3+} (x%) samples with different Yb^{3+} concentrations and NIR photoluminescence spectra of YBO_3:Ce^{3+} (1%), Yb^{3+} (3%) sample (λ_{ex} = 358 nm, Ce^{3+}: 4f → 5d transition). (Reproduced with permission from Ref. [44], © 2010 Elsevier B.V.)

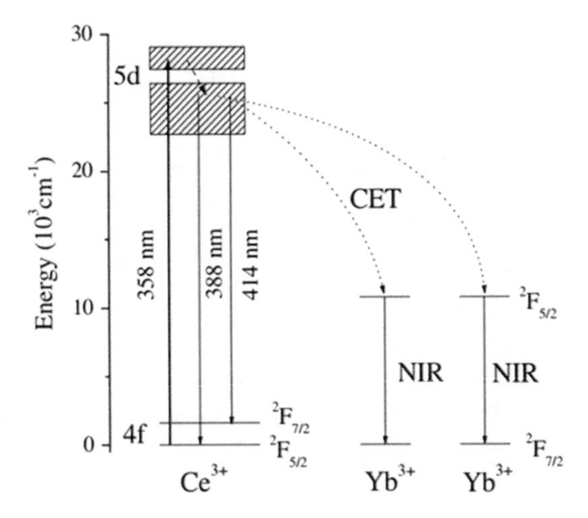

FIGURE 4.6 Schematic energy level diagram and cooperative energy transfer mechanism of Ce^{3+} and Yb^{3+} in YBO_3:Ce^{3+}, Yb^{3+}. (Reproduced with permission from Ref. [44], © 2010 Elsevier B.V.)

Simultaneous Ce^{3+} and Yb^{3+} ion emissions were produced in response to 358 nm light excitation, as depicted in Figure 4.5(b). The shift from the lower 5d level to $^2F_{5/2}$ and $^2F_{7/2}$ of Ce^{3+} has been attributed to the broad bands centering at 388 nm and 414 nm in the visible range of 370–480 nm. Ce^{3+} has a $^5d_1$ excited electrical arrangement. The nearby anion ligands in the compounds engage strongly with the 5d electron of Ce^{3+} because it is not protected from its environment. Because of this, Ce^{3+} has a broad excitation band and the host materials can modify its excitation wavelength. It is also important to point out that a definite emission with a center at 973 nm was also seen, which corresponds to the Yb^{3+} ions $^2F_{5/2}$→ $^2F_{7/2}$ transition. In order to demonstrate the ET from Ce^{3+} to Yb^{3+}, the Yb^{3+} single doped YBO_3 sample was likewise stimulated by 358 nm; nevertheless, J. Chen et al. [44] were unable to detect any peaks in the range from 900 nm to 1100 nm.

Their findings suggest that the Ce^{3+}-Yb^{3+} dual ions can emit two NIR photons for every UV photon that is absorbed when using a CET procedure. In Figure 4.6, transitions from one Ce^{3+} ion to two Yb^{3+} ions are schematically depicted at various energy levels. The transitions from the lower 5d level to $^2F_{5/2}$ and $^2F_{7/2}$ of Ce^{3+}, correspondingly, can be attributed to the Ce^{3+} emissions that follow excitation at 358 nm, which happen at 388 nm and 414 nm (shown in Figure 4.5). Owing to the $^2F_{5/2}$ → $^2F_{7/2}$ transition of Yb^{3+} ions, two NIR photons (about 973 nm) are produced from one absorbing UV photon.

4.5.2 Lu_2GeO_5:Bi^{3+}, Yb^{3+}

Figure 4.7 illustrates the CET process for Yb^{3+} NIR emission and the diagrammatic representation for energy transfer from Bi^{3+} to Yb^{3+} in Lu_2GeO_5 with transition $^2F_{5/2}$ → $^2F_{7/2}$. The electrons of Bi^{3+} are promoted from the ground state $^1S_0$ to the excited state $^3P_1$ during the excitation of UV light, and the excited energy of Bi^{3+} is more than double that of the Yb^{3+} ion. As a result, there's a CET from Bi^{3+} $^3P_1$ level to Yb^{3+} $^2F_{5/2}$ level. According to Figure 4.7, the blue-violet emission of Bi^{3+} ions result from electrons hopping from ground levels $^1S_0$ to $^3P_1$. The energy from the cooperative $^3P_1$ states of Bi^{3+} is transmitted to the $^2F_{5/2}$ levels of

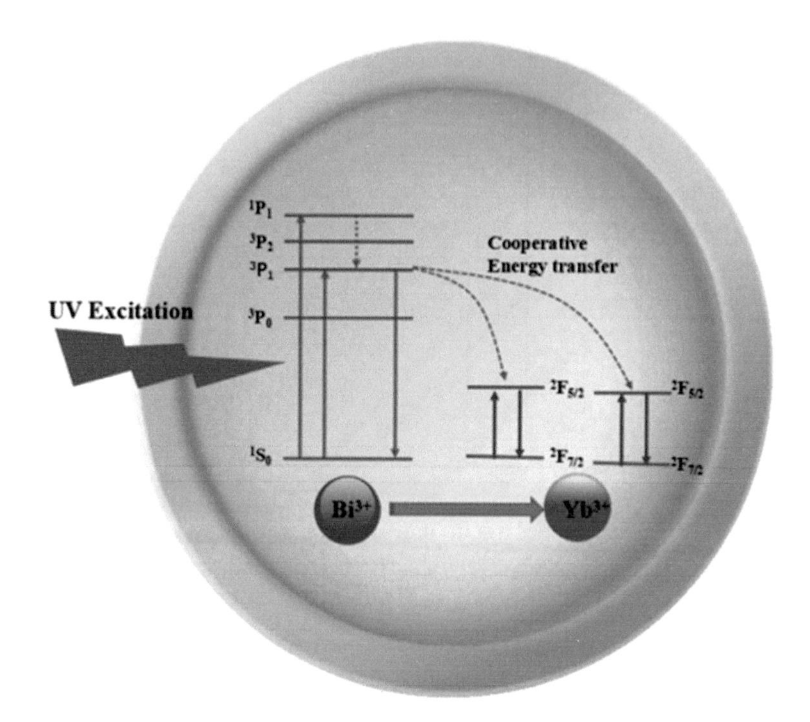

FIGURE 4.7 Schematic diagram of Bi^{3+} and Yb^{3+} ions in the Lu_2GeO_5 host, presenting the cooperative energy transfer mechanism for the NIR emission from Yb^{3+}: $^2F_{5/2} \rightarrow ^2F_{7/2}$ upon direct excitation into the $^3P_1$ level of Bi^{3+} by UV light. (Reproduced with the permission from Ref. [45], © 2019 Elsevier B.V.)

two Yb^{3+} ions while the $^3P_1$ states of Bi^{3+} are in resonance with the $^2F_{5/2}$ levels of Yb^{3+} ions. One UV photon splits into two NIR photons as a result of the Yb^{3+} ions two $^2F_{5/2} \rightarrow ^2F_{7/2}$ transitions, according to the results of the quantum cutting process [45].

4.5.3 $NaBaPO_4$: Bi^{3+}, Er^{3+}

One of the best ways to increase QC luminescence is by doping Bi^{3+}. Bi^{3+} ion was, thus, doped into the $NaBaPO_4$: Er^{3+} phosphors. Figure 4.8(a) displays the NIR emission spectra of $NaBaPO_4$: 2.5% Er^{3+} co-doped with various concentrations of Bi^{3+} under 377 nm excitation. NIR emission intensity increases till Bi^{3+} concentration reaches 4%, after which it starts to decline. It is wonderful to see that when Bi^{3+} was doped, the NIR emission intensity was significantly increased at 377 nm excitation. The ratio of $NaBaPO_4$:Bi^{3+}, Er^{3+}, and $NaBaPO_4$: Er^{3+} NIR emission intensities is used here to establish the emission enhancement factor. The maximal NIR emission enhancement factor is 10.64. Figure 4.8(c) displays the NIR emission spectra and excitation spectra of $NaBaPO_4$ with 1% Er^{3+} co-doped with various concentrations of Bi^{3+} under 377 nm excitation. With varying doses of Bi^{3+}, $NaBaPO_4$:1% Er^{3+} also exhibits a significant boost of NIR emission. The maximal NIR emission enhancement factor is 8.79.

Figure 4.8(b) and Figure 4.8(d) depict equivalent excitation spectra of $NaBaPO_4$:Er^{3+} co-doped with various concentrations of Bi^{3+}. Following is a corresponding mechanism: Bi^{3+} ion is excited to $^3P_1$ under 377 nm excitation. Er^{3+} is excited to interact with $^4G_{11/2}$ at the

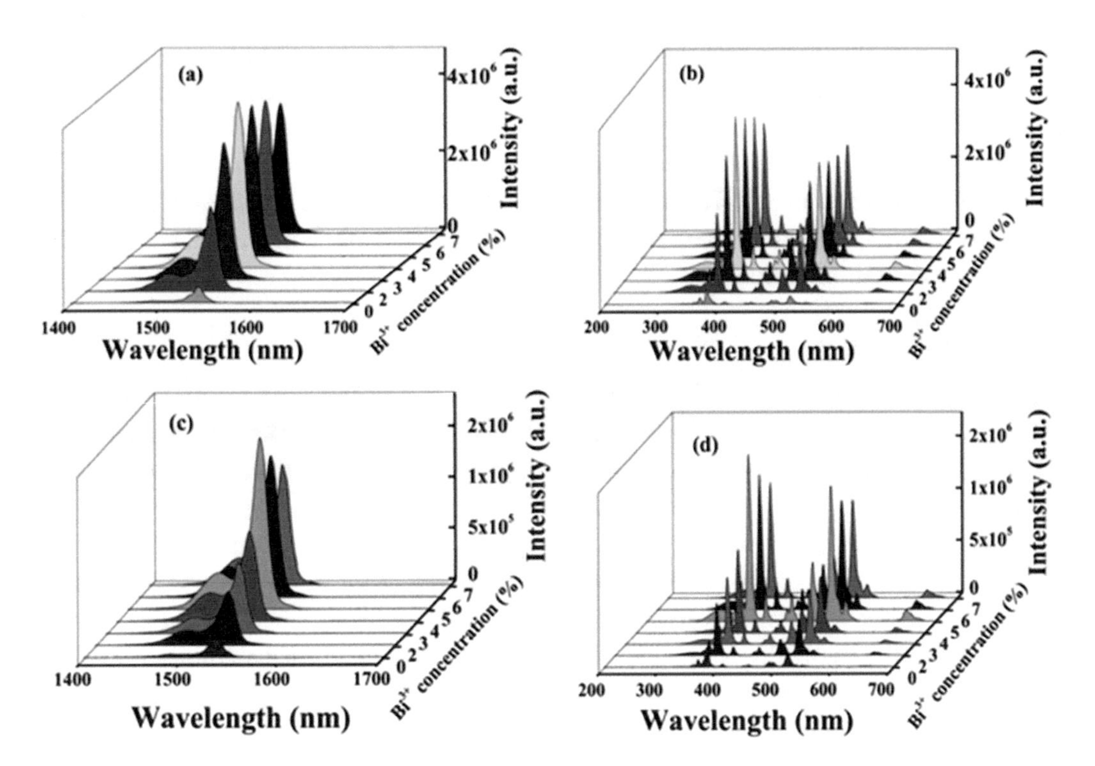

FIGURE 4.8 (a) NIR emission spectra of $NaBaPO_4$: x% Bi^{3+}, 2.5% Er^{3+} (x = 0, 2, 3, 4, 5, 6, 7) (λ_{ex} = 377 nm); (b) excitation spectra of $NaBaPO_4$: x% Bi^{3+}, 2.5% Er^{3+} (x = 0, 2, 3, 4, 5, 6, 7) (λ_{em} = 1534 nm); (c) NIR emission spectra of $NaBaPO_4$: x% Bi^{3+}, 1.0% Er^{3+} (x = 0, 2, 3, 4, 5, 6, 7) (λ_{ex} = 377 nm); (d) excitation spectra of $NaBaPO_4$: x% Bi^{3+}, 1.0% Er^{3+} (x = 0, 2, 3, 4, 5, 6, 7) (λ_{em} = 1534 nm). (Reproduced with permission from Ref. [46], © 2019 Elsevier B.V.)

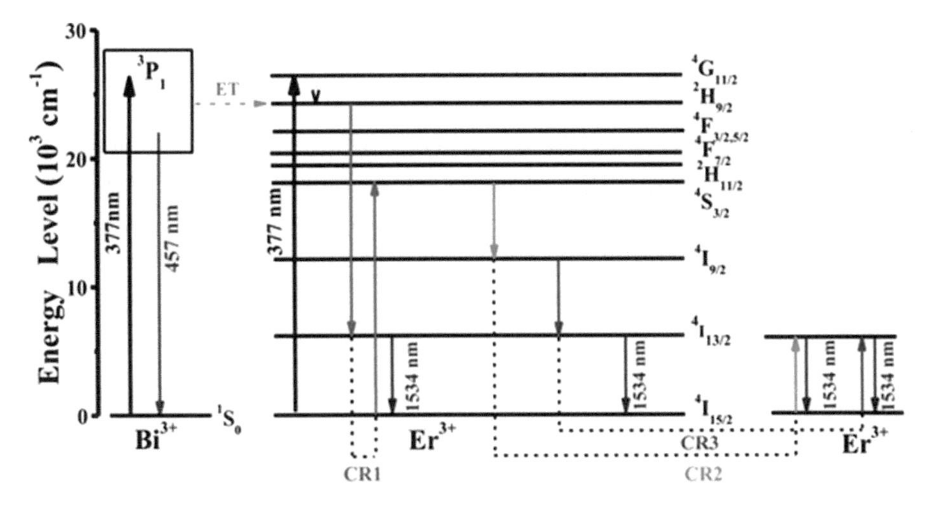

FIGURE 4.9 Schematic QC mechanisms of $NaBaPO_4$: Er^{3+} and $NaBaPO_4$: Bi^{3+}, Er^{3+}. (Reproduced with the permission from Ref. [46], © 2019 Elsevier B.V.)

same moment. Due to the fact that the broad band emission band of Bi^{3+} ions match the excitation peaks of Er^{3+} NIR emission nicely. The ET procedure from Bi^{3+} to Er^{3+} takes place, which is beneficial for Er^{3+} NIR emission. Figure 4.9 depicts the $NaBaPO_4$: Bi^{3+} and Er^{3+} conceptual QC method. It is important to note that ET is not the primary source of the significant amplification of NIR emission for two reasons: first, the broad excitation band's intensity is low, and second, there is a considerable increase in the intensity of each of Er^{3+} distinctive excitation peaks.

4.6 CONCLUSION

Based on the idea that a QC phosphor can generate two visible photons for every VUV photon absorbed, QC materials have an efficient gain. In order to achieve the requisite red-shift of the absorbed radiation without sacrificing energy efficiency, the excitation energy is split between the two photons. Research into QC systems has begun with regard to single ions like Pr^{3+}, Tm^{3+}, Gd^{3+}, and Er^{3+} that are capable of cascade emission. The emphasis is currently on the pairing of two ions, in which the energy of the donor ion might be incrementally transferred to two acceptor ions through a DC process. Since that it has become generally known, many RE-based phosphors, particularly those based on fluoride, contain QC via DC. Fluoride compounds having trivalent RE ions may offer greater opportunities for the creation of innovative optical materials that are appropriate for a variety of applications with VUV radiation as the excitation source due to their wide-band gap and low photon energy. Nonetheless, a thorough examination of oxide-based QC phosphors is required due to the possibility of their use in numerous high-performing displays and devices.

However, QC materials could also be used in solar cells. The idea of NIR QC, which involves emitting two NIR photons in exchange for an absorbed visible photon, has been put out for this application. Energy loss resulting from the thermalization of electron-hole pairs could be considerably decreased using NIR QC. With the energy of their emission just above silicon's band gap, QC phosphors could be extremely useful in the construction of solar cells. In order to produce high efficiency Si solar cells, RE^{3+}-Yb^{3+} dual-ions driven QC phosphors have the ability to convert the green to UV portion of the solar spectrum to 1000 nm photons with almost twice as many photons.

Several more original ideas and fresher materials are still in the research phase. In the ensuing decades, some of them might result in significantly greater efficiency and cheaper costs. Reaching the objective with effective QC phosphors still requires a lot of work. Nonetheless, given the emergence of new and exciting materials and concepts, it has been determined that the prospects of achieving this goal are satisfactory and even more intriguing.

REFERENCES

[1] Z. Liu, J. Li, L. Yang, Q. Chen, Y. Chu, N. Dai, Efficient near-infrared quantum cutting in Ce^{3+}-Yb^{3+} codoped glass for solar photovoltaic, *Sol. Energy Mater. Sol. Cells.* 122 (2014) 46–50. https://doi.org/10.1016/j.solmat.2013.10.030

[2] J. Chen, H. Zhang, F. Li, H. Guo, High efficient near-infrared quantum cutting in Ce^{3+},Yb^{3+} co-doped $LuBO_3$ phosphors, *Mater. Chem. Phys.* 128 (2011) 191–194. https://doi.org/10.1016/j.matchemphys.2011.02.057

[3] F. Zhang, J. Xie, G. Li, W. Zhang, Y. Wang, Y. Huang, Y. Tao, Cation composition sensitive visible quantum cutting behavior of high efficiency green phosphors $Ca_9Ln(PO_4)_7$:Tb^{3+} (Ln = Y, La, Gd), *J. Mater. Chem. C.* 5 (2017) 872–881. https://doi.org/10.1039/c6tc03982k

[4] D. Abeysinghe, M.D. Smith, J. Yeon, G. Morrison, H.-C. zur Loye, Observation of multiple crystal-to-crystal transitions in a new reduced vanadium oxalate hybrid material, $Ba_3[(VO)_2(C_2O_4)_5(H_2O)_6]\cdot(H_2O)_3$, prepared via a mild, two-step hydrothermal method, *Cryst. Growth Des.* 14 (2014) 4749–4758. https://doi.org/10.1021/cg500888u

[5] X. Chen, X. Liu, Y. Feng, X. Li, H. Chen, T. Xie, H. Kou, R. Kučerková, A. Beitlerová, E. Mihóková, M. Nikl, J. Li, Microstructure evolution in two-step-sintering process toward transparent Ce:$(Y,Gd)_3(Ga,Al)_5O_{12}$ scintillation ceramics, *J. Alloys Compd.* 846 (2020). https://doi.org/10.1016/j.jallcom.2020.156377

[6] Q.Y. Zhang, X.Y. Huang, Recent progress in quantum cutting phosphors, *Prog. Mater. Sci.* 55 (2010) 353–427. https://doi.org/10.1016/j.pmatsci.2009.10.001

[7] N. Soudi, S. Nanayakkara, N.M.S. Jahed, S. Naahidi, Rise of nature-inspired solar photovoltaic energy convertors, *Sol. Energy.* 208 (2020) 31–45. https://doi.org/10.1016/j.solener.2020.07.048

[8] Suman, P. Sharma, P. Goyal, Evolution of PV technology from conventional to nano-materials, *Mater. Today Proc.* 28 (2020) 1593–1597. https://doi.org/10.1016/j.matpr.2020.04.846

[9] B. Roose, E.M. Tennyson, G. Meheretu, A. Kassaw, S.A. Tilahun, L. Allen, S.D. Stranks, Local manufacturing of perovskite solar cells, a game-changer for low- and lower-middle income countries?, *Energy Environ. Sci.* 15 (2022) 3571–3582. https://doi.org/10.1039/d2ee01343f

[10] H.J. Snaith, Perovskites: The emergence of a new era for low-cost, high-efficiency solar cells, *J. Phys. Chem. Lett.* 4 (2013) 3623–3630. https://doi.org/10.1021/jz4020162

[11] G. Zhang, Q. Cui, G. Liu, Efficient near-infrared quantum cutting and downshift in Ce^{3+}-Pr^{3+} codoped $SrLaGa_3S_6O$ suitable for solar spectral converter, *Opt. Mater. (Amst).* 53 (2016) 214–217. https://doi.org/10.1016/j.optmat.2016.01.042

[12] A. Dwivedi, K. Mishra, S.B. Rai, Multi-modal luminescence properties of RE^{3+} (Tm^{3+}, Yb^{3+}) and Bi^{3+} activated $GdNbO_4$ phosphors - Upconversion, downshifting and quantum cutting for spectral conversion, *J. Phys. D. Appl. Phys.* 48 (2015) 435103. https://doi.org/10.1088/0022-3727/48/43/435103

[13] J. Xu, A. Gulzar, P. Yang, H. Bi, D. Yang, S. Gai, F. He, J. Lin, B. Xing, D. Jin, Recent advances in near-infrared emitting lanthanide-doped nanoconstructs: Mechanism, design and application for bioimaging, *Coord. Chem. Rev.* 381 (2019) 104–134. https://doi.org/10.1016/j.ccr.2018.11.014

[14] J.K. Swabeck, S. Fischer, N.D. Bronstein, A.P. Alivisatos, Broadband sensitization of lanthanide emission with indium phosphide quantum dots for visible to near-infrared downshifting, *J. Am. Chem. Soc.* 140 (2018) 9120–9126. https://doi.org/10.1021/jacs.8b02612

[15] P. Babu, I.R. Martín, V. Lavín, U.R. Rodríguez-Mendoza, H.J. Seo, K.V. Krishanaiah, V. Venkatramu, Quantum cutting and near-infrared emissions in Ho^{3+}/Yb^{3+} codoped transparent glass-ceramics, *J. Lumin.* 226 (2020) 117424. https://doi.org/10.1016/j.jlumin.2020.117424

[16] J.C.G. Bünzli, S.V. Eliseeva, Lanthanide NIR luminescence for telecommunications, bioanalyses and solar energy conversion, *J. Rare Earths.* 28 (2010) 824–842. https://doi.org/10.1016/S1002-0721(09)60208-8

[17] A.K. Singh, S.K. Singh, S.B. Rai, Role of Li^{+} ion in the luminescence enhancement of lanthanide ions: Favorable modifications in host matrices, *RSC Adv.* 4 (2014) 27039–27061. https://doi.org/10.1039/c4ra01055h

[18] C. Lorbeer, F. Behrends, J. Cybinska, H. Eckert, A.V. Mudring, Charge compensation in RE^{3+} (RE = Eu, Gd) and M^{+} (M = Li, Na, K) co-doped alkaline earth nanofluorides obtained by microwave reaction with reactive ionic liquids leading to improved optical properties, *J. Mater. Chem. C.* 2 (2014) 9439–9450. https://doi.org/10.1039/c4tc01214c

[19] L. Li, X. Wei, Y. Chen, C. Guo, M. Yin, Energy transfer in Tb^{3+},Yb^{3+} codoped Lu_2O_3 near-infrared downconversion nanophosphors, *J. Rare Earths.* 30 (2012) 197–201. https://doi.org/10.1016/S1002-0721(12)60022-2

[20] F. Enrichi, E. Cattaruzza, P. Riello, G.C. Righini, A. Vomiero, Ag-sensitized Tb^{3+}/Yb^{3+} codoped silica-zirconia glasses and glass-ceramics: Systematic and detailed investigation of the broadband energy-transfer and downconversion processes, *Ceram. Int.* 47 (2021) 17939–17949. https://doi.org/10.1016/j.ceramint.2021.03.107

[21] P. Vergeer, T.J.H. Vlugt, M.H.F. Kox, M.I. Den Hertog, J.P.J.M. Van Der Herden, A. Meijerink, Quantum cutting by cooperative energy transfer in $Yb_xY1\text{-}_xPO_4$:Tb^{3+}, *Phys. Rev. B - Condens. Matter Mater. Phys.* 71 (2005) 1–11. https://doi.org/10.1103/PhysRevB.71.014119

[22] A. Bahadur, R.S. Yadav, R.V. Yadav, S.B. Rai, Multimodal emissions from Tb^{3+}/Yb^{3+} co-doped lithium borate glass: Upconversion, downshifting and quantum cutting, *J. Solid State Chem.* 246 (2017) 81–86. https://doi.org/10.1016/j.jssc.2016.11.004

[23] F. Zhao, Y. Liang, J.B. Lee, S.J. Hwang, Applications of rare earth Tb^{3+}-Yb^{3+} co-doped down-conversion materials for solar cells, *Mater. Sci. Eng. B.* 248 (2019) 114404. https://doi.org/10.1016/j.mseb.2019.114404

[24] S. Ye, B. Zhu, J. Chen, J. Luo, J.R. Qiu, Infrared quantum cutting in Tb^{3+}, Yb^{3+} codoped transparent glass ceramics containing CaF_2 nanocrystals, *Appl. Phys. Lett.* 92 (2008) 2011–2014. https://doi.org/10.1063/1.2907496

[25] K. Mishra, S.K. Singh, A.K. Singh, M. Rai, B.K. Gupta, S.B. Rai, New perspective in garnet phosphor: Low temperature synthesis, nanostructures, and observation of multimodal luminescence, *Inorg. Chem.* 53 (2014) 9561–9569. https://doi.org/10.1021/ic500854k

[26] J.L. Yuan, X.Y. Zeng, J.T. Zhao, Z.J. Zhang, H.H. Chen, X.X. Yang, Energy transfer mechanisms in Tb^{3+}, Yb^{3+} codoped Y_2O_3 downconversion phosphor, *J. Phys. D. Appl. Phys.* 41 (2008). https://doi.org/10.1088/0022-3727/41/10/105406

[27] G. Alombert-Goget, C. Armellini, S. Berneschi, A. Chiappini, A. Chiasera, M. Ferrari, S. Guddala, E. Moser, S. Pelli, D.N. Rao, G.C. Righini, Tb^{3+}/Yb^{3+} co-activated Silica-Hafnia glass ceramic waveguides, *Opt. Mater. (Amst).* 33 (2010) 227–230. https://doi.org/10.1016/j.optmat.2010.09.030

[28] M. Singh, P.D. Sahare, Nuclear instruments and methods in Physics Research B Redox reactions in Cu-activated nanocrystalline LiF TLD phosphor, *Nucl. Inst. Methods Phys. Res. B.* 289 (2012) 59–67. https://doi.org/10.1016/j.nimb.2012.08.003

[29] A.R. Kadam, S.J. Dhoble, Synthesis and luminescence study of Eu^{3+}-doped $SrYAl_3O_7$ phosphor, *Luminescence.* 34 (2019) 846–853. https://doi.org/10.1002/bio.3681

[30] J. Malleshappa, H. Nagabhushana, B.D. Prasad, S.C. Sharma, Y.S. Vidya, K.S. Anantharaju, Structural, photoluminescence and thermoluminescence properties of CeO_2 nanoparticles, *Optik - Int. J. Light Electron Opt.* (2015). https://doi.org/10.1016/j.ijleo.2015.10.114

[31] V. Ramón, O. Barrón, F. María, E. Ochoa, C. Cruz, R. Bernal, Thermoluminescence of novel MgO – CeO_2 obtained by a glycine-based solution combustion method, *Appl. Radiat. Isot.* (2016) 1–5. https://doi.org/10.1016/j.apradiso.2016.02.002

[32] G. Sahu, A.S. Gour, R.K. Chandrakar, Thermoluminescence and optical properties of Dy^{3+} doped MgO nanoparticles, prepared by solution combustion synthesis method, *J. Pure Appl. Ind. Phys.* 7 (2017) 115–127.

[33] N.J. Shivaramu, B.N. Lakshminarasappa, K.R. Nagabhushana, F. Singh, Synthesis characterization and luminescence studies of gamma irradiated nanocrystalline yttrium oxide, Spectrochim. *Acta Part A Mol. Biomol. Spectrosc.* 154 (2016) 220–231.

[34] J.J. Eilers, D. Biner, J.T. van Wijngaarden, K. K. Kramer, H.U. Gudel, A. Meijerink, Efficient visible to infrared quantum cutting through downconversion with the Er^{3+}–Yb^{3+} couple in $C_s3Y_2Br_9$, *Appl. Phys. Lett.* 96 (2010) 151106. https://doi.org/10.1063/1.3377909

[35] P.S. Peijzel, A. Meijerink, Visible photon cascade emission from the high energy levels of Er^{3+}, *Chem. Phys. Lett.* 401 (2005) 241–245. https://doi.org/10.1016/j.cplett.2004.11.049

[36] Y.Z. Wang, D.C. Yu, H.H. Lin, S. Ye, M.Y. Peng, Q.Y. Zhang, Broadband three-photon near-infrared quantum cutting in Tm^{3+} singly doped YVO_4, *J. Appl. Phys.* 114 (2013). https://doi.org/10.1063/1.4836897

[37] D. Yu, T. Yu, A.J. van Bunningen, Q. Zhang, A. Meijerink, F.T. Rabouw, Understanding and tuning blue-to-near-infrared photon cutting by the Tm^{3+}/Yb^{3+} couple, *Light Sci. Appl.* 9 (2020). https://doi.org/10.1038/s41377-020-00346-z

[38] X. Chen, G.J. Salamo, S. Li, J. Wang, Y. Guo, Y. Gao, L. He, H. Ma, J. Tao, P. Sun, W. Lin, Q. Liu, Two-photon, three-photon, and four-photon excellent near-infrared quantum cutting luminescence of Tm^{3+} ion activator emerged in Tm^{3+}:$YNbO_4$ powder phosphor one material simultaneously, *Phys. B Condens. Matter.* 479 (2015) 159–164. https://doi.org/10.1016/j.physb.2015.10.009

[39] J.A. Skarulis, P. Kissinger, The systems M_2SiF_6-$(NH_4)_{2SiF6}$-H_2O at 25° (M = Na, K, Rb, Cs), *J. Phys. Chem.* 70 (1966) 186–192. https://doi.org/10.1021/j100873a030

[40] G. Blasse, H.S. Kiliaan, A.J. De Vries, A study of the energy transfer processes in sensitized gadolinium phosphors, *J. Less-Common Met.* 126 (1986) 139–146. https://doi.org/10.1016/0022-5088(86)90272-9

[41] P.S. Jaiswal, M. Ubale, National Conference on Multidisciplinary Research in Science and Technology for Healthy Lifestyle Management (NCMRST- 2020) Organizer:- Shri R.L.T. College of Science, Akola 24, *Innov. Libr. Auetom. Inf. Sci.* 293 (2020) 473–476. https://www.rltsc.edu.in/wp-content/uploads/2022/07/27-Paper-Mr.-Ubale-2019-20.pdf

[42] B. Zheng, J. Hong, B. Chen, Y. Chen, R. Lin, C. Huang, C. Zhang, J. Wang, L. Lin, Z. Zheng, Quantum cutting properties in KYF_4:Tb^{3+}, Yb^{3+} phosphors: Judd-Ofelt analysis, rate equation models and dynamic processes, *Results Phys.* 28 (2021) 104595. https://doi.org/10.1016/j.rinp.2021.104595

[43] X.J. Wang, S. Huang, L. Lu, W.M. Yen, A.M. Srivastava, A.A. Setlur, Energy transfer in Pr^{3+}- and Er^{3+} -codoped CaAl12O19 crystal, *Opt. Commun.* 195 (2001) 405–410. https://doi.org/10.1016/S0030-4018(01)01344-X

[44] J.D. Chen, H. Guo, Z.Q. Li, H. Zhang, Y.X. Zhuang, Near-infrared quantum cutting in Ce^{3+}, Yb^{3+} co-doped YBO_3 phosphors by cooperative energy transfer, *Opt. Mater. (Amst).* 32 (2010) 998–1001. https://doi.org/10.1016/j.optmat.2010.01.040

[45] H. Luo, S. Zhang, Z. Mu, F. Wu, Z. Nie, D. Zhu, X. Feng, Q. Zhang, Near-infrared quantum cutting via energy transfer in Bi^{3+}, Yb^{3+} co-doped Lu_2GeO_5 down-converting phosphor, *J. Alloys Compd.* 784 (2019) 611–619. https://doi.org/10.1016/j.jallcom.2019.01.060

[46] J. Hong, L. Lin, X. Li, J. Xie, Q. Qin, Z. Zheng, Enhancement of near-infrared quantum-cutting luminescence in $NaBaPO_4$: Er^{3+} phosphors by Bi^{3+}, *Opt. Mater. (Amst).* 98 (2019) 109471. https://doi.org/10.1016/j.optmat.2019.109471

CHAPTER 5

Recent Developments in Rare-Earth Doped Phosphors for Eco-Friendly and Energy-Saving Lighting Applications

Yatish R. Parauha

Shri Ramdeobaba College of Engineering and Management, Nagpur, India

Marta Michalska-Domanska

Military University of Technology, Warszawa, Poland

Sanjay J. Dhoble

Rashtrasant Tukadoji Maharaj Nagpur University, Nagpur, India

5.1 ENERGY-SAVING LIGHTING SYSTEMS

Energy is a key indicator of a country's success. Energy has been essential for the ongoing progress of human civilization. As global energy consumption has expanded since the start of the Industrial Revolution nearly 200 years ago, the improvement of human living standards has accelerated, particularly in the world's industrialized countries. According to estimates, 20% of the world's total electricity production goes into lighting [1, 2]. Our health, comfort, safety, and productivity are all affected by the type and amount of light in our immediate environment. Lighting accounts for a significant portion of the electricity budget in many countries. Energy consumption in the form of lighting accounts for more than 35% depending on the activities that take place in buildings [3]. Lighting management is one of the most popular energy-saving strategies because it is one of the simplest ways to achieve significant energy savings for very little investment. Standards, rules, and building codes control lighting systems for commercial buildings. Lighting must, in addition to

DOI: 10.1201/9781003315261-7

being necessary, be suitable for its intended use and comply with occupational health and safety regulations. In many situations, workplace lighting is excessive, and there is significant opportunity for ideal energy reduction. Devices with obsolete lighting and inefficient luminaires can be replaced with high-performance/low-consumption alternatives, and higher quality light can be achieved. These strategies are particularly suited to locations where illumination is required continuously or for extended periods of time and savings cannot be realized simply by turning off the lights. The time it takes to pay back the investment varies from situation to situation, although projects typically require around two years. The industrialized world consumes much more energy per capita than the developing world and there is a finite amount of energy resources available. Therefore, it is imperative to find new energy sources and use the currently available energy sources wisely, so that they can be sustained for a long time. Energy conservation is the right step towards energy saving. In short, energy conservation is an effort towards reducing the amount of energy used. This can be done using less energy while achieving the same results, which can be done through efficient energy use or using low-energy services. Energy saving can promote financial capital, environmental value, national security, personal safety, and people's comfort in the future [4]. Energy conservation is desired by people and businesses that use energy directly both to reduce energy prices and advance economic security. Users in the industrial and commercial sectors want to boost productivity and maximize profits as a result. In reality, the industrial sector is where most of the energy is consumed, thus, there is a huge opportunity for business to implement energy-saving practices. Energy efficiency has become even more important now, as it is the most reliable and cost-effective way to mitigate global climate change. Nations have made a steady shift to energy-efficient lighting as the most cost-effective and reliable means of energy-saving. Lighting devices have improved, and new technologies have been developed to reduce energy consumption while improving the quality of light.

Energy efficiency is achieved when a product's energy consumption is minimized without affecting its output, final response, or the user's level of comfort. Compared to the same product with higher energy consumption, an energy-efficient product uses less energy to accomplish the same function. The energy efficiency of a lighting area provides the required illumination level of a lighting scheme for the purpose it is designed for, using the least amount of energy possible. Simply put, energy-efficient lighting can save electricity while retaining high light quality and quantity. In energy-efficient lighting, traditional lamps (such as incandescent lamps) can be replaced with energy efficient bulbs such as fluorescent lamps, compact fluorescent (CF) lamps, and light-emitting diode (LED) lamps. Depending on the lighting requirements, type and age, more energy efficient lighting may be an option. For example, new fluorescent lighting is easily accessible, but replacing the lighting also requires replacing the ballasts. The development of LED technology has great potential, especially for smart control, which has only recently been introduced. To meet energy conservation goals in the lighting industry, LEDs are considered a sustainable alternative solution. It is the first lighting technology that offers excellent energy efficiency and smart-management capability, making it ideal for a variety of sectors (residential, service sector buildings, infrastructure, etc.) [3]. LED light luminaires provide considerable energy efficiency advantages over high-pressure sodium lamps, fluorescent lighting, and other light sources.

5.1.1 History of Lighting

The history of the light-emitting diode (LED) goes back more than a century. Light revolutionized human society, illuminating the evening, making life comfortable, and electrifying the whole world. Thomas Alva Edison developed the first electric light bulb in 1879 [5]. However, his incandescent light was made from carbonized sewing thread, while commercial incandescent lamps were made from carbonized bamboo fiber. These lamps had an efficiency of about 1.4 lm/W and ran at 60 W for 100 hours. With additional improvements, the performance was increased to 120 V, 60 W, 15 lm/W efficiency, and an average lifespan of 1,000 hours. American Peter Cooper Hewitt received a patent for the first mercury vapor discharge lamp in 1901. It was the forerunner of today's fluorescent lamp. George Inman, a General Electric employee, enhanced Hewitt's design and increased fluorescent light efficiency to between 65 and 100 lm/W. However, it depended on the wattage and the type of bulb. Over the past 60–120 years, both incandescent and fluorescent light sources have increased in performance. H. J. Round reported the first research on the light-emitting diode (LED) in 1907 after studying the light emission from a silicon carbide junction diode. Losev observed emission independently from ZnO and SiC diodes, as documented in 1927. The potential of the technology was not appreciated at the time, and the inventions were mostly ignored until 1962 [5–7]. Nick Holonyak was the first to develop an efficient visible-spectrum LED. In subsequent decades, LEDs became widely employed in numerical display and signaling applications. However, it was not until 1995 or so that high brightness and blue LEDs could be produced, or LEDs could be used for general lighting. The progress of lighting technology is shown in Figure 5.1.

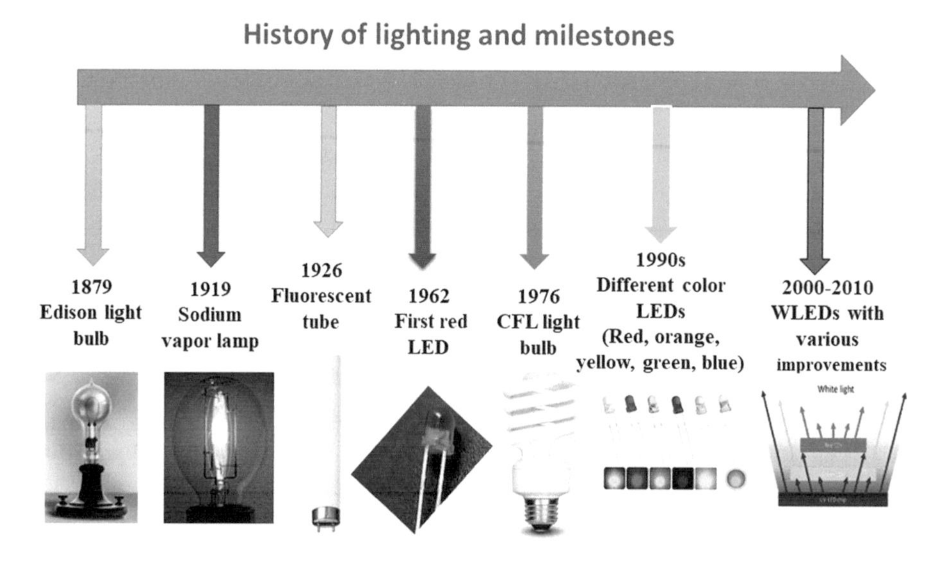

FIGURE 5.1 Progress of lighting technology.

(Source: Author.)

5.2 PHOSPHOR-CONVERTED WHITE LIGHT EMITTING DIODES (PC-WLEDS)

Phosphor-Converted White Light-Emitting Diodes (pc-WLEDs) have gained more popularity as market leaders in smart-lighting technologies in economically developed countries due to their amazing features such as long-term stability, simple circuitry, high efficiency, eco-friendly behavior, high quantum efficiency, and so on. However, this dominance has only just come into existence. Prior to this, various lighting techniques were used over the past 20 years. During the last 40 years incandescent light bulbs have been replaced by effective discharge lamps, which use the glow of an excited gas followed by photoluminescence of a phosphor to convert light. As a result, with the exception of specialized applications, the lighting market in the preceding decade was dominated by two discharge devices, the fluorescent tube and the high-pressure sodium lamp used for indoor and outdoor lighting, respectively. The lighting industry and our daily lives have seen a significant changes due to the recent development of energy-efficient LED bulbs. The pc-LEDs have already been important in a wide range of applications, including indoor and outdoor lighting, automobile lighting, plant illumination, medical equipment, and future visible light communication applications. The pc-WLEDs are widely regarded as the most influential new generation solid-state light sources among a wide variety of applications. Many factors are taken into account to create a highly efficient WLED device, including the luminaire, phosphor-conversion components, semiconductor chips (indium gallium nitride (InGaN) semiconductor chips), and packaging techniques. Phosphors are essential components of pc-WLED devices. One of the most important and urgent tasks that can be solved by leading-edge science and technology is the discovery and development of pc-WLED phosphors, which can be used as lighting and display backlight sources. Generally, prepared phosphors are connected with blue/near ultraviolet chip for the generation of white light. There are two basic tactics employed in this process [8–12]:

(i) In the first combination, yellow-emitting cerium doped yttrium aluminate garnet (YAG:Ce^{3+}) phosphor is combined with InGaN blue LED chip (460 nm). This method is the most popular and widely used method and is still in use. However, fabricated LEDs using this technique represent some serious drawbacks like deficiency of red color component, low color rendering index (R_a = 70–80), and high correlated color temperature (CCT = 4000–7500 K). To solve these problems, researchers and scholars are trying to combine the combination of red/green-emitting phosphor and yellow-emitting phosphor with InGaN blue LED chip [8–12].

(ii) In the second approach, a combination of three-color-emitting (red, green, blue) single-phased phosphors is coated with a near UV LED chip (380–420 nm). This approach also has some common problems such as low efficiency as a result of reabsorption, as well as sedimentation and uniform phosphor dispersion in the silicone resin. The exceptionally high color rendering index (R_a > 90), the wide color gamut coverage, and the consistent light color output at various driving currents are some amazing advantages of this technique [8–12].

However, previous studies and literature surveys have pointed out that researchers and scientists are continuously focusing on the development of new phosphor materials with high efficiency and tunable emission because phosphor is the essential component for WLEDs lighting technology. So far, a number of research reviews, book chapters, and books have been published on the topic of phosphors. This chapter intended to provide a comprehensive discussion of aspects of WLEDs and a summarized review of rare-earth doped phosphors that show potential for WLEDs applications.

5.3 RARE-EARTH DOPED PHOSPHORS

Generally, phosphors are white colored solid materials, which exhibit the phenomena of luminescence. They are light-emitting substances that are composed from host material and very small amounts of rare-earth ions or transition metal ions. Through electronic transitions, these materials may convert higher-frequency light to lower-frequency light. Research on luminescence materials has proved to be a milestone in lighting technology or, in other words, luminescent materials are of paramount importance to the development of solid-state lighting technologies. Due to growing worldwide energy consumption, rare-earth doped phosphors are extremely advantageous and essential for clean and energy-saving technologies. These materials are needed to enhance the efficiency of light sources with low voltage consumption. The rare-earth element plays an important role for tuning the various important properties – optical, magnetic, electrical, and electronic – when it dopes with inorganic host materials. It is doped in a very small quantity so that it does not disturb the internal structure of the lattice. Recently, there are different rare-earth activated host materials that have been studied such as oxides, aluminates, silicates, tungstate, vanadate, phosphate, borates, and titanates and they find potential applications in LED and display devices [13]. Rare-earth activated phosphors have piqued the interest of researchers due to significant advances in lumen output, energy efficiency, thermal and chemical stability, and color rendering index. In general, trivalent rare-earth ions emit line emissions attributable to 4f electrons that are suitably shielded from disturbances caused by the host matrix's environment. As a result, rare-earth activated phosphors have a high luminous quantum yield when compared to other phosphors, with quenching occurring only at higher temperatures or activator concentrations. Rare-earth doped phosphors find applications in various fields such as solid-state lighting, field emission displays (FEDs), biomedical applications, luminescent paints, high and low dose dosimetric applications, 3D display technology, interior decoration, solar energy, fingerprint detection, phototherapy, safety indications, and plant growth, among others [7, 12, 14–17]. These phosphors are prepared using a variety of methods, including electrospinning, coprecipitation, sol-gel synthesis, combustion synthesis, hydrothermal synthesis, solid-state synthesis, and laser ablation [11, 16, 18–21]. The luminescent characteristics of rare-earth doped materials are significantly affected by the synthesis process. Therefore, it is essential to understand the synthesis pathway and its implications for the production of phosphors with practical uses.

5.4 FUNDAMENTAL ASPECTS OF PC-WLEDS

5.4.1 Low-Cost Synthesis

Producing white light by combining red, green, blue, and yellow light is a very complicated and expensive process. Therefore, researchers have found a way to produce white light at a lower cost. For this, multiple rare ions are doped in a single host material. In addition, the selection of synthesis approach is another aspect to help reduce the cost of synthesis.

5.4.2 Color Rendering Index (CRI)

A color rendering index (CRI) is a quantitative assessment of a light source's ability to correctly show the colors of diverse objects when compared to a natural or conventional light source. It is represented by R_a and its values vary from 0–100.

CRI 95–100 → Phenomenal color rendering. Colors appear as they should, subtle tones pop out and are accented, skin tones look beautiful, art comes alive, backsplashes and paint show their true colors.

CRI 90–95 → Great color rendering. Almost all colors 'pop' and are easily distinguishable. Noticeably great lighting starts at a CRI of 90.

CRI 80–90 → Good color rendering, where most colors are rendered well. Acceptable for most commercial uses.

CRI Below 80 → Lighting with a CRI below 80 would be considered to have poor color rendering. Under this light, items and colors may look desaturated, drab, and at times unidentifiable. It would be difficult to distinguish between similar colors.

5.4.3 Correlated Color Temperature (CCT)

Correlated color temperature (CCT) is one measure of the quality of white light. It is measured in kelvin (K) and is most commonly found between 2,000 and 6,500 K. The color temperature of a light source can be classified as warm or cool, depending on the visible emission color. Cool light is closer to blue light emission whereas warm light is closer to infrared emission. Cool colors have temperatures above 6,000 K, whereas warm colors have temperatures ranging between 2,000 and 3,500 K.

5.4.4 Thermal and Chemical Stability

In order to tolerate the temperature difference, the phosphor material must be thermally stable. There should be no difference in its qualities. The phosphor material must also be chemically stable in order to improve synthesis results and produce the required chemicals.

5.4.5 Quantum Yield (QY)

Quantum yield (QY) is one of the most significant parameters in the study of phosphor characteristics. The quantum yield of a luminescent material is defined as the ratio of the number of photons emitted to the number of photons absorbed by the irradiated sample.

It characterizes a radiative transition in combination with the luminescence lifetime, luminescence spectrum, and photo stability of the phosphor. Quantum yield is a selection criterion for luminescent materials used in solid-state lighting applications, as shown in Equation (5.1):

$$QY = \frac{\eta_{\text{emitted}}}{\eta_{\text{absorbed}}} = \frac{\eta_{\text{emitted}}}{\left(\eta_{\text{directed}} - \eta_{\text{reflected}}\right)} \tag{5.1}$$

Where, η_{emitted}, η_{absorbed}, η_{directed} and $\eta_{\text{reflected}}$ are the released photon number, absorbed photon number, incoming photons number, and reflected photon number, respectively.

5.4.6 Lumen Depreciation

Lumen depreciation is the process through which all electric light sources lose some of their brightness over time. Over time, tungsten particles accumulate on the wall of the bulb as the evaporation of the incandescent filament leaves them behind. During the 1,000-hour lifetime of an incandescent lamp, this causes a 10–15% drop relative to the initial lumen output.

5.4.7 Lifetime

The electrical and thermal architecture of the LED system or fixture influence how long LEDs last and how much light they emit. Driving the LED at a current greater than the rated current increases relative light output but decreases usable life. Operating the LED at temperatures greater than the design temperature will dramatically reduce useful life. Most manufacturers of high-power white LEDs estimate a lifetime of roughly 30,000 hours at 70% lumen maintenance, assuming continuous current operating at 350 milliamps (mA) and junction temperature no greater than 90°C. LED durability, on the other hand, is improving, allowing for larger driving currents and operating temperatures. Specific manufacturer data should be consulted because some LEDs available today are rated for 50,000 hours at 1,000 mA with junction temperature up to 120°C.

5.5 LITERATURE SURVEY OF PC-WLEDS PHOSPHORS

At present, new efforts are being made to develop phosphors of high quality and high efficiency that can meet the objectives of lighting technology. Most academicians and researchers are now working to develop efficient phosphors by spectrum tuning, which can enhance the luminous efficiency, color rendering index and color gamut of pc-LEDs [22]. These color-tunable phosphors currently have the potential to make a significant difference in the lighting field. According to the literature, several methods are often available to adjust the emission color of synthesized phosphors, including designing the energy transfer mechanism and changing the host structure through cationic, anionic, and cationic-anionic co-substitution. The color-tunable strategies which may be successfully employed for tuning the photoluminescence (PL) properties, so as to achieve the properties desirable for efficient working of pc-LEDs are explained in detail in a review reported by Baiqi Shaho et al. [23]. Moreover, Li et al. [24] have expounded on tunable strategies in their review paper.

5.5.1 Spectral Tuning by Host Substitution

It is important to note that multi-site emission phosphors are one of the most promising phosphors that can significantly aid the development of WLEDs [25]. Such materials can generate light at different wavelengths, increase luminous efficiency, conserve raw materials, and promote material utilization [26]. It has been demonstrated that cation or anion substitution is a successful method to modify the luminous colors and characteristics of phosphor materials [25, 27–29]. In recent years, various host material-based phosphors have been investigated, but limited materials are available for commercial use. Therefore, researchers are trying to vary the host structure by doping of anions and cations and their result showed that the spectral tunning and efficiency enhancement of sample. Some of these materials are reviewed and briefly discussed below.

Wu et al. [30] investigated $Ba_{2.94-2x}La_xNa_x P_4O_{13}$:0.06Eu ($x = 0–0.50$) [BLNPO] phosphors with cation substitution [La^{3+}-Na^+] for [Ba^{2+}-Ba^{2+}] pairs. Authors reported minor phase change in the X-ray diffraction (XRD) pattern of host material with cationic substitution of [La^{3+}-Na^+] for [Ba^{2+}-Ba^{2+}]. By adjusting the [La^{3+}-Na^+] concentration, the emission colors of BLNPO:Eu phosphors may be tailored from blue-green to white and finally to orange. This method successfully achieves and optimizes color-tunable white light emission in single-phase BLNPO hosts, making the BLNPO:Eu phosphors suitable color-tunable phosphor candidates for WLEDs.

Kadam et al. [31] investigated anion, cation substitution and co-substitution in $Na_2Sr_2Al_2PO_4Cl_9:Eu^{3+}$ phosphor. In this investigation proposed materials were synthesized by solid-state reaction method. Authors investigated the effect on PL properties after doping of anion and cation in $Na_2Sr_2Al_2PO_4Cl_9:Eu^{3+}$ phosphor. The PL emission spectra of $Na_2Sr_2Al_2PO_4Cl_9:Eu^{3+}$ phosphor exhibits orange and strong red emission under 395 nm and 466 nm excitation. In this investigation, the authors substituted various atoms of the host matrix in the given way: $Cl^- \rightarrow F^-$; $PO_4^{3-} \rightarrow WO_4^{2-}$, MoO_4^{2-}, VO_4^{3-}; $Al^{3+} \rightarrow La^{3+}$, Y^{3+}. From the investigation, it was found that with the doping of Y^{3+} ions highest emission intensity was observed. Figure 5.2(a–d) shows variation in emission intensity with anion, cation substitution and co- substitution in $Na_2Sr_2Al_2PO_4Cl_9:Eu^{3+}$ phosphor. The shift in PL emission intensity is caused by the ions changing the local crystal structure. Surprisingly, the PL intensity of phosphor rises several times after the inclusion of a (Y^{3+}) ion excited at 317 nm. The overall results of this study represent a potential red-emitting candidate for WLEDs application.

$ABZn_2Ga_2O_7:Bi^{3+}$ (ABZGO, A = Ca, Sr; B = Ba, Sr) phosphor is a yellow-orange emitting material reported by Liu et al. with tunable optical properties [32]. The technique of cation substitution may be used to study phosphors with outstanding luminous performance. The excitation wavelength shifts to the red from 325 to 363 nm when Sr^{2+} replaces Ca^{2+} and Ba^{2+}, which works well with n-UV chip-based WLEDs. $CaBaZn_2Ga_2O_7:0.01Bi^{3+}$ phosphor optical tuning is accomplished by varying the Bi^{3+} doping level. On the other hand, $SrBaZn_2Ga_2O_7:0.01Bi^{3+}$ (SBZGO:0.01Bi^{3+}) and $Sr_2Zn_2Ga_2O_7:0.01Bi^{3+}$ (SZGO:0.01Bi^{3+}) phosphors display a single wide emission band, peaking at 600 and 577 nm. Furthermore, the temperature-dependent PL spectra demonstrate that SZGO:0.01Bi^{3+} has the best thermal stability compared to other ABZGO:0.01Bi^{3+} phosphors. According to the above results, SZGO:0.01Bi^{3+} phosphors that emit yellow-orange light and have better thermal

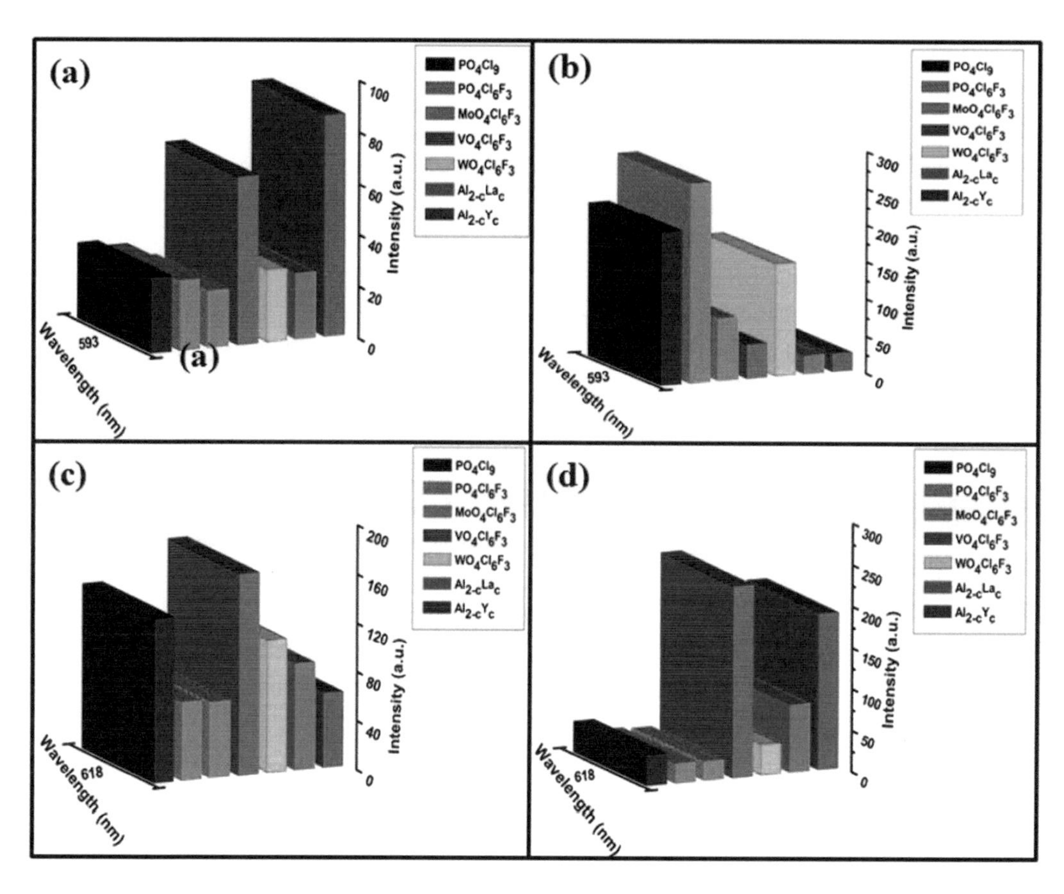

FIGURE 5.2 Variation in PL emission intensity under (a) and (c) 395 nm and (b) and (d) 467 nm for 593 nm and 618 nm emission wavelength. (Reproduced with permission from Ref. [31], © 2019 Elsevier Ltd.)

stability with strong n-UV excitation were fabricated as WLEDs. The EL properties of proposed sample shows high CRI (97.9) and low CCT (3932 K) values that indicate SZGO:0.01Bi^{3+} phosphor has huge potential for high-quality n-UV WLED devices.

The PL properties of $La_2(MoO_4)_3$:Eu^{3+} phosphors were investigated by Parauha et al. when molybdate ions were replaced by phosphate, sulphate, and vanadate ions [33]. A solid-state reaction method was employed for synthesis of materials. Under 467 nm excitation, the PL emission spectra of the $La_2(MoO_4)_3$:Eu^{3+} phosphor emits a red-colored emission band at 614 nm due to the $^5D_0 \rightarrow {}^7F_2$ transition. With the inclusion of phosphate, sulfate, and vanadate ions the emission intensity increased in the $La_2(MoO_4)_3$:Eu^{3+} phosphor. Figure 5.3 shows the higher emission intensity when vanadate ion is doped in $La_2(MoO_4)_3$:Eu^{3+} phosphor. The emission intensity is enhanced as the local crystal structure and particle size are increased by the doping of anions (phosphate, sulfate, and vanadate). Thus, phosphate, sulfate, and vanadate doped $La_2(MoO_4)_3$:Eu^{3+} phosphors can be employed as blue-light excited white LEDs.

Tamboli et al. investigated color tunable properties by anion substation in $Ba_{1.96}Mg(PO_4)_2$: 0.04Eu^{2+} phosphor [34]. In this investigation, PO_4^{3-} was replaced by BO_3^{3-} ions and

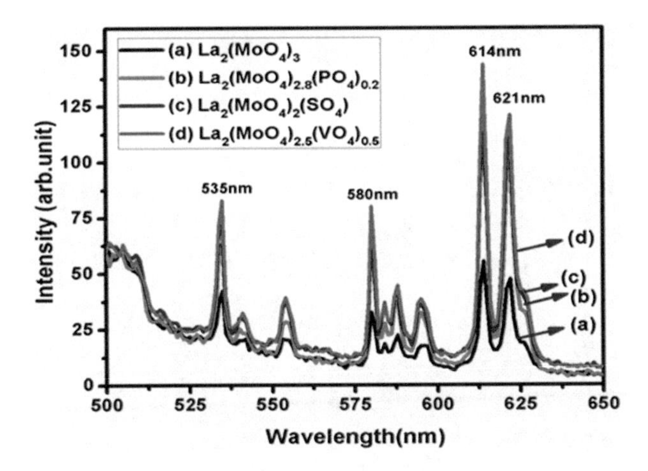

FIGURE 5.3 Comparative PL emission spectra of phosphate, sulfate, and vanadate doped $La_2(MoO_4)_3:Eu^{3+}$ phosphors under 467 nm excitation. (Reproduced with permission from Ref. [33], © 2019 Elsevier Ltd.)

materials were synthesized by the sol-gel/Pechini method. According to PL investigations, changing the composition of the $(BO_3)^{3-}$ group allowed for the color tuning of blue-white-orange light when it was excited by near-UV light. A higher-energy emission (420 nm) and a lower-energy emission (585 nm) resulted from the coexistence of the PO_4 and BO_3 groups, respectively. To calculate the energy transfer effectiveness between higher- and lower-energy emission sites of the phosphor, the PL decay curves were examined. In this study, a new method is presented to adjust the color emission of single-doped phosphors by host lattice structure modification. Figure 5.4 demonstrates color tuning of samples with variation of BO_3^{3-} ions.

Ming et al. [35] investigate spectra tuning in PL properties of $Lu_2MgAl_4SiO_{12}:Eu^{2+}$ phosphor by using cation substitution and chemical unit co-substitution method. The PL properties of $Lu_2MgAl_4SiO_{12}$:1%Eu^{2+} phosphor show asymmetric blue emission band under 365 nm UV excitation. Furthermore, by replacing Mg^{2+} with Ca^{2+} in the $Lu_2MgAl_4SiO_{12}$:1%Eu^{2+} system, the redshift behavior from 463 to 503 nm was achieved with broad emission band. Chemical unit co-substitution of [Ca^{2+}-Ge^{4+}] for [Lu^{3+}-Al^{3+}] in $Lu_2MgAl_4SiO_{12}$:1%Eu^{2+} phosphor was used to achieve the spectral tuning behavior. This effectively shifted the PL largest peak from 463 to 523 nm and also increased the full-width at half maxima (FWHM) value of the emission spectra. In order to achieve color-tunable emission, this work showed that it is effective to construct innovative Eu^{2+}-doped garnet-type phosphors by controlling the chemical composition of the hosts by cation substitution and the chemical unit co-substitution approach.

5.5.2 Spectral Tuning by Energy transfer

Over the past decades, the use of inorganic phosphor materials has increased rapidly due to their wide spectrum of applications. As discussed earlier, there are some problems encountered with the commercial approach to generating white light. Therefore, researchers are investigating another approach to generating white light. According to this approach, a

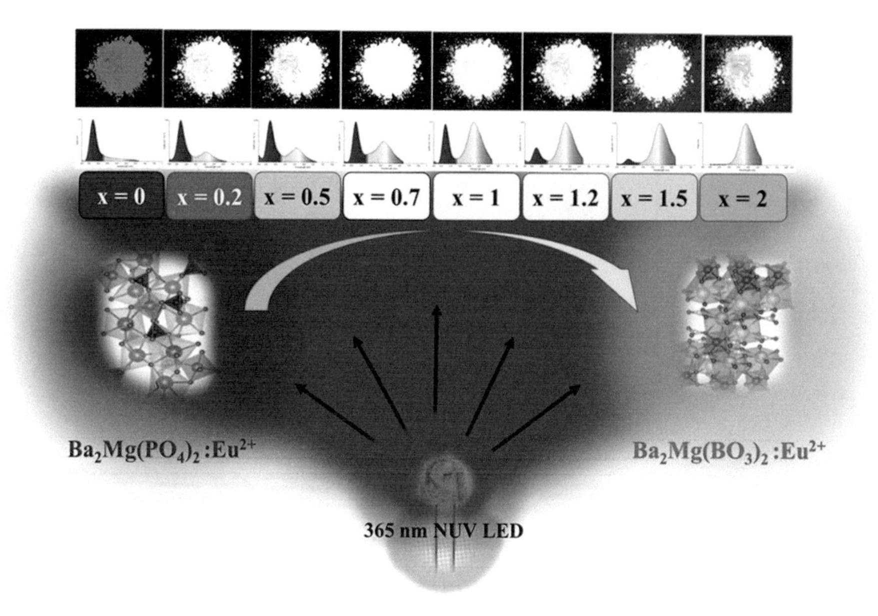

FIGURE 5.4 Graphical representation of color-tunable emission in $Ba_{1.96}Mg(PO_4)_{2-x}(BO_3)_x$:$0.04Eu^{2+}$ ($x = 0.2$, 0.5, 0.7, 1, 1.2 and 1.5) phosphors by host modification. (Reproduced with permission from Ref. [34]. © 2020 Elsevier Ltd.)

single-phase RGB phosphor would be synthesized. In this approach an energy transfer mechanism is mainly working. In the energy transfer process two or more rare-earth ions are doped in single host material. One is known as the sensitizer (S), and the other is the activator (A). The S absorbs most of the energy and transfers it to the A for the process of emission. When a single doping of activators injected into the host structure is not effective at all in absorbing the incoming radiation, a new dopant ion or sensitizer can be added to the current activator to improve absorption. Additionally, energy transfer makes it easier to adjust the photoluminescence. Energy transfer is the process of combining a sensitizer and activator inside the same host structure. Nevertheless, sensitizer must have a higher energy barrier than activators. As a result, after first absorbing the incoming radiation, the sensitizer finally transmits this energy directly to the activator, which then emits light. Prior to the development of energy transfer, white light was produced using the multi-phosphor method, which had significant drawbacks. Consequently, energy transfer is another simple method with which white light can be generated by a single host. Energy transfer is one of the most effective strategies as it often enhances the emission intensity. At the present time, the development of color-tunable phosphors through the doping of multiple activators in a host material and the study of energy transfer mechanisms is the most effective and popular strategy [36]. Generally, by altering the doping ratio of these activators, one can control the ET efficiency among them and then achieve multi-color luminescence in a single-phase phosphor. According to the literature, multi-color luminescence can be achieved by various energy transfer systems right now such as $Eu^{2+} \rightarrow Tb^{3+} \rightarrow Eu^{3+}$ [37], $Tb^{3+} \rightarrow Eu^{3+}/Sm^{3+}$ [38], $Ce^{3+} \rightarrow Tb^{3+} \rightarrow Eu^{3+}$ [39], $Bi^{3+} \rightarrow Eu^{3+}/Sm^{3+}/Mn^{4+}$ [40, 41], $Dy^{3+} \rightarrow Eu^{3+}$ [42, 43], $Tm^{3+} \rightarrow Dy^{3+}$ [44] etc. In this chapter, we have briefly discussed color-tunable phosphors, and we believe that a detailed description could be introduced in this topic, but in this chapter our focus is only on the capable and compatible phosphors for pc-LEDs.

Chen et al. [45] reported green-emitting $Sr_2LiScB_4O_{10}:Ce^{3+}/Tb^{3+}$ phosphor, synthesized by solid-state reaction method. $Sr_2LiScB_4O_{10}:Ce^{3+}/Tb^{3+}$ phosphor shows broad strong absorption band from 235 to 375 nm due to presence of Ce^{3+} ions. Ce^{3+} act as a sensitizer and transfer their energy to Tb^{3+}, therefore, the emission intensity of the green color is increased. The investigation revealed effective energy transfer from Ce^{3+} to Tb^{3+}. The prepared sample shows excellent color stability and thermal stability with increasing temperature.

Parauha et al. [46] reported the color-tunable behavior of Eu^{3+}/Dy^{3+} doped/co-doped $LaPO_4$ phosphors by the energy transfer process. In this investigation, Eu^{3+}/Dy^{3+} doped/co-doped $LaPO_4$ phosphors were prepared by solid-state reaction method. They observed effective energy transfer from $Dy^{3+} \rightarrow Eu^{3+}$ and color tuning from near-white to white color region. The proposed material emits blue, yellow and red emission under near-UV excitation. In addition, this material could be used in a solar cell as a downconversion layer to improve the efficiency of the solar cell. The I-V characteristics performance shows 17.34% and 12.76% solar-cell efficiency enhancement under direct sunlight or a solar simulator, respectively. Therefore, it can be said that the proposed phosphor sample has great potential for WLED and solar cell applications.

Zhang et al. [47] investigated the color-tunable properties of $Ca_9LiMn(PO_4)_7:Eu^{2+},Tb^{3+}$ phosphor and energy transfer between Eu^{2+} and Tb^{3+}. The authors synthesized single and co-doped (Eu^{2+},Tb^{3+}) phosphors by solid-state reaction method. With doping of Eu^{2+} ions, the red color emission observed may be due to the presence of Mn ions. Here, Eu^{2+} acts as the sensitizer; it absorbs the energy and transfers it to Mn^{2+}. Mn^{2+} converts to red emission and Eu^{2+} escapes to blue emission. Figure 5.5(a) represents an energy level diagram, which explains energy transfer from Eu^{2+} to Mn^{2+} and Tb^{3+} and Mn^{2+}. Prepared CLMP:$0.02Eu^{2+}$, $0.90Tb^{3+}$ phosphor also shows good performance, demonstrating high color purity (~93.9%), thermal stability (~70% intensity at 373 K), and quantum efficiency (IQE = 51.2%). The pc-WLED device is fabricated with CLMP:Eu^{2+},Tb^{3+} phosphor. Figure 5.5(b) shows the EL spectra of the fabricated device that exhibits a warm white light with color coordinates of (0.368, 0.350), a high CRI (~90.6), and suitable CCT (~4196 K). These findings suggest that a structural confinement effect is important to be established as a general method for increasing ET efficiency for net and effective emission.

Khan et al. [48] investigated color-tunable $Ca_3YAl_3B_4O_{15}:Ce^{3+},Tb^{3+},Sm^{3+}$ phosphor, synthesized by solid-state reaction method. In this investigation, authors studied color tunable behavior and energy transfer between Ce^{3+},$Tb^{3+}$and Sm^{3+}. The prepared sample exhibits green, orange emission under near UV excitation. Therefore, authors blended this sample singly and with commercial blue phosphor (BAM:Eu^{2+}) for fabrication of White LEDs. The EL performance of fabricated devices shows low CCT $\approx$ 4543 K or 3913 K and high color rendering index around 86.0 or 84.4, respectively. These results proved that the prepared sample has the suitability for the fabrication of warm WLEDs.

Li et al. [49] investigated $KBaY(MoO_4)_3$ (KBYMO):Dy^{3+}, Eu^{3+} phosphors that are synthesized by sol-gel/Pechini-type method. The Dy^{3+} doped samples show blue and yellow emission under UV and near-UV excitation, whereas, Eu^{3+} doped samples show orange red

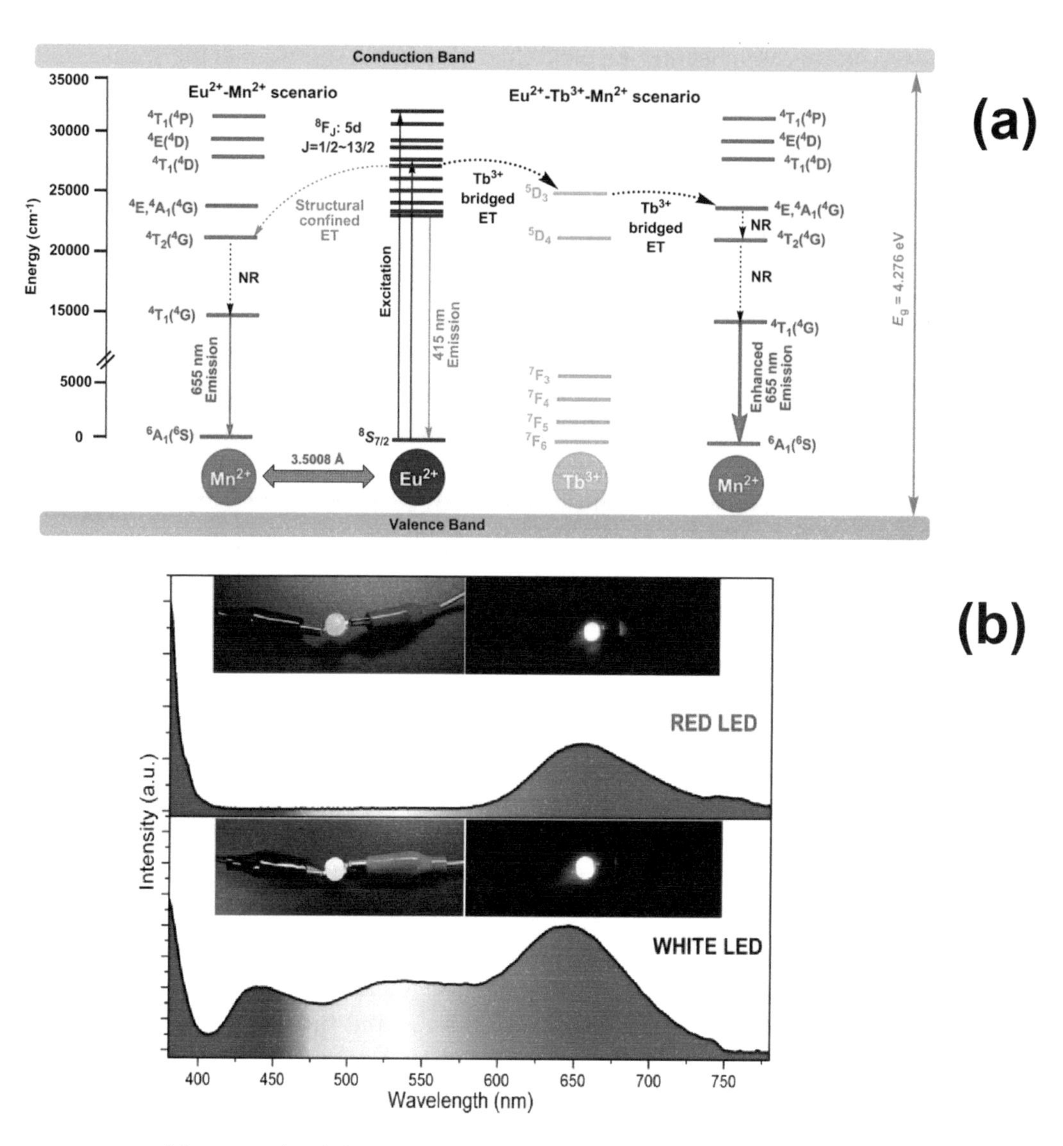

FIGURE 5.5 (a) Energy level diagram of the CLMP:$0.02Eu^{2+}$,yTb^{3+} samples (b) EL spectrum of fabricated red and white LEDs. (Reproduced with permission from Ref. [47], © 2020, American Chemical Society.)

color emission under near UV and blue excitation. To investigate the energy transfer between Dy^{3+} and Eu^{3+}, both rare-earths were co-doped into the host material. The prepared material shows three prominent emission bands around 488, 573, and 615 nm: their respective transitions can be seen in Figure 5.6(a). It was also observed from this investigation that energy is transferred from $Dy^{3+} \rightarrow Eu^{3+}$. The thermal stability of the sample shows a stability of 82.8% emission intensity at 423 K from room temperature. Figure 5.6(b) shows a Commission Internationale de l'Eclairage (CIE) chromaticity diagram of KBYMO:$0.05Dy^{3+}$,yEu^{3+} (y = 0, 0.02, 0.04, 0.06, 0.10, 0.14 and 0.18) phosphor and their respective color emission. The prepared materials show orange-red emission, and their overall results suggest their suitability for WLEDs.

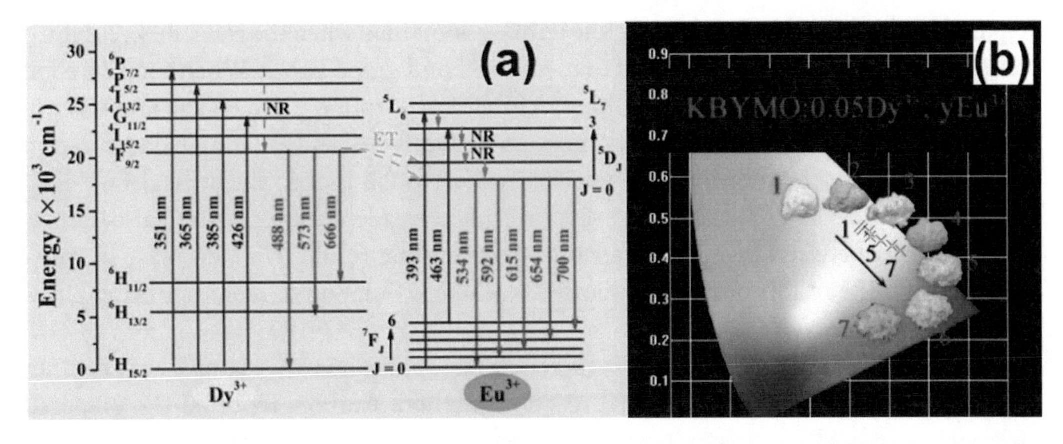

FIGURE 5.6 (a) Energy level diagram of KBYMO:Dy^{3+},yEu^{3+} phosphor (b) CIE chromaticity diagram of KBYMO:0.05Dy^{3+},yEu^{3+} (y = 0, 0.02, 0.04, 0.06, 0.10, 0.14 and 0.18) phosphor and their respective color emission. (Reproduced with permission from Ref. [49], © 2020, American Chemical Society.)

Parauha et al. [50] studied color-tunable behavior and energy transfer mechanism in Na_2SrPO_4F:Eu^{2+},Tb^{3+} phosphor. The proposed materials were synthesized by solid-state reaction method. The PL emission and excitation properties of singly and co-doped Na_2SrPO_4F:Eu^{2+},Tb^{3+} phosphors were investigated. With the co-doping of Tb^{3+} ions, strong blue and green emission was observed under near-UV excitation. This investigation represents energy transfer from $Eu^{2+} \rightarrow Tb^{3+}$. In addition, I-V characteristics of prepared phosphor were investigated. They showed enhancement in solar-cell efficiency under solar simulator and direct sunlight.

5.5.3 Some Other Rare-Earth Doped Phosphors

In addition to spectrally adjustable phosphors, numerous additional materials with promise for illuminating applications have been studied in recent years. Some rare-earth phosphors can be combined with other phosphors and used to make LEDs.

Han and co-workers [51] reported $NaLa_{1-x}MgWO_6$:xPr^{3+} (x being number of moles of Pr^{3+}) prepared following the solid-state technique. The excitation spectrum comprises a band from 250 nm (4.96 eV) to 350 nm (3.54 eV) that is related to the host and bands of Pr^{3+} between 440 nm (2.81 eV) and 500 nm (2.48 eV). The latter matches well with that of blue-emitting LEDs that are available commercially. When this phosphor was excited with 268 nm (4.62 eV) and 451 nm (2.75 eV) wavelengths corresponding to UV and blue lights, respectively, it displayed red emission owing to the $^3P_0 \rightarrow {}^3F_2$ (654 nm, 1.8598 eV) transition in Pr^{3+}. The optimum concentration of Pr^{3+} was verified to be 7%. Beyond this, concentration quenching dominated and reduced luminosity. The luminescence decay lifetime is in the microsecond range for $^3P_0 \rightarrow {}^3F_2$ transition in Pr^{3+} and the lifetime decreased with increased Pr^{3+} ion concentration. The authors have also fabricated a WLED device to explore the possibility of using the synthesized $NaLaMgWO_6$:Pr^{3+} red phosphor in SSL [51].

In the work of Maggay and co-workers [52], new blue-emitting garnet $NaCaBeSi_2O_6F$:Eu^{2+} phosphors that exhibited a broad NUV excitation band between 300 and 400 nm were prepared by the solid-state method. White light was obtained when the $NaCaBeSi_2O_6F:Eu^{2+}$ phosphor was mixed with phosphors that emit red and green lights. When monitored at 467 nm, the $NaCaBeSi_2O_6F:Eu^{2+}$ phosphor exhibited absorption between 250 and 450 nm and on excitation at 397 nm displayed emission between 400 and 600 nm. Concentration quenching was observed beyond Eu^{2+} concentration of 0.8 mol%. The critical Eu^{3+}-Eu^{3+} distance was 31.96 Å. When the Eu^{2+}-Eu^{2+} distance was less than the critical distance, energy transfer between Eu^{2+} ions overcame emission that resulted in a decrease in emission intensity. The results obtained indicated $NaCaBeSi_2O_6F:Eu^{2+}$ as a blue emitting phosphor that can improve WLEDs [52].

Wu et al. [53] reported $Na_5W_3O_9F_5$ (NWOF):Eu^{3+} phosphor, this sample was synthesized using conventional solvothermal reaction method. Authors reported the prepared material has monoclinic crystal structure. The PL emission properties show novel red color emission at 607 nm under 466 nm blue light excitation. The color purity of red color emission is around 97.44%. In addition, a pc-WLED was fabricated by combining a synthesized NWOF:0.25Eu^{3+} red phosphor and a commercial YAG:Ce^{3+} phosphor in a 460 nm blue chip. The overall outcomes suggested that the NWOF:0.25Eu^{3+} phosphor is a suitable candidate as a red component for the preparation of pc-WLEDs.

Zhang and his co-workers investigated PL properties of $Na_2YMg_2V_3O_{12}:Eu^{3+}$ phosphors, which were synthesized by conventional solid-state reaction method [53]. The authors reported that the prepared material was luminophores host matrix, which shows self-emission in the blue-green region under UV light excitation. The PL emission spectrum of $Na_2YMg_2V_3O_{12}:Eu^{3+}$ phosphors show a broad emission band due to d-d transition of $(VO_4)^{3-}$ groups with some sharp Eu^{3+} characteristic bands ranging from 400 to 700 nm. The obtained powder was mixed with commercial green- [$(Ba, Sr)_2SiO_4:Eu^{2+}$] and blue- [$BaMgAl_{10}O_{17}:Eu^{2+}$] emitting phosphors in appropriate proportions to fabricate a white LED with a UV LED chip. The fabricated device has CIE coordinates around (0.3068, 0.3491), CRI ~ 88.20 and CCT ~ 4460 K. All these results represent that the obtained powder $Na_2YMg_2V_3O_{12}:Eu^{3+}$ can be used as a red-emitting candidate for WLEDs devices. Table 5.1 summarizes the list of rare-earth activated phosphors that can be potentially used for solid-state lighting.

5.6 CHALLENGES AND FUTURE ADVANCES

The advancement of LED technology prompts an increasing reliance on this technology to deliver new and innovative products to replace not only existing lighting, but lighting that would not have been possible previously due to the intelligent control of color, power, spatial distribution, and dimmer functionality. This, combined with the reality that the world needs affordable, energy-efficient lighting, means that LEDs have opened the door to a wide range of different applications that can result in lucrative economic and financial scenarios where everyone ultimately succeeds. Over the last decade, solid-state lighting (SSL) technology has made significant breakthroughs and has become the technology of choice for replacing both outdated incandescent and fluorescent lights. Furthermore, LED efficiency has been improved and the cost of LEDs has been steadily lowered. However, several issues remain constant in the development of LEDs like thermal, electrical, optical,

TABLE 5.1 List of Rare-Earth Activated Phosphors with Brief Description

Ref. No	Compound Name	Activator	Synthesis Method	Photoluminescence Properties		Additional Features, If Any
				Excitation Wavelength	Emission Wavelength	
[54]	$LaTiSbO_6$	Sm^{3+}	Solid-state reaction route	$\lambda_{exc} = 403$ nm	$\lambda_{em} = 567, 598, 645$ and 711 nm	Activation energy $E_a = 0.308$ eV, $R_a = 90$, CCT = 5536 K.
[55]	$Ba_3In_2WO_9$	Eu^{3+}	Solid-state reaction	$\lambda_{exc} = 464$ nm	$\lambda_{em} = 613$ nm	Color Purity = 97.92%
[56]	$Li_7La_3Zr_2O_{12}$	Dy^{3+}	Solid-state reaction	$\lambda_{exc} = 351$ nm	$\lambda_{em} = 481, 579, 670$ nm	
[57]	Ca_2YNbO_6	Eu^{3+}	Solid-phase method	$\lambda_{exc} = 465$ nm	$\lambda_{em} = 612$ nm	Color Purity = 99.90%, CCT = 2656 K
[58]	$Cs_3GdGe_3O_9$	Eu^{3+}	Solid-state reaction	$\lambda_{exc} = 464$ nm	$\lambda_{em} = 611$ nm	Color Purity = 95.07%, IQE= 94%, CRI = 89.7, CCT = 4508 K
[59]	$BaYAlZn_3O_7$	Dy^{3+}	Solution-combustion synthesis	$\lambda_{exc} = 355$ nm	$\lambda_{em} = 485, 575$ nm	
[60]	$AlPO_4$	Eu^{3+}	Co-precipitation method	$\lambda_{exc} = 394$ nm	$\lambda_{em} = 588, 594, 685$ and 700 nm	Good thermal stability (84% at 160°C),CIE coordinate = (0.5573, 0.3253) and IQE = 38.7%
[61]	Ca_2LuSbO_6	Eu^{3+}	Solid-state reaction	$\lambda_{exc} = 396, 466$ nm	$\lambda_{em} = 594, 613$ and 700 nm	IQE = 82%, CIE coordinate = (0.661, 0.339)
[62]	$Ba_2Y_5B_5O_{17}$	Bi^{3+}	Solid-state reaction	$\lambda_{exc} = 370$ nm	Blue emission band (380–460 nm)	QE = 46%
[63]	$(NH_4)_2SnCl_6$	Te	Chemical synthesis route		Broad orange emission peaked at 590 nm	QY = 83.51%, CIE coordinate = (0.39, 0.38), CCT = 3855 K, CRI = 83
[64]	$NaBaB_9O_{15}$	Tb^{3+}	Solid-state reaction	$\lambda_{exc} = 370$ nm	Green emission	
[65]	$CaAlSiN_3$	Eu^{2+}	Solid-state reaction	Broad excitation band from 250 to 600 nm (centered at 460 nm)	Broad emission band from 550 to 800 nm (centered at 655 nm)	Suitable EL performance for WLEDs with two other phosphors, CRI – ($R_a = 95.9$ and R9 = 92)
[66]	$Ca_3Al_2O_6$	Eu^{2+}	Solid-state reaction	$\lambda_{exc} = 334$ nm (Asymmetric broad band)	$\lambda_{em} = 445$ nm (Broad emission band)	CIE Coordinate ~(0.212, 0.110) and Color Purity ~84.21%
[67]	$CaNa_2(SO_4)_2$	Eu^{3+}	Solid-state reaction	$\lambda_{exc} = 396, 466$ nm	$\lambda_{em} = 595, 618$ nm	
[68]	$NaCaPO_4$	Dy^{3+}	Solution-combustion synthesis	$\lambda_{exc} = 325, 355, 365, 388$ nm	$\lambda_{em} = 482, 576$ nm	7.6% Solar cell efficiency enhancement under solar simulator
[69]	Zn_2SnO_4	Mn^{2+}	Solid-state reaction	Broad excitation band with five distinct peaks at 359, 380, 424, 435, and 444 nm	Green emission band peaking at 523 nm	CCT ~ 3858 K, CRI ~ 91
[70]	$K_3TaO_2F_4$	Mn^{4+}	Modified solid-state method	$\lambda_{exc} = 460$ nm	$\lambda_{em} = 630$ nm	CCT = 3488 K, $R_a = 93.0$, R9 = 90.0

and packaging implementations. Currently, pc-LEDs are the most energy-efficient option. They are currently the most readily available. However, pc-LEDs have some serious issues, as discussed earlier, the most important being low CRI, high CCT and lack of red components. Higher CRI requirements are more restrictive of spectral content, and in general require a broader spectrum. Hybrid approaches, where more than one spectral LED is combined with a phosphor emission (e.g., blue, red, and phosphor), are gaining momentum and promise increased efficacy with favorable color quality attributes. Temperature and other external factors have an effect on the efficiency and lifetime of LEDs. Additionally, there are some difficulties with cost control during manufacturing and thermal management. The ambient temperature of the operating environment is ideal for the performance of the LEDs. This characteristic is important for high-power LEDs. The energy efficiency of LEDs has increased significantly in recent years. However, too much heat can damage the LED package. If LEDs are exposed to too much heat, they can be permanently damaged. Subsequently, the performance of the LED may be affected, its quality and light output may be reduced, or the device may break. In order to achieve proper heat dissipation, it is necessary to equip the luminaires with suitable heat sinks or cooling systems [4]. As a result, the LEDs will have a longer lifetime and be able to operate at different temperatures. Aiming high and having strategies to deal with the new challenges in the lighting business will be key in the new decade. The SSL industry will reach its highest point in the next ten years due to the rapid digitization and dazzling transition of the sector, which is unleashing some significant technological advancements. Manufacturers are trying to take advantage of every opportunity by incorporating the most recent technology into the design of LED products. The LED industry has gone through some ups and downs in recent times. However, in the long run, LED has proven to be a profitable track and will continue to do so.

5.7 SUMMARY

Solid state lighting (SSL) is defined as the use of semiconductor electroluminescence to generate visible light for illumination, resulting in more environmentally friendly structures that consume less electricity and reduce dependence on fossil fuels. Let's deduce that SSL will be essential over the next ten years to produce low-cost, high-performance structures that use less energy and emit fewer greenhouse gases than their counterparts. Academic and industry research groups are increasingly focusing on the research of SSL sources based on WLEDs as next-generation light sources. WLEDs offer exceptional light weight, flexibility, light-uniform, and glare-free displays, and thin profiles that are more comfortable to human perception than current light sources such as incandescent light bulbs and fluorescent lamps. In addition, WLEDs for SSL have numerous essential performance criteria, including efficiency, longevity, and color quality. We have discussed the current improvements in LED materials in detail in this chapter. However, it is worth noting that considerable technical hurdles must be addressed before LEDs can be considered a success in the future. Despite the substantial obstacles facing LEDs, remarkable progress in the development of LEDs in recent years has provided significant confidence in this potential technology for future lighting and displays. We think the scientific research in

LEDs is active and prosperous, and in the near future LEDs will infiltrate and enhance our daily lives in many exciting ways.

REFERENCES

[1] G. Li, Y. Tian, Y. Zhao, J. Lin, Recent progress in luminescence tuning of Ce^{3+} and Eu^{2+}-activated phosphors for pc-WLEDs, *Chem. Soc. Rev.* 44 (2015) 8688–8713. https://doi.org/10.1039/c4cs00446a

[2] J. Qiao, J. Zhao, Q. Liu, Z. Xia, Recent advances in solid-state LED phosphors with thermally stable luminescence, *J. Rare Earths.* 37 (2018) 565–572. https://doi.org/10.1016/j.jre.2018.11.001

[3] Ç. Iris, J. Siu, L. Lam, A review of energy efficiency in ports: Operational strategies, technologies and energy management systems, *Renew. Sustain. Energy Rev.* 112 (2019) 170–182. https://doi.org/10.1016/j.rser.2019.04.069

[4] G.B. Nair, S.J. Dhoble, Current trends and innovations, in: *Fundam. Appl. Light. Diodes, Elsevier,* 2020: pp. 253–270. https://doi.org/10.1016/B978-0-12-819605-2.00010-0

[5] D. Dilaura, A brief history of lighting, *Opt. Photonics News.* 19 (2008) 22–28.

[6] R.U. Ayres, The history of artificial light, in: *The History and Future of Technology*, Springer, Cham, 2021.

[7] J. Cho, J.H. Park, J.K. Kim, E.F. Schubert, White light-emitting diodes: History, progress, and future, *Laser Photonics Rev.* 11 (2017) 974. https://doi.org/10.1002/lpor.201600147

[8] M.S. Khan, M.D. Mehare, Y.R. Parauha, S.A. Dhale, S.J. Dhoble, Synthesis and novel emission properties of Bi^{3+}-doped Ca_2BO_3Cl phosphor for plant cultivation, *Luminescence.* (2022). https://doi.org/10.1002/bio.4418

[9] Y.R. Parauha, D.K. Halwar, S.J. Dhoble, Photoluminescence properties of Eu^{3+}-doped Na_2CaSiO_4 phosphor prepared by wet-chemical synthesis route, *Displays.* 75 (2022). https://doi.org/10.1016/j.displa.2022.102304

[10] D.J. Dhiman, Y.R. Parauha, A.B. Chourasia, S.J. Dhoble, Tunable luminescence from Eu^{3+} and Ce^{3+} doped/co-doped color tunable $Na_4Ca(PO_3)_6$ phosphors for white LEDs and solar cell applications, *J. Mater. Sci. Mater. Electron.* 33 (2022) 11106–11123. https://doi.org/10.1007/s10854-022-08087-y

[11] C. Gandate, Y.R. Parauha, S.G.M. Mushtaque, S.J. Dhoble, Current progress and comparative study of performance of the energy saving lighting devices: A review, in: *J. Phys. Conf. Ser.*, IOP Publishing Ltd, 2021. https://doi.org/10.1088/1742-6596/1913/1/012018

[12] S. Khan, Y.R. Parauha, D.K. Halwar, S.J. Dhoble, Rare Earth (RE) doped color tunable phosphors for white light emitting diodes, in: *J. Phys. Conf. Ser.*, IOP Publishing Ltd, 2021. https://doi.org/10.1088/1742-6596/1913/1/012017

[13] A.M. Bhake, Y.R. Parauha, S.J. Dhoble, Synthesis and photoluminescence study of Ce^{3+} ion-activated $Na_2ZnP_2O_7$ and $Na_4P_2O_7$ pyrophosphate phosphors, *J. Mater. Sci. Mater. Electron.* (2019). https://doi.org/10.1007/s10854-019-02559-4

[14] H. Zhang, H. Zhang, Rare earth luminescent materials, *Light Sci. Appl.* 11 (2022) 260. https://doi.org/10.1038/s41377-022-00956-9

[15] R. Rajeswari, N. Islavath, M. Raghavender, L. Giribabu, Recent progress and emerging applications of rare earth doped phosphor materials for dye-sensitized and perovskite solar cells: A review, *Chem. Rec.* 19 (2019) 1–25. https://doi.org/10.1002/tcr.201900008

[16] X. Wanga, J. Xu, J. Yu, Y. Bu, J. Marques-Hueso, X. Yana, Morphology control, spectrum modification and extended optical applications of rare earth ions doped phosphors, *Phys. Chem. Chem. Phys.* 22 (2020) 15120–15162. https://doi.org/10.1039/D0CP01412E

[17] I. Gupta, S. Singh, S. Bhagwan, D. Singh, Rare earth (RE) doped phosphors and their emerging applications: A review, *Ceram. Int.* 47 (2021) 19282–19303. https://doi.org/10.1016/j.ceramint.2021.03.308

[18] G.B. Nair, H.C. Swart, S.J. Dhoble, A review on the advancements in phosphor-converted light emitting diodes (pc-LEDs): Phosphor synthesis, device fabrication and characterization, *Prog. Mater. Sci.* (2020) 100622.
[19] R. Xie, N. Hirosaki, Y. Li, T. Takeda, Rare-earth activated nitride phosphors: Synthesis, luminescence and applications, *Materials (Basel).* 3 (2010) 3777–3793. https://doi.org/10.3390/ma3063777
[20] R.S. Yadav, S.B. Rai, S.J. Dhoble, Recent advances on morphological changes in chemically engineered rare earth doped phosphor materials, *Prog. Solid State Chem.* (2019) 100267. https://doi.org/10.1016/j.progsolidstchem.2019.100267
[21] Y.R. Parauha, V. Sahu, S.J. Dhoble, Prospective of combustion method for preparation of nanomaterials: A challenge, *Mater. Sci. Eng. B Solid-State Mater. Adv. Technol.* 267 (2021) 115054. https://doi.org/10.1016/j.mseb.2021.115054
[22] D.V. Deyneko, I.V. Nikiforov, D.A. Spassky, P.S. Berdonosov, P.B. Dzhevakov, B.I. Lazoryak, $Sr_8MSm1\text{-}_xEu_x(PO_4)_7$ phosphors derived by different synthesis routes: Solid state, sol-gel and hydrothermal, the comparison of properties, *J. Alloys Compd.* 887 (2021) 161340. https://doi.org/10.1016/j.jallcom.2021.161340
[23] B. Shao, J. Huo, H. You, Prevailing strategies to tune emission color of lanthanide-activated phosphors for WLED applications, *Adv. Opt. Mater.* 7 (2019) 1–23. https://doi.org/10.1002/adom.201900319
[24] G. Li, Y. Tian, Y. Zhao, J. Lin, Recent progress in luminescence tuning of Ce^{3+} and Eu^{2+}-activated phosphors for pc-WLEDs, *Chem. Soc. Rev.* 44 (2015) 8688–8713. https://doi.org/10.1039/C4CS00446A
[25] Y. Guo, S.H. Park, B.C. Choi, J.H. Jeong, J.H. Kim, Effect of La3+ ion doping on the performance of Eu^{2+} ions in novel $Sr_3CeNa(PO_4)_2SiO_4$ phosphors, *J. Alloys Compd.* 724 (2017) 763–773. https://doi.org/10.1016/j.jallcom.2017.07.114
[26] P.P. Dai, C. Li, X.T. Zhang, J. Xu, X. Chen, X.L. Wang, Y. Jia, X. Wang, Y.C. Liu, A single Eu^{2+}-activated high-color-rendering oxychloride white-light phosphor for white-light-emitting diodes, *Light Sci. Appl.* 5 (2016) 1–9. https://doi.org/10.1038/lsa.2016.24
[27] Y. Sato, H. Kato, M. Kobayashi, T. Masaki, D.H. Yoon, M. Kakihana, Tailoring of deep-red luminescence in Ca_2SiO_4:Eu^{2+}, *Angew. Chemie - Int. Ed.* 53 (2014) 7756–7759. https://doi.org/10.1002/anie.201402520
[28] L. He, L. Bi, X. He, C. Xu, Y. Liu, D. Lin, Chemical unit cosubstitution and tuning of photoluminescence in Eu^{2+} doped $Ca_{1.65\text{-}x}Sc_xSr_{0.35Si1\text{-}x}Ga_xO_4$ phosphors, *J. Lumin.* 198 (2018) 84–91. https://doi.org/10.1016/j.jlumin.2018.02.030
[29] M. Seibald, T. Rosenthal, O. Oeckler, W. Schnick, Highly efficient pc-LED phosphors $Sr1\text{-}_xBa_xSi_2O_2N_2$:$Eu^{2+}$ ($0 \leq x \leq 1$) – Crystal structures and luminescence properties revisited, *Crit. Rev. Solid State Mater. Sci.* 39 (2014) 215–229. https://doi.org/10.1080/10408436.2013.863175
[30] Z.C. Wu, L.L. Cui, X. Zhang, X.X. Zhang, J. Liu, L. Ma, X.J. Wang, J.L. Zhang, Cationic substitution induced tuning of photoluminescence in $Ba_{2.94\text{-}2x}La_xNa_xP_4O_{13}$: 0.06Eu phosphors for WLEDs, *J. Alloys Compd.* 835 (2020) 2–7. https://doi.org/10.1016/j.jallcom.2020.155109
[31] A.R. Kadam, R.S. Yadav, G.C. Mishra, S.J. Dhoble, Effect of singly, doubly and triply ionized ions on downconversion photoluminescence in Eu^{3+} doped $Na_2Sr_2Al_2PO_4Cl_9$ phosphor: A comparative study, *Ceram. Int.* 46 (2020) 3264–3274. https://doi.org/10.1016/j.ceramint.2019.10.032
[32] D. Liu, X. Yun, P. Dang, H. Lian, M. Shang, G. Li, J. Lin, D. Liu, X. Yun, P. Dang, H. Lian, M. Shang, Phosphors: Optical temperature sensing and WLED Applications yellow / orange-emitting $ABZn_2Ga_2O_7$:Bi^{3+} (ABZGO, A = Ca, Sr; B = Ba, Sr) phosphors: Optical temperature sensing and WLED applications, *Chem. Mater.* 32 (2020) 3065–3077. https://doi.org/10.1021/acs.chemmater.0c00054
[33] Y.R. Parauha, R.S. Yadav, S.J. Dhoble, Enhanced photoluminescence via doping of phosphate, sulphate and vanadate ions in Eu^{3+} doped $La_2(MoO_4)_3$ downconversion phosphors for white LEDs, *Opt. Laser Technol.* 124 (2019) 105974. https://doi.org/10.1016/J.OPTLASTEC.2019.105974

[34] S. Tamboli, G.B. Nair, R.E. Kroon, S.J. Dhoble, H.C. Swart, Color tuning of the $Ba_{1.96}Mg(PO_4)_2$:0.04Eu^{2+} phosphor induced by the chemical unit co-substitution of the (BO3)3– anion group, *J. Alloys Compd.* 864 (2021) 158124. https://doi.org/10.1016/j.jallcom.2020.158124

[35] Z. Ming, J. Qiao, M.S. Molokeev, J. Zhao, H.C. Swart, Z. Xia, Multiple substitution strategies toward tunable luminescence in $Lu_2MgAl_4SiO_{12}$:Eu^{2+} phosphors, *Inorg. Chem.* 59 (2019) 1405–1413. https://doi.org/10.1021/acs.inorgchem.9b03142

[36] M. Song, W. Zhao, J. Xue, L. Wang, J. Wang, Color-tunable luminescence and temperature sensing properties of a single-phase dual-emitting La_2LiSbO_6:Bi^{3+}, Sm^{3+} phosphor, *J. Lumin.* 235 (2021) 2–4. https://doi.org/10.1016/j.jlumin.2021.118014

[37] M. Xie, G. Zhu, D. Li, R. Pan, X. Fu, $Eu^{2+} \rightarrow Tb^{3+} \rightarrow Eu^{3+}$ energy transfer in $Ca_6La_2Na_2(PO_4)_6F_2$:Eu, *Tb Phosphors*, 2016. https://doi.org/10.1039/c6ra03154d

[38] K. Li, R. Van Deun, Photoluminescence and energy transfer properties of a novel molybdate $KBaY(MoO_4)_3$:Ln^{3+} (Ln^{3+} = Tb^{3+}, Eu^{3+}, Sm^{3+}, Tb^{3+}/Eu^{3+}, Tb^{3+}/Sm^{3+}) as a multi-color emitting phosphor for UV w-LEDs, *Dalt. Trans.* 47 (2018) 6995–7004. https://doi.org/10.1039/c8dt01011k

[39] Y. Yang, X. Lv, L. Wei, J. Xu, H. Yu, Y. Hu, H. Zhang, B. Liu, X. Wang, Q. Li, Energy transfer from Ce^{3+} to Eu^{3+} through Tb^{3+} chain in YPO_4:Ce^{3+}/Tb^{3+}/Eu^{3+} phosphors, *Solid State Commun.* 269 (2018) 35–38. https://doi.org/10.1016/j.ssc.2017.10.002

[40] Y. Zhang, X. Zhang, L. Zheng, Y. Zeng, Y. Lin, Y. Liu, B. Lei, H. Zhang, Energy transfer and tunable emission of $Ca_{14}Al_{10}Zn_6O_{35}$:Bi^{3+},Sm^{3+} phosphor, *Mater. Res. Bull.* 100 (2018) 56–61. https://doi.org/10.1016/j.materresbull.2017.12.003

[41] D. Huang, P. Dang, H. Lian, Q. Zeng, J. Lin, Luminescence and energy-transfer properties in Bi^{3+}/Mn^{4+}-co-doped Ba_2GdNbO_6 double-perovskite phosphors for white-light-emitting diodes, *Inorg. Chem.* 58 (2019) 15507–15519. https://doi.org/10.1021/acs.inorgchem.9b02558

[42] W. Zhou, M. Song, Y. Zhang, Z. Xie, W. Zhao, Color tunable luminescence and optical temperature sensing performance in a single-phased $KBaGd(WO_4)_3$:Dy^{3+}, Eu^{3+} phosphor, *Opt. Mater. (Amst).* 109 (2020) 110271. https://doi.org/10.1016/j.optmat.2020.110271

[43] J. Wang, Y. Bu, X. Wang, H.J. Seo, A novel optical thermometry based on the energy transfer from charge transfer band to Eu^{3+}-Dy^{3+} ions, *Sci. Rep.* 7 (2017) 1–11. https://doi.org/10.1038/s41598-017-06421-7

[44] S.P.N. Tm, Multiple energy transfer in luminescence-tunable single-phased phosphor $NaGdTiO_4$: Tm^{3+}, Dy^{3+}, Sm^{3+}, *Nanomaterials.* 10 (2020) 1249.

[45] H. Chen, Y. Wang, $Sr_2LiScB_4O_{10}$:Ce^{3+}/Tb^{3+}: A green-emitting phosphor with high energy transfer efficiency and stability for LEDs and FEDs, *Inorg. Chem.* 58 (2019) 7440–7452. https://doi.org/10.1021/acs.inorgchem.9b00639

[46] Y.R. Parauha, S.J. Dhoble, Color-tunable luminescence, energy transfer behavior and I-V characteristics of Dy^{3+},Eu^{3+} co-doped $La(PO_4)$ phosphors for WLEDs and solar applications, *New J. Chem.* 46 (2022) 6230–6243. https://doi.org/10.1039/d2nj00232a

[47] X. Zhang, J. Luo, Z. Sun, W. Zhou, Z.C. Wu, J. Zhang, Ultrahigh-energy-transfer efficiency and efficient Mn^{2+} red emission realized by structural confinement in $Ca_9LiMn(PO_4)_7$:Eu^{2+} ,Tb^{3+} phosphor, *Inorg. Chem.* 59 (2020) 15050–15060. https://doi.org/10.1021/acs.inorgchem.0c01991

[48] W. Ullah, L. Zhou, X. Li, W. Zhou, D. Khan, S. Niaz, Single phase white LED phosphor Ca_3Y $Al_3B_4O_{15}$:Ce^{3+},Tb^{3+}, Sm^{3+} with superior performance : Color-tunable and energy transfer tudy, *Chem. Eng. J.* 410 (2021) 128455. https://doi.org/10.1016/j.cej.2021.128455

[49] K. Li, R. Van Deun, Color tuning from greenish-yellow to orange-red in thermal-stable $KBaY(MoO_4)_3$:Dy^{3+}, Eu^{3+}phosphors via energy transfer for UV W-LEDs, *ACS Appl. Electron. Mater.* 2 (2020) 1735–1744. https://doi.org/10.1021/acsaelm.0c00307

[50] Y.R. Parauha, N.S. Shirbhate, S.J. Dhoble, Color-tunable luminescence by energy transfer mechanism in RE (RE = Eu^{2+}, Tb^{3+})–doped Na_2SrPO_4F phosphors, *J. Mater. Sci. Mater. Electron.* 33 (2022) 15333–15345. https://doi.org/10.1007/s10854-022-08423-2

[51] B. Han, Y. Dai, J. Zhang, X. Wang, W. Shi, H. Shi, $NaLaMgWO_6:Pr^{3+}$: A novel blue-light excitable red-emitting phosphor for white light-emitting diodes, *J. Lumin.* 196 (2018) 275–280. https://doi.org/10.1016/j.jlumin.2017.12.050

[52] I.V.B. Maggay, K.Y. Yeh, B. Lei, M.G. Brik, M. Piasecki, W.R. Liu, Luminescence properties of Eu^{2+}-activated $NaCaBeSi_2O_6F$ for white light-emitting diode applications, *Mater. Res. Bull.* 100 (2018) 26–31. https://doi.org/10.1016/j.materresbull.2017.11.059

[53] G. Wu, J. Xue, X. Li, Q. Bi, M. Sheng, Z. Leng, A novel red-emitting $Na_5W_3O_9F_5$: Eu^{3+} phosphor with high color purity for blue-based WLEDs, *Ceram. Int.* 49 (2023) 10615–10624.

[54] Y. Xie, X. Geng, Z. Hu, Z. Zhou, L. Zhou, Y. Zhao, J. Chen, W. Chen, N. Gong, B. Deng, R. Yu, Synthesis and photoluminescence properties of novel orange-emitting Sm^{3+}-activated $LaTiSbO_6$ phosphors for WLEDs, *Opt. Laser Technol.* 147 (2022) 107605. https://doi.org/10.1016/j.optlastec.2021.107605

[55] P. Yu, R. Cui, X. Gong, X. Zhang, C. Deng, Photoluminescence properties of $Ba_3In_2WO_9:Eu^{3+}$ a novel red-emitting phosphor for WLEDs, *J. Mater. Sci. Mater. Electron.* 33 (2022) 14882–14893. https://doi.org/10.1007/s10854-022-08406-3

[56] R.A. Ganesh Kumar, K. Balaji Bhargav, P. Aravinth, K. Ramasamy, P. Shashwati Sen, Tunable photoluminescence properties of Dy^{3+} doped LLZO phosphors for WLED and dosimetry applications, *Ceram. Int.* 48 (2022) 1402–1407.

[57] C.D. Youzhen Shi, Ruirui Cui, Xinyong Gong, A novel red phosphor $Ca_2YNbO_6:Eu^{3+}$ for WLEDs, *Luminescence.* 37 (2022) 1343–1351.

[58] P. Dang, G. Li, X. Yun, Q. Zhang, D. Liu, H. Lian, M. Shang, J. Lin, Thermally stable and highly efficient red-emitting Eu^{3+}-doped $Cs_3GdGe_3O_9$ phosphors for WLEDs: non-concentration quenching and negative thermal expansion, *Light Sci. Appl.* 10 (2021). https://doi.org/10.1038/s41377-021-00469-x

[59] V.B.T. Priyanka Sehrawat, R.K. Malik, R. Punia, S.P. Khatkar, Augmenting the photoluminescence efficiency via enhanced energy-relocation of new white-emanating $BaYAlZn_3O_7:Dy^{3+}$ nano-crystalline phosphors for WLEDs, *J. Alloys Compd.* 879 (2021) 160371.

[60] P.T.H.M.T. Tran, Nguyen Tu, N.V. Quang, D.H., Nguyen, L.T.H., Thu, D.Q., Trung PhD b c d, excellent thermal stability and high quantum efficiency orange-red-emitting $AlPO_4:Eu^{3+}$ phosphors for WLED application, *J. Alloys Compd.* 853 (2021) 156941.

[61] J. Liang, B. Devakumar, L. Sun, G. Annadurai, S. Wang, Q. Sun, X. Huang, Synthesis and photoluminescence properties of a novel high-efficiency red-emitting $Ca_2LuSbO_6:Eu^{3+}$ phosphor for WLEDs, *J. Lumin.* 214 (2019) 116605. https://doi.org/10.1016/j.jlumin.2019.116605

[62] X. Wang, J. Wang, X. Li, H. Luo, M. Peng, Novel bismuth activated blue-emitting phosphor $Ba_2Y_5B_5O_{17}:Bi^{3+}$ with strong NUV excitation for WLEDs, *J. Mater. Chem. C.* 7 (2019) 11227–11233. https://doi.org/10.1039/c9tc03729b

[63] Z. Li, C. Zhang, B. Li, C. Lin, Y. Li, L. Wang, R. Xie, Large-scale room-temperature synthesis of high-efficiency lead-free perovskite derivative $(NH_4)_2SnCl_6$: Te phosphor for warm wLEDs, *Chem. Eng. J.* 420 (2021) 129740. https://doi.org/10.1016/j.cej.2021.129740

[64] Y.Z. Qinling Li, C. Chen, Y. Qiao, B. Yu, B. Shen, High thermal stability of green-emitting phosphor $NaBaB_9O_{15}:Tb^{3+}$ via energy compensation, *J. Alloys Compd.* 897 (2022) 163131.

[65] Y. Zhang, Z. Zhang, X. Liu, G. Shao, L. Shen, J. Liu, W. Xiang, X. Liang, A high quantum efficiency $CaAlSiN_3$: Eu^{2+} phosphor-in-glass with excellent optical performance for white light-emitting diodes and blue laser diodes, *Chem. Eng. J.* 401 (2020) 125983. https://doi.org/10.1016/j.cej.2020.125983

[66] Y.R. Parauha, S.J. Dhoble, Photoluminescence and electron-vibrational interaction in 5d state of Eu^{2+} ion in $Ca_3Al_2O_6$ down-conversion phosphor, *Opt. Laser Technol.* 142 (2021) 107191. https://doi.org/10.1016/j.optlastec.2021.107191

[67] S.R. Bargat, Y.R. Parauha, G.C. Mishra, S.J. Dhoble, Combustion synthesis and spectroscopic investigation of $CaNa_2(SO_4)_2:Eu^{3+}$ phosphor, *J. Mol. Struct.* 1221 (2020) 128838. https://doi.org/10.1016/j.molstruc.2020.128838

[68] Y.R. Parauha, S.J. Dhoble, Structural and photoluminescence properties of Dy^{3+}-activated $NaCaPO_4$ phosphor derived from solution combustion, *Luminescence.* 37 (2022) 141–152. https://doi.org/10.1002/bio.4155

[69] L.T.T. Vien, N. Tu, D.X. Viet, D.D. Anh, D.H. Nguyen, P.T. Huy, Mn^{2+}- doped Zn_2SnO_4 green phosphor for WLED applications, *J. Lumin.* 227 (2020) 117522. https://doi.org/10.1016/j.jlumin.2020.117522

[70] Y. Zhou, S. Zhang, X. Wang, H. Jiao, Structure and luminescence properties of Mn^{4+} -activated $K_3TaO_2F_4$ red phosphor for white LEDs, *Inorg. Chem.* 58 (2019) 4412–4419. https://doi.org/10.1021/acs.inorgchem.8b03577

CHAPTER 6

Spectroscopic Properties of Rare-Earth Activated Energy-Saving LED Phosphors

M. D. Mehare
Priyadarshini College of Engineering, Nagpur, India

Chaitali M. Mehare
R.T.M. Nagpur University, Nagpur, India

Vijay Singh
Konkuk University, Seoul, Republic of Korea

Sanjay J. Dhoble
R.T.M. Nagpur University, Nagpur, India

6.1 INTRODUCTION

Solid-state lighting is an evolving technology in the recent era of development, in which energy efficiency and low-power consumption become important criteria for light and illumination sources and other pervasive products. In general, the average home energy budget requires approximately 25% electric lighting, and in the world nearly 20% of energy is required for lighting and displays. Hence, major research is inclined towards the development of eco-friendly, cost-effective, low-energy consumables along with high operational life, light-emitting devices (W). At present it is virtually certain these will displace all traditional lighting technologies such as incandescent, fluorescent, and high-intensity discharge lighting [1, 2]. The light-emitting diode works on the principle of electroluminescence that was first reported by Henry Joseph in 1906, when the first LED was fabricated using carborundum (SIC). Continuous progress followed over the next 50 years until concerted research into semiconductors began. In the year 1950, the journey of light emission was carried out at the p-n junction using different semiconductor materials, but the obtained results revealed low intensity light that was very expensive and only available

DOI: 10.1201/9781003315261-8

in very few colors [3]. Nonetheless, this motivated continuous research into the evolution of improved performance of LEDs. In the year 1960, Texas Instruments developed the LED to exhibit emission in the region of 870 nm, but it was very expensive and, hence, sold in very low volumes. In 1964, IBM reported GaAsP LEDs used as on-off indicator lights in computers and this was further perceived as a significant breakthrough. Further, these LEDs were mounted directly onto the circuit board. Their low power consumption as well as high operational stability reduced maintenance, which further surpassed other traditional light sources. In the year 1990, Nakamura and co-workers reported the synthesis of low resistive p-type GaN and AlGaN using Mg-doping produced after growth and thermal annealing in N_2 atmosphere [4], which further supported the development of bright blue-emitting GaInN/AlGaN LEDs, along with blue-green LEDs [5, 6]. The progressive innovations by Nakamura and co-workers resulted in advancements in InGaN LED technology, which initiated the wide-scale commercialization of blue and green solid-state sources, as well as the development of white LEDs. Around 1997, a prototype white solid-state lighting was reported using red, green, and blue LEDs. Recently the combination of blue InGaN LED chip with yellow YAG: Ce^{3+}phosphors is the most common approach adopted for the development of the white light-emitting diode (WLED). Unfortunately, their use is greatly limiting their commercialization owing to their poor color rendering index, and high correlated color temperature (CCT), which occur due to deficiency of a red-light-color-emitting component [7–10]. An alternative way to overcome this problem and obtain a WLED is the combination of UV or near-UV(n-UV) LED chips with tri-color (red, green, and blue) phosphors, which can provide superior color uniformity with a high color-rendering index and excellent quality of white light. However, owing to the strong re-absorption of blue light by red/green-emitting phosphors and large Stokes shift, the luminous efficiency is low in this approach [11–13]. Hence, there is significant research interest in looking for appropriate alternatives with high thermal stability and high color-rendering properties. The use of single compositional phosphors (instead of multi-component phosphors) could be an efficient way to improve the performance of white LEDs because it may reduce some variations in a phosphors blend [14–16]. This approach generally allows a controllable adjustment of the emission in complementary color regions, to obtain (by suitable mixing) white light. In actuality, lanthanide ions, commonly known as rare-earth ions, reveal wide use in various optical applications and play a very important role in modern lighting, displays, biological labels, fluorescent falsification-prevention, and other optical-electronic devices, owing to their excellent luminescence properties based on their 4f–4f or 5d–4f transitions [17, 18]. With an ever-growing demand for phosphor and other optical materials based on lanthanide, research interest is inclined towards systematic studies of their optical properties. Widely used rare-earths are: Eu^{3+}, Tb^{3+}, Tm^{3+}, Sm^{3+}, Dy^{3+}, among others, which normally exhibit intense emission in the red, green, and blue region in the host phosphor. The red emission in Eu^{3+} ions is attributed to $^5D_0 \rightarrow {}^7F_J$ transitions (J = 0, 1, 2, 3, 4), however, the green emission in Tb^{3+} ions is owing to $^5D_3 \rightarrow {}^7F_J$ (J = 3, 4, 5, 6) and blue emission from Tm^{3+} ions originating from the transitions of $^1D_2 \rightarrow {}^3F_4$ and $^1G_4 \rightarrow {}^3H_6$

transition. The desired luminescence can be achieved by the appropriate doping of the rare-earth in a suitable host. The present chapter focuses on the spectroscopic and electronic structure of rare-earths along with the effect of their doping on the luminescence characteristics of the phosphor, and also briefly on the impact of energy transfer via singly, doubly doping and enlightening its luminescent mechanism.

6.2 FUNDAMENTAL AND ELECTRONIC STRUCTURE OF RARE-EARTH IONS

The rare-earth metals by definition are the Group IIIb elements including Sc, Y, La, and the 14 lanthanides Ce–Lu in the periodic table. However, in actuality, the term 'rare-earth' has often been more prominently applied to lanthanides with the exclusion of Sc, Y, and La. Moreover, they belong to a unique family of elements characterized by a trivalent state with high symmetry among their physical and chemical properties. Currently, these metals have gained a prime position in modern technologies ranging from cell phones and televisions to LED light bulbs and wind turbines [19]. In general, the lanthanide elements are classified into two groups namely: (1) light rare-earth elements (LREEs), including lanthanum (57) to europium (63), and (2) heavy rare-earth elements (HREEs), including gadolinium (64) to lutetium (71). The HREEs differ from the LREEs in that they have 'paired electrons' (clockwise and counter-clockwise spinning electrons). However, yttrium (39) is the lightest REE but is usually included in the HREEs owing to similar physical and chemical properties. Scandium (21) shows sufficiently different properties and, hence, is not classified as LREE or HREE [20]. The quantum number (n) and angular quantum number (*L*) describe the electronic configuration of an atom in the ground state. In general, lanthanide elements show two types of electronic configurations including $[Xe]4f^n6s^2$ and $[Xe]4f^{n-15}d^16s^2$, where *n* represents the principal quantum number that varies from 1 to 14, and [Xe] indicates the electronic configuration of xenon given by $1s^2\ 2s^2\ 2p^6\ 3s^2\ 3p^6\ 3d^{10}\ 4s^2\ 4p^6\ 4d^{10}\ 5s^2\ 5p^6$. In the series of rare-earths, lanthanum, cerium, and gadolinium correspond to electronic configuration $[Xe]4f^n6s^2$, while other such as praseodymium, neodymium, promethium, samarium, europium, terbium, dysprosium, holmium, erbium, thulium, ytterbium, and lutetium ascribe to $[Xe]4f^{n-15}d^16s^2$ electronic configuration, Moreover, though 4f electrons are not present in scandium and yttrium, owing to similar physical and chemical properties, they are considered as lanthanide elements [21, 22]. In addition to this, one of the most important characteristics associated with lanthanide is lanthanide contraction, which describes a progressive decrease in atomic and ionic radii with an increase in nuclear charge from La to Lu owing to the enhancement of electrostatic repulsion among the electrons [23]. This phenomenon corresponds to a large increase in the effective positive charge over $5s^2$ and $5p^6$ electrons with increasing atomic numbers, as an effect of the low shielding power of 4f electrons. Additionally, the lanthanide contraction is usually ascribed to relativistic effects due to the high atomic numbers of these elements [24, 25]. In the first approach, $4f^n$ configurations are described by three quantum numbers of angular momentum, namely spin angular momentum (S), orbital angular momentum (L), and total angular momentum (J), in which the spin-orbit coupling (LS) is assumed to obey the Russell-Saunders scheme. In free ions, energy levels are initially affected by the central field which depends upon the principal and azimuthal quantum numbers and separates the different configurations by the order of 10^5 cm^{-1}. However, Coulomb interaction removes the degeneracy of the 4f configuration which is the configuration into $^{(2S+1)}L$ level separated by 10^4 cm^{-1}. Each of these levels is further affected by spin-orbit interaction which results in the splitting of the term into $^{(2S+1)}L_J$ multiplets exhibiting the typical separation of 10^3 cm^{-1}. Finally, the crystal-field interaction results in the loss of $^{(2S+1)}L_J$ levels, which can split into a maximum of $(2J + 1)$ Stark components with energy separations of 10^2 cm^{-1} [21]. The crystal-field parameters depend upon site symmetry and the nature of other ions and, hence, reveal the variation between crystals. However, the free-ion interactions and, therefore, the positions of the multiplets, are almost independent of the crystal. The basic properties regarding 4f configurations can be represented for each trivalent lanthanide in a practically universal scheme applicable to any chemical environment. The crystal-field splitting is small owing to the lack of sensitivity of the f orbital to the environment of the REE ions, the width of the levels in the Dieke diagram indicates the magnitude of the crystal-field splitting.

6.3 PRINCIPLE OF SELECTION RULES

In the electronic system, the atomic orbital is represented by a wave function. In a non-degenerate state, it must be orthogonal, known as a Hermitian condition. The different electronic states, say i and j represented by a wave function ψ_i and ψ_j in the particular electronic configuration, satisf they condition $\int \psi_i \psi_j d\tau = 0$, where $d\tau$ represents all the coordinates of the system. The photons can be absorbed or emitted by the system followed by the transition from state i to state j and can be possible if this integral is assumed to be non-zero owing to perturbation operating over the two non-degenerate electronic states. The amount of energy absorbed or emitted is the energy difference between states i and j. However, the transition between the electronic states can be possible only when the perturbation can obey the specific requirement termed as the selection rule. The allowed transition from state i to state j is attributed to $\int \psi_i^* \alpha \o_j \, d\tau \neq 0$, where ψ_i^* is the complex conjugate of wave function and α is the perturbation operator, also known as the transition momentum operator over the system obtained as a result of the interaction of the system with electric or magnetic components of the electromagnetic radiation. The selection rule is normally applied to angular momenta and parities of photon and atomic/molecular wave functions. The integral of allowed transition is assigned by Dirac notation represented as $<i |\alpha| j>$ in which bra $<i|$ is the complex conjugate of the state i and ket $|j>$ is the wave function of state j. The zero values of the integral assigned to the forbidden transition represent a low probability of existence is very low. There are three mechanisms that must be considered for the interpretation of the observed transitions: (1) electric dipole transition (2) magnetic dipole transition, and (3) electric quadrupole transition [26].

1) **Electric dipole transition**

   The majority of observed optical transitions in rare-earth ions are electric dipole transitions, which arise due to the angular momentum of the photon involved in each case. In actuality, the photons are boson that obey Bose-Einstein Statistics with unitary spin ($s = 1$) and intrinsic odd parity ($\pi = -1$). The total angular momentum of the photon is not zero($J \neq 0$), as the spin (s) and orbital angular momentum (L) are not independent of each other. The electronic transition associated with photons having $s = 1$, $L = 0$, $J = 1$, and $\pi = -1$ in interaction with electric components of radiation is known as electric dipole transition. In the electric dipole transition the perturbation operator has odd parity, the matrix element for transitions $r_{ij} = \langle i|er|j\rangle$ is necessarily zero unless state i and j have opposite parity. According to Laporte's selection rule, the intra-configurational electric dipole transition is not allowed and hence are forbidden transition. In the case of rare-earths, all states of a single $4f^n$ configuration have the same parity, and hence all optical transitions within the configuration are strictly forbidden. But owing to several considerations Laporte's selection rule is relaxed. First, if the crystalline environment lacks inversion symmetry, the crystal field admixes a small fraction of the excited configurations of opposite parity into the ground configuration, which makes such transitions weakly allowed. Such transitions, their probabilities and intensities are further evaluated with *Judd-Ofelt* theory [27, 28]. They are strongly linked to the symmetry of the site where the rare-earth is located.

2) **Magnetic dipole transition**
A magnetic dipole transition is owing to the interaction of the spectroscopic active ion with the magnetic field component of the light through a magnetic dipole. In this transition, the photons which exhibit absorption or emission phenomena possess orbital angular momentum of unit value ($L = 1$) which reveals even parity ($\pi = 1$). The total angular momentum (J), which is a combination of angular momentum ($L = 1$) and spin angular momentum ($s = 1$), possesses two states denoted as $J = 1$ and $J = 2$ since $J = 0$ is not possible for photons. When the photon with unit orbital angular momentum ($L = 1$) is characterized by unitary total angular momentum ($J = 1$), and intraconfigurational electronic transitions, it is not the oscillating electric vector of radiation that interacts with the 4f electrons of Ln^{3+}, but the oscillating magnetic vector of the radiation and hence called magnetic dipole transition. The intensity associated with magnetic dipole transition is very weak; about 10^{-6} times that of allowed electronic transition [29].

3) **Electric quadrupole transition**
If the photon corresponds to unitary orbital angular momentum ($L = 1$) and with total angular momentum 2 ($J = 2$), the transition associated with such electronic component of radiation is termed as electric quadrupole transition. In actuality, the electric quadrupole transition arises from a displacement of charge that has a quadrupole nature. An electric quadrupole has even parity and such transitions are very much weaker of the order of 10^{-6} as compared to magnetic dipole and electric dipole transitions. The S, L, and J selection rule for the various type of transition is depicted in Table 6.1.

TABLE 6.1 Selection Rule of Different Electric and Magnetic Transitions of Lanthanide Ions

Interaction Type	Mode of Interaction	Photon Parity (π)	Selection Rule
Electric	Dipole	−1 (odd) $\pi_a = -\pi_b$	$\Delta s = 0$, $\Delta L = 0, \pm1$, $\Delta J = 0, \pm1$; ($J = 0 \leftrightarrow J' = 0$ and $L = 0 \leftrightarrow L' = 0$ are not allowed)
	Quadrupole	+1 (even) $\pi_a = \pi_b$	$\Delta s = 0$, $\Delta L = 0, \pm1, \pm2$ $\Delta J = 0, \pm1; \pm2$ ($J = 0 \leftrightarrow J' = 0, 1$ and $L = 0 \leftrightarrow L' = 0, 1$ are forbidden)
Magnetic	Dipole	+1 (even) $\pi_a = \pi_b$	$\Delta s = 0$, $\Delta L = 0$, $\Delta J = 0, \pm1$; ($J = 0 \leftrightarrow J' = 0$ are not allowed)
	Quadrupole	−1(odd) $\pi_a = -\pi_b$	$\Delta s = 0, \pm1$ $\Delta L = 0, \pm1$ $\Delta J = 0, \pm1; \pm2$ ($J = 0 \leftrightarrow J' = 0, 1$ and $L = 0 \leftrightarrow L' = 0$ are forbidden)

6.4 BASIC ASPECT OF LIGHT EMISSION BY RARE-EARTH ACTIVATED PHOSPHORS

6.4.1 Light Emission by 4f-4f Transition

The 4f-4f transition is also termed as an intra-configurational transition which describes the transition of electron between the different energy levels of the 4f orbital of the same rare-earth ions. In actuality, according to the parity selection rule, these transitions are forbidden as an electronic transition between the energy levels of the same parity does not exist [30]. Though the 4f-4f electronic transitions are forbidden, still this transition occurs due to the relaxation of the parity rule owing to perturbation such as electron vibration coupling and uneven crystal field terms from the host [31]. The crystal-field effect in the host lattice is less effective as the 4f electrons of Ln^{3+} lanthanides ions are protected by the external field of 5*s* and 5*p* electrons. Therefore, each lanthanides ion (Ln^{3+}) is uniquely characterized by its energy levels, which are independent of host materials revealing a narrow *f-f* absorption emission band. The energy levels along with their possible energy transitions of 4fn levels for 4f-4f transitions are well discussed by Dieke [32] and Wybourne [33]. The various Ln^{3+} lanthanides ions like Eu^{3+}, Sm^{3+}, Dy^{3+}, Tb^{3+}, Gd^{3+}, and Yb^{3+} possess 4f-4f transition.

6.4.2 Light Emission by 4f-5d Transition

The rare-earth activated phosphor consists of rare-earth as an active center and inorganic crystalline phosphor as a host material. In actuality, there are various rare-earths like Ce^{3+}, Tb^{3+}, Er^{3+}, Pr^{3+}, Eu^{2+} that can be excited via suitable sources with a desired excitation wavelength and excite from the state of $4f^n$ configuration to the excited state $4f^{n-1}d^1$ configuration and exhibit emission via an intraband 4f→4f transition or 5d→4f transition. As per the selection rule, the transition corresponding to multiple states of $4f^n$ configuration is forbidden and that of 4f→5d transition is a dipole-allowed transition [34]. Moreover, 4f→5d transition is strongly influenced by host materials and exhibits the emission range from deep UV to far IR region as 4f-state and 5d states are diffuse and overlap with ligand orbitals. Hence, it is observed that various rare-earth activated host materials compose of 4f→5d transition reveal the absorption in UV or blue region with the corresponding emission in the visible region. The Ce^{3+} and Eu^{2+} ions widely show 4f→5d transition and hence inorganic phosphor doped with Ce^{3+} and Eu^{2+} reveals wide application for WLEDs. The electronic configuration of Ce^{3+} ions consists of one electron in a 4f shell. This ion exhibits absorption and emission via 4f→5d transition. Under UV excitation Ce^{3+} activation phosphor exhibits emission from UV to red region depending upon the host materials and hence shows applications in the field of scintillation and UV absorption filters. The Eu^{2+} ions also exhibit absorption and emission in the same manner and $4f^7$ excited state levels are higher in energy than that of the $4f^65d^1$ configuration. Its emission varies from host to host, as in a fluoride host it shows emission in the UV region, blue in a phosphate, and red in a sulfate host, attributed to $4f^n$-$4f^{n-1}5d$ transition. Photonic energies are associated with these transitions, which are highly sensitive and are affected by the crystalline environment of the host.

6.4.3 Concentration Quenching

The pure inorganic crystalline phosphor does not exhibit the desired emission under UV or visible absorption and hence it is co-doped with an activator to achieve the expected fluorescent yield (η), which is described as the number of photons emitted per photon absorbed by the activator. Experimentally, it is observed that an increase in dopant concentration in the host results in enhancement in luminescent intensity up to a certain concentration, beyond which luminescence intensity goes on decreasing with further increase in dopant concentration. This phenomenon is termed as concentration quenching. The resonant energy transfer between some ion species is responsible for the phenomenon of concentration quenching. When the ions are close to each other there is a possibility of an increase in energy transfer. This energy migration enhances the possibility of excitation energy trapping at defect or impurity sites showing non-radiative energy losses, resulting in a decrease in emission intensity [35].

6.5 RARE-EARTH ACTIVATED PHOSPHORS

6.5.1 $SrAl_{12}O_{19}:Dy^{3+}$ Phosphor

Solid-state lighting is becoming an encouraging lighting technique, revealing the primary focus of attention on the development of highly efficient white light-emitting diodes, and it is in a position to replace traditional lighting sources soon. In actuality, the Dy^{3+} activator plays a very crucial role in white light generation as its emission spectrum shows two dominant bands centered at the blue and yellow regions, and the position of the spectrum greatly relies on the effect produced by the crystal field of the host lattice [36]. These two emission bands correspond to $^4F_{9/2}$→$^6H_{15/2}$ and $^4F_{9/2}$→$^6H_{13/2}$ and exhibit dominance of emission in blue (480 nm) and yellow (580 nm) respectively [37]. Recently Dev et al. reported the development of Dy^{3+} singly doped $SrAl_{12}O_{19}$ phosphor via the combustion method [38]. All the diffraction peaks show well, indexing with standard data confirmed from X-ray diffraction (XRD) analysis. Validating the formation of prepared phosphor also exhibits its uniform substitution of Dy^{3+} ions into the Sr sites. The surface morphology analyzed from 'field-emission scanning electron microscopy (FE-SEM) micrographs exhibits foamy, non-uniform shapes, agglomerated nature occurs owing to irregular mass flow, and non-uniform temperature distribution during the combustion process along the elemental detection is confirmed from energy-dispersive spectra (EDS) analysis. The antisymmetric stretching vibrations corresponding to a particular wavelength are explained from Fourier-Transform Infrared (FTIR) analysis.

The photoluminescence excitation spectrum monitor corresponds to an absorption wavelength of 478 nm, and 573 nm exhibits the several excitation peaks positioned at 324, 350, 364, and 387 nm corresponding to the transition of Dy^{3+} ions from ground state $^6H_{15/2}$ to higher energy levels $^6P_{3/2}$, $^6P_{7/2}$, $^6P_{5/2}$ and $^4F_{7/2}$ respectively [39]. The intense excitation peaks are centered at 350 nm and 387 nm. Upon the excitation of 350 nm, it shows emission spectrum centered at two regions at wavelength positions at 478 nm and the other at 573 nm corresponding to $^4F_{9/2} \rightarrow {}^6H_{15/2}$ and $^4F_{9/2} \rightarrow {}^6H_{13/2}$ transitions of Dy^{3+} ions in the host. A similar emission can be observed at an excitation of 387 nm as shown in Figure 6.1(a) and (b).

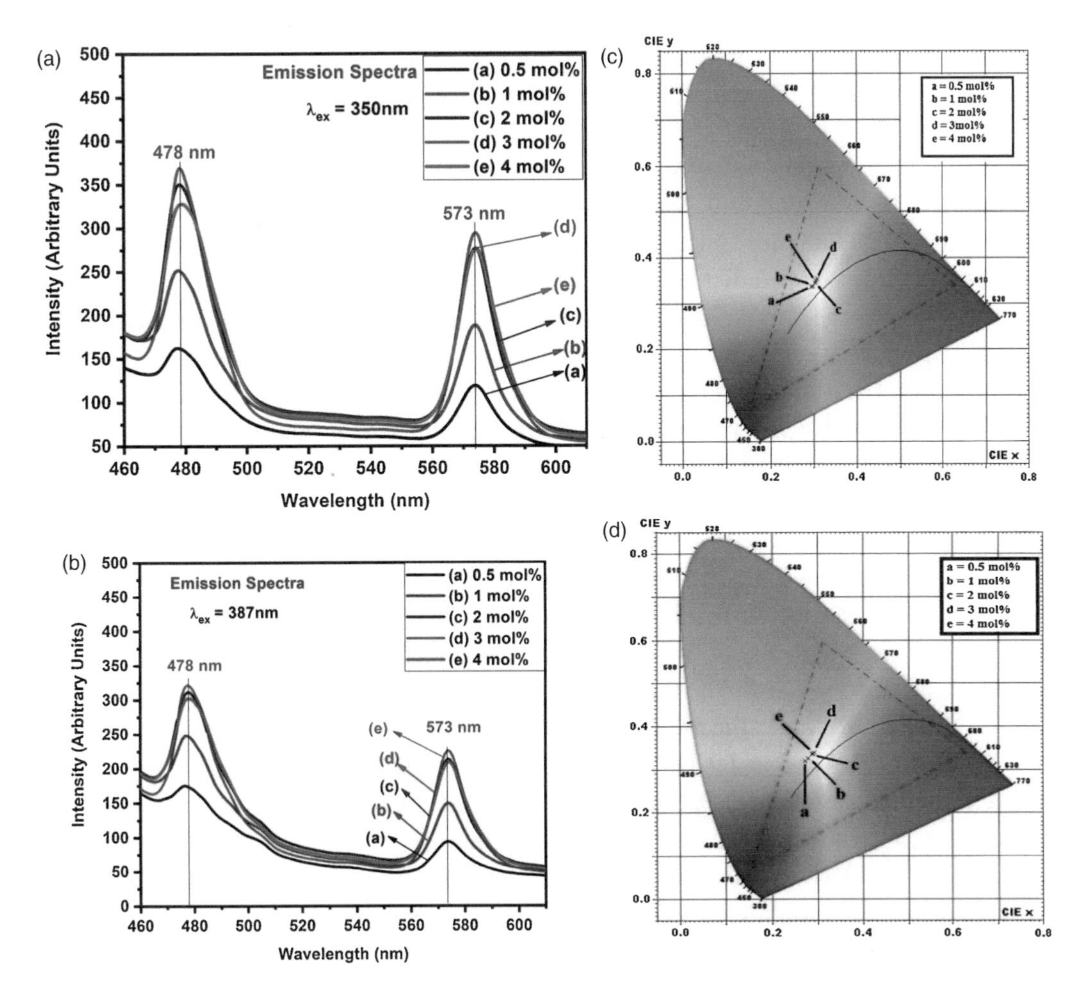

FIGURE 6.1 (a) Photoluminescence emission spectra of $SrAl_{12}O_{19}:Dy^{3+}$ phosphor monitored at 350 nm excitation; (b) Photoluminescence emission spectra of $SrAl_{12}O_{19}:Dy^{3+}$phosphor monitored at 387 nm excitation; (c) CIE Chromaticity Diagram for $SrAl_{12}O_{19}:nDy^{3+}$ ($n = 0.5$, 1, 2, 3 and 4 mol%) phosphor at $\lambda_{ex} = 350$ nm; (d) CIE Chromaticity Diagram for $SrAl_{12}O_{19}:nDy^{3+}$ ($n = 0.5$, 1, 2, 3 and 4 mol%) phosphor $\lambda_{ex} = 387$ nm. (Reproduced with permission from Ref. [38], © 2019 Elsevier GmbH.)

It shows prepared phosphor is suitable for blue and yellow emissions. The CIE chromaticity corresponding to an excitation wavelength of 350 nm and 387 nm is depicted in Figure 6.1 (c) and (d), which show that CIE coordinates vary with the activation of Dy^{3+} ions in the host and lie in the white region followed by CCT = 6542 K. The obtained outcomes reveal reported phosphor is favorable for single-component white-light-emitting, UV convertible phosphor for WLEDs.

6.5.2 $LiBaB_9O_{15}$:Eu^{3+} Phosphor

In recent years, borate-based phosphors have been extensively studied followed by doping of various concentrations of Eu^{3+} ions with excellent luminescent intensity, and have been widely reported. Zhang et al. reported the series of Eu^{3+} singly doped $LiBaB_9O_{15}$ borate-based phosphor, synthesized via high-temperature solid reaction method, and also investigated the effect of excess addition of H_3BO_3 on the phase purity and luminescent characteristics of prepared phosphor [40]. The crystal structure of $LiBaB_9O_{15}$ material belongs to a trigonal system with the space group R-3c. The emission spectra of $LiBaB_9O_{15}$:xEu^{3+} (x = 0.07–0.95), with the 15% excess H_3BO_3, showed emission at 612 nm under UV excitation followed by gradual enhancement with an increase in the concentration of Eu^{3+} ions up to 0.59mol%. However, with further increase in the concentration of Eu^{3+} ions, emission intensity decreases owing to the quenching effect normally associated with the non-radiative energy transfer, which may occur due to radiation re-absorption, exchange interaction, or multipole-multipole interaction [41]. The exclusion of overlapping between excitation and emission spectra revealed a non-existence of radiation re-absorption. The critical distance (R_c) between Eu^{3+} in the host obtained to be 9.86 Å representing the existence of exchange interaction. From the Dexter theory, the electric multipolar character (θ) obtained to be 11.31 A° (which is close to 10) indicates the dominance of quadrupole-quadrupole interaction for energy transfer responsible for the concentration quenching mechanism. The PLE spectra monitor at 614 nm gives several excitation peaks in the range of 200–500 nm. The existence of a broad band at 200–290 nm is attributed to $O^{2-}\rightarrow Eu^{3+}$ charge transfer band (CTB); moreover, several excitation peaks are observed at 319, 361, 381, 394, 413, and 464 nm corresponding to $^7F_0\rightarrow{}^5H_6$, $^5D_4$, $^5G_2$, $^5L_6$, $^5D_3$, $^5D_2$ transition of Eu^{2+} ions in the host represent the suitability of prepared phosphor for UV or n-UV excitation. Under the excitation wavelength of 394 nm it exhibits a narrow emission band as located at 586, 597, 612, 650 and 697 nm, attributed to $^5D_0\rightarrow{}^7F_J$ (J = 1, 2, 3, 4) respectively, and transition of Eu^{3+} ions in the host. The intense peak is observed at 612 nm. The obtained CIE coordinates for reported phosphor are (0.6415, 0.3411) and are close to standard red light CIE coordinates (0.67, 0.33). The results are clear: the prepared phosphor will be a potential candidate for WLEDs applications in solid-state lighting.

6.5.3 Y_2O_2S:Eu^{3+} Phosphor

In recent years, research has been predominantly focused on the development of novel UV excitable RGB- (red-green-blue) emitting phosphor, specially blended in different weight ratios that exhibit its potential applicability in the field of displays and light-generating systems [42]. In the process of blending, the red component is greater than the blue or green emission component. On this concept recently Sundarakannan et al. synthesized UV excitable Eu^{3+} singly doped Y_2O_2S with ZnO entrapped white-light-emitting phosphor via the sol-gel combustion technique [43]. The reported phosphor revealed excellent luminescent efficiency, followed by intense emission in the red region in comparison with the $Y_2O_3:Eu^{3+}$ phosphor [44]. The XRD analysis exhibited that as-synthesized $Y_2O_2S:Eu^{3+}$ and $Y_2O_2S:Eu^{3+}$ with ZnO entrapped show crystalline hexagonal structure followed by good consistency with standard JCPDS card no. 24-1424. Moreover, the retention peak positions, even with the addition of ZnO, indicate uniform distribution of $Y_2O_2S:Eu^{3+}$ in the amorphous ZnO matrix. The clear formation of the amorphous phase of ZnO with a uniform fusion of crystalline Y_2O_3 in the ZnO matrix exhibits from SEM analysis with confirmation of elements by the EDAX spectrum. Under UV excitation (λ_{ex} = 395 nm) at room temperature, the $Y_2O_3S:Eu^{3+}$ phosphor shows several emission peaks corresponding to $^5D_I \rightarrow {}^7F_J$ (I = 0, 1, 2, 3; J = 0–6) transition of Eu^{3+}, however, the intense emission peak at 625 nm is attributed to$^5D_0 \rightarrow {}^7F_2$ transition of trivalent europium ions in the host. The PL emission of $Y_2O_3S:Eu^{3+}$ trapped with ZnO matrix shows the emission in the visible region in the range of 420–650 nm. The broad band emission at 420–590 nm is owing to an interstitial oxygen defect developed in ZnO, along with sharp peak at 625 nm assigned to Eu^{3+} ions in the host. Remarkable emission intensity is observed for $Y_2O_3S:Eu^{3+}$-ZnO (70:30). The electroluminescence spectrum of $Y_2O_3S:Eu^{3+}$-ZnO (70:30) phosphor under UV (λ_{ex} = 375 nm) excitation exhibits the emission in the wavelength range of 400–700 nm, covering the entire visible region. The blue and orange emissions are due to the creation of a defect in ZnO, however, the red emission is attributed to transition of Eu^{3+} ions in the host Y_2O_3S phosphor. Under applied power of 150mW, fabricated LED shows warm white light emission with CIE color coordinates of (0.44, 0.35), and with color rendering index (CRI) (75) followed by correlated color temperature (CCT) of 2420 K. The obtained results revealed the advantageous utility of the reported phosphor for the development of WLED in mass production.

6.5.4 $Gd_2O_2SO_4:Tb^{3+}$ Phosphor

Luminescent efficiency is greatly influenced by the choice of host material as well as its composition, structural and morphological behavior, and size as well as co-dopant in the host materials [45]. Among the various phosphors, rare-earth doped oxysulfates-based phosphor ($Ln_2O_2SO_4$) exhibits outstanding performance. Various methods are reported for the development of $Ln_2O_2SO_4$-based phosphor. Some of them involve the thermal decomposition of hydrate lanthanide sulfates, but the synthesis process is very slow and time-consuming [46]. Shoji et al. reported high-energy ball milling for the synthesis of $Y_2O_2SO_4:Eu^{3+}$ phosphor but the as-synthesized phosphor is bulky and results in weak luminescent intensity in comparison with a nano-scale phosphor [47]. In addition, co-precipitation, as well as a template-assisted method, are also employed for the synthesis of nanostructure $Ln_2O_2SO_4$ phosphor, but the requirement of complicated post-treatment for the separation and also the purity of material limit its utilization. Song and co-workers reported the synthesis of Tb^{3+} singly doped $Gd_2O_2SO_4$ nanostructure phosphor by the electro-spinning method and also studied the effect of alkali metal ions (Na^+, K^+, Li^+)

co-doping on the luminescence properties of prepared phosphor [48]. The formation of the compound is confirmed from the XRD analysis and exhibits good indexing with JCPDS card nos. 29-0613 and 41-0683 with the exclusion of additional peaks representing effective mixing of Tb^{3+} and M^+ (Li^+, K^+, Na^+) in the host matrix. The PL emission, corresponding to UV excitation ($\lambda_{ex} = 230$ nm) for $Gd_2O_2SO_4$:Tb^{3+} in bulk and nano pieces, exhibits similar emission peaks with the maximum intensity centered at 545 nm in the green region attributed to $^5D_4 \rightarrow ^7F_5$ transition of Tb^{3+} ions in the host along with several emission peaks. It is more clearly observed that the emission intensity of $Gd_2O_2SO_4$:Tb^{3+} nano pieces phosphor is very dominant as compared to that of bulk phosphor attributed to well-exchanged interaction between Tb^{3+} ions in the nano pieces host, as well as direct electron transfer from excited state to $^5D_4$ level, owing to confinement of Tb^{3+} in the host. Also, the effect of co-doping of alkali metal ions (Li^+, Na^+, K^+) exhibits the enhancement of luminescent efficiency by 34% for Li^+ ions as compared to host, demonstrating its potential application in the field of display devices and in light-emitting diodes.

6.6 ENERGY TRANSFER FROM DIFFERENT RARE-EARTH IONS IN ECO-FRIENDLY LED PHOSPHORS

6.6.1 $Ca_8ZnGd(PO_4)_7$: Eu^{2+}, Mn^{2+} Phosphor

Recently, more than the conventional light sources like incandescent and fluorescent lamps, white-light-emitting diodes (WLEDs) have attracted crucial attention owing to their numerous characteristics like excellent brightness, long lifetime, small size, and eco-friendliness [49]. The first commercial WLED was invented around 1996 followed by the combination of blue InGaN chip with yellow $Y_3Al_5O_{12}$:Ce^{3+} phosphor, but its emission in the cold white light region due to deficiency in red light contribution, along with poor CRI and high CCT, limited its commercialization [50]. With a suitable selection of host and activator the above frailties minimize and achieve a single-component full-color emitting phosphor in order to enhance the luminescence efficiency of WLEDs [51, 52]. The crystal structure and phase purity confirmed from the XRD pattern reveal all diffraction peaks are well indexed with standard data (JCPDS card no. 49-502) without impurity peaks, exhibiting uniform incorporation of Eu^{2+}/Mn^{2+} ions in the host. The prepared host shows a rhombohedral crystal structure with space group R3c (161). The photoluminescence characteristic of host and co-doped phosphor was analyzed using PL and PLE spectra. The PLE spectrum of CZGP:0.06Eu^{2+} phosphor exhibits broad excitation in the region from 200 to 430 nm ascribed to $4f^7 - 4f^65d^1$ transition of Eu^{2+}, which shows clear n-UV excitability of prepared phosphor [53]. Upon the excitation wavelength 284 nm and 365 nm it shows emission in the wavelength range of 400–700 nm with green emission centered at 487 nm corresponding to electric-dipole allowed transition from lower level 5d excited state to 4f ground state of Eu^{2+} ions [54]. The PL and PLE spectra of CZGP:0.15Mn^{2+} phosphor exhibit a wide range of emissions in the region of 575–730 nm centered at 641 nm at the excitation wavelength ascribed to spin-forbidden $4T_1(^4G)–^6A_1(^6S)$ transition of Mn^{2+} ions. Furthermore, the overlapping of emission spectra of Eu^{2+} with excitation spectra of Mn^{2+} ions indicate that effective energy transfer (ET) from Eu^{2+} to Mn^{2+} ions in the host phosphor [53]. The emission of Mn^{2+} doped luminescent material corresponds to crystal-field strength and coordination number. Further emission spectra of CZCP:$0.06Eu^{2+}$,yMn^{2+} ($0 \leq y \leq 0.3$) were analyzed at the excitation wavelength 365 nm showing that the emission intensity of Mn^{2+} ions increases gradually with an increase in the concentration of Mn^{2+} ions, followed by a monotonical decrease in emission intensity of Eu^{2+} ions, demonstrating the ET from Eu^{2+} to Mn^{2+} ions in the host. However, beyond $y = 0.15$, a decrease in emission intensity of Mn^{2+} ions is due to the quenching effect [55]. The obtained energy transfer crystal distance (R_c) reveals the dominance of electric multipolar interaction. The plot $(I_{so}/I_s) - C^{n/3}$ shows linear relation is observed only when $n = 8$ and implies that the ET from Eu^{2+} to Mn^{2+} in CZGP host is mainly due to dipole-quadrupole (d-q) mechanism. This was further supported from fluorescence decay curves of Eu^{2+} at excitation wavelength 365 nm exhibiting a decrease of the lifetime of Eu^{2+} from 1.825 μs to 1.021 μs. In addition to this, energy transfer efficiency is enhanced from 0 to 44.03% with increase in the content of Mn^{2+}. Eu^{2+} as the sensitizer absorbs energy and jumps from ground state $^8S_{7/2}$ to $4f^65d$ excited state. Some Eu^{2+} ions made radiative transition, exhibiting green emission and the rest of energy at $4f^65d$ state effectively transfers to Mn^{2+} ions via resonant energy transfer, which results in enhancement of the emission efficiency of Mn^{2+} ions [56]. With an increase in Mn^{2+} ions, the CIE chromaticity diagram shows shifting of CIE coordinates from green to white light with co-ordinates (0.297, 0.298), as well as a decrease in CCT parameters of the prepared phosphor. Co-doping of Eu^{2+}/Mn^{2+} in the host shows color tunable from green to white light reveals its applicability for n-UV excitable WLED devices.

6.6.2 $Ca_6Y_2Na_2(PO_4)_6F_2$:Eu^{2+}, Mn^{2+} Phosphor

Compared to multi-component emitter systems in WLEDs, single component white-emitting phosphors under UV excitation have received more attention owing to their higher stability, remarkable efficiency, excellent color rendering index, and ease of fabrication process [57]. Guo et al. reported apatite type Eu^{2+}, Mn^{2+} co-doped fluorophosphates $Ca_6Y_2Na_2(PO_4)_6F_2$ (CYNP) phosphor synthesized via high temperature conventional solid-state reaction method [58]. The prepared phosphor reveals a hexagonal crystal structure with space group $P6_3/m$ similar to the reported crystal structure $Ca_6Eu_2Na_2(PO_4)_6F_2$ [59, 60]. The XRD analysis exhibits the formation of a single crystal structure with good incorporation of Eu^{2+}/Mn^{2+} ions in the host. The PLE and PL emission spectra of Eu^{2+}, Mn^{2+} co-doped CNYP host, represent overlapping between the emission band of CYNP:$0.01Eu^{2+}$ and the excitation band of CYNP:$0.05Mn^{2+}$ and exhibits effective resonant energy transfer from Eu^{2+} to Mn^{2+} ions in CYNP, as the PLE spectrum which monitors emission of Mn^{2+} ions is well consistent with that emission of Eu^{2+}[61, 62]. In actuality, optimum results can be obtained by exciting the sensitizer (Eu^{2+}) with allowed electronic transition that transfers the excitation energy to the activator (Mn^{2+}), which is expected to reveal white light emission owing to a spectral combination of blue emission of Eu^{2+} ions with yellow emission of Mn^{2+} ions. The PL spectrum exhibits that with an increase in Mn^{2+} ions concentration, there is a progressive increase in the emission intensity of Mn^{2+} with a monotonical decrease in intensity of Eu^{2+} ions revealing the phenomenon of ET from Eu^{2+} to Mn^{2+} ions, which is also supported by the decay curve that represents a continuous depletion in the lifetime of Eu^{2+} from 0.382 to 0.258μs with an increase in Mn^{2+} ion concentration.

The obtained values of $s = 8$ represent dominance of d-q interaction with efficiency of 32.5% as well as excellent thermal stability. The CIE chromaticity diagram of CYNP:0.01Eu^{2+}, xMn^{2+} phosphor, shows coordinates of prepared phosphor are (0.335, 0.337) which are very close to coordinates of pure white light (0.333, 0.333), representing standard white light emission from CYNP:0.01Eu^{2+},0.025Mn^{2+} phosphor and demonstrates that reported phosphor is a potential candidate for UV-based white LEDs.

6.6.3 $Ca_9Mg(PO_4)_6F_2$:Eu^{2+}, Mn^{2+} Phosphor

The apatite structure composes the general formula $M_{10}(AO_4)_6X_2$ where M stands for univalent to trivalent cations including Na^+, K^+, Ca^{2+}, Sr^{2+}, Cd^{2+}, La^{3+}, Cd^{3+}, Y^{3+}, A represent P^{5+}, V^{5+}, Si^{4+}, Ge^{4+} and so on, and X represents F^-, Cl^- and O^{2-}. This structure belongs to space group $P6_3/m$, comprising 9-fold coordinated 4f sites with c_3 point symmetry and 7-fold coordinated 6h sites with c_s point symmetry, revealing a suitable host for rare-earth ions [63]. Li et al. reported the synthesis of Eu^{2+}, Mn^{2+} co-doped $Ca_9Mg(PO_4)_6F_2$(CMPF:Eu^{2+},Mn^{2+}) phosphor via high-temperature solid-state reaction [64]. Rietveld refinement carried out at room temperature of the reported phosphor is depicted in Figure 6.2(a). It exhibits a hexagonal crystal structure with space group ($P6_3/m$) and lattice parameters as $a = 9.350702$ Å, $b = 9.350702$ Å, $c = 6.845183$ Å, $V = 518.327$ Å^3.

All the parameters including atom coordinates, fraction factors, and thermal vibration were refined with convergence and well fitted the reflection condition [i.e. $R_p = 3.99\%$, $R_{wp} = 5.60\%$, and $\chi^2 = 5.06$]. This exhibits that Mg^{2+} ions substitute Ca^{2+} sites with two different crystallographic lattices involving 9-fold coordinated 4f sites with C_3 point symmetry [denoted as Ca^{2+}(1)] and 7-fold coordinated 6h sites with Cs point symmetry [denoted asCa^{2+}(2)] [65] which can be clearly observed from the structure depicted in Figure 6.2(b). The PLE and PL spectra of singly doped Eu^{2+}, Mn^{2+} and co-doped Eu^{2+}/Mn^{2+} ions in CMPF host is shown in Figure 6.3.

With the excitation wavelength 294 nm CMPF:0.18Eu^{2+} exhibits strong emission in the region of 380–580 nm concentrated at 454 nm corresponding to $4f^65d^1 \rightarrow 4f^7$ transition of Eu^{2+} ions shown in Figure 6.3(a). However, at the excitation wavelength of 404 nm CMPF:0.18Mn^{2+} gives broad emission range from 500 to 600 nm centered at 565 nm and ascribed to $4T_{1g}(G) \rightarrow 6A_{1g}(S)$ spin-forbidden transition of Mn^{2+} ions shown in Figure 6.3(b). The overlapping of excitation and emission peaks reveals energy transfer, further validated from Figure 6.3(c).

Further emission spectra of CMPF:0.18Eu^{2+},$y$$Mn^{2+}$ ($0 \le y \le 0.38$) phosphor were monitored at the excitation wavelength $\lambda_{ex} = 365$ nm, depicted in Figure 6.4(a) representing a gradual increase in emission intensity of Mn^{2+} ions with an increase in its ions concentration, followed by a monotonical decrease in the emission intensity of Eu^{2+} becoming maximum at $y = 0.34$, with further increase in the concentration of Mn^{2+} ions emission decreasing due to quenching effect [66]. This represents energy transfer from Eu^{2+} to Mn^{2+} ions in the host which was further verified by the decay curve exhibiting a decrease in the lifetime of Eu^{2+} ions from 495.93 to 274.36ns with an increase in the concentration of Mn^{2+} ions. The occurrence of color-tunability with an increase in the concentration of Mn^{2+} ions is also evidence for energy transfer from Eu^{2+} to Mn^{2+} in the prepared host. The CIE

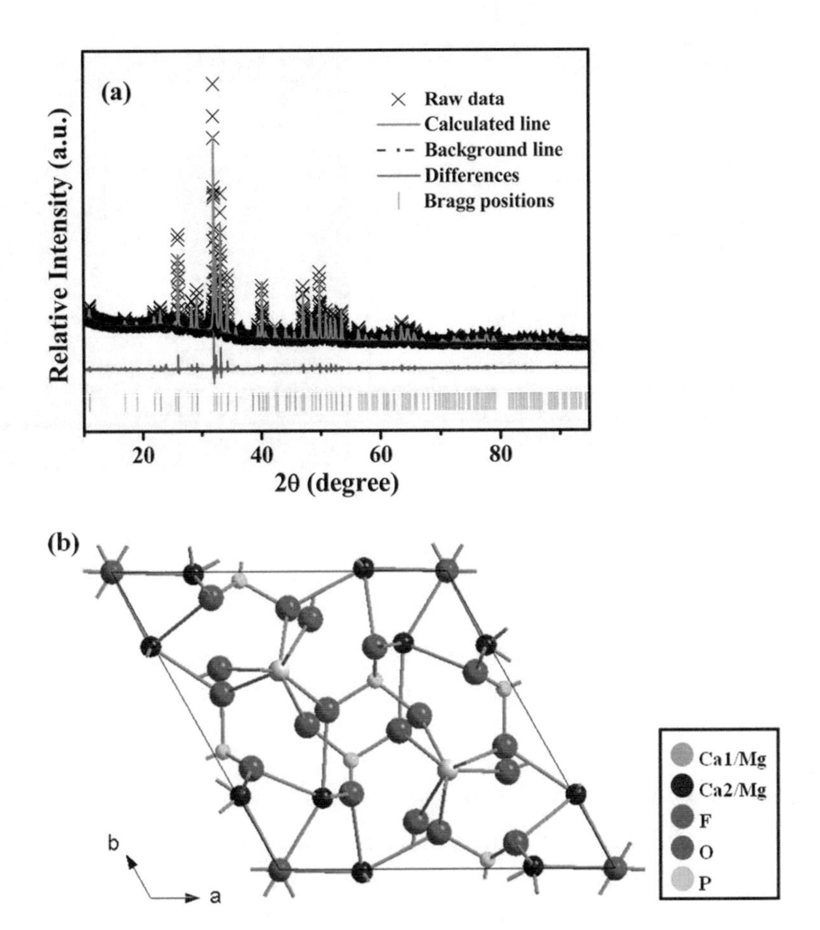

FIGURE 6.2 (a) Rietveld refinement of powder XRD profile of $Ca_9Mg(PO_4)_6F_2$. (b) Crystal structure of Ca_9Mg $(PO_4)_6F_2$ host. (Reproduced with permission from Ref. [64], © 2014, American Chemical Society.)

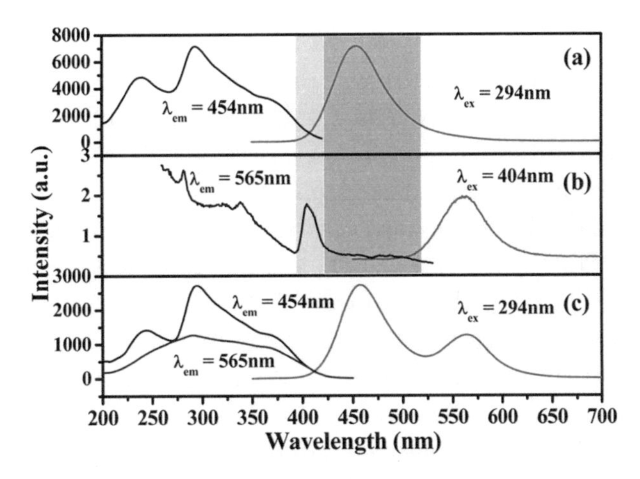

FIGURE 6.3 Excitation and emission spectra of: (a) CMPF:0.18Eu^{2+}, (b) CMPF:0.18Mn^{2+}, and (c) CMPF:0.18Eu^{2+},0.18Mn^{2+}phosphors (Reproduced with permission from Ref. [64], © 2014, American Chemical Society.)

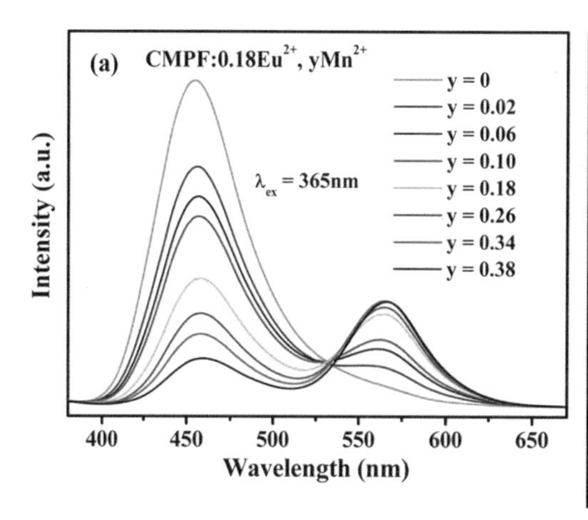

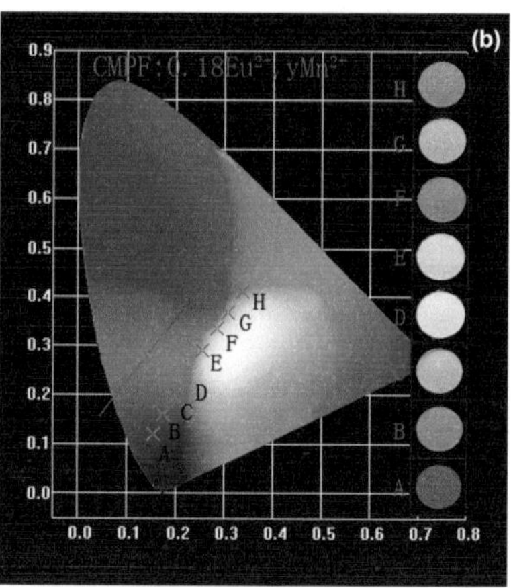

FIGURE 6.4 (a) Variation of emission intensity. (b) CIE chromaticity for CMPF:$0.18Eu^{2+}$,yMn^{2+}($0 \leq y \leq 0.38$) phosphors at excitation wavelength 365 nm. (Reproduced with permission from Ref. [64]. © 2014, American Chemical Society.).

chromaticity coordinate shown in Figure 6.4(b) reveals the variation of color-tunable at an excitation wavelength of 365 nm, representing that the prepared phosphor is suitable for UV-pumped WLEDs.

6.6.4 $Sr_3NaSc(PO_4)_3F$:Eu^{2+},Tb^{3+} Phosphor

In next-generation solid-state lighting, in order to achieve optimum energy saving n-UV excitable WLEDs received close attention owing to their numerous merits such as high luminescence efficiency, power-saving, eco-friendliness, compared to that of conventional light sources such as fluorescent and incandescent lamps. The luminescence efficiency depends upon a combination of phosphor co-doped with rare-earth. In actuality, Tb^{3+} act as promising green-emitting activators and exhibit sharp emission at about 488, 543, 582 and 625 nm corresponding to 4f-4f transition and hence are mainly used to provide green-emitting light in fluorescent materials [67]. Normally, a sensitizer plays a very crucial role in order to enhance the emission efficiency of the activator. As Eu^{2+} shows, a broad absorption band in the n-UV region, owing to its $^4f_7 \rightarrow {}^4f_6{}^5d_1$ electric dipole, allowed transition and hence become appropriate for pumping by UV-LEDs. In addition, the blue emission band of Eu^{2+} is also suitable for sensitizing the f–d spin in allowed as well as forbidden transitions of Tb^{3+} in the UV region via energy transfer, which have been extensively studied in many phosphors, such as $NaMg_4(PO_4)_3$, $CaSi_2O_2N_2$, and $BaCa_2MgSi_2O_8$ [68, 69]. Recently, Yang et al. reported Eu^{2+},Tb^{3+} co-doped color-tunable $Sr_3NaSc(PO_4)_3F$ phosphor by conventional high-temperature solid-state reaction [70]. The XRD analysis revealed prepared phosphor shows a hexagonal structure with space group P_{-6} along with all diffraction peaks beingwell matched with standard JCPDS card no. 50-1595, without impurity peaks representing the formation of single-phase phosphor as well as indicating proper incorporation of Eu^{2+} and Mn^{2+} ions in the host phosphor. The surface morphology of prepared phosphor at different magnifications reveals the uniform distribution of elements over the whole range of particles indicating suitability for the fabrication of phosphor. Blue-emitting phosphor $Sr_3NaSc(PO_4)_3F:Eu^{2+}$ shows an absorption band in the region 200–400 nm reveals well matching with commercial near-UV chips. However, all the absorption bands in this region tend to be intensified and extended with the increase of Eu^{2+} doping content. The overlapping of emission peak of SNScPF:0.03Eu^{2+} and excitation peak of SNScPF:0.1Tb^{3+} reveals the energy transfer from Eu^{2+} to Tb^{3+}. The PL spectrum of SNScPF:0.03Eu^{2+},yTb^{3+} ($0 \leq y \leq 0.5$) phosphor at excitation wavelength 356 nm exhibits that an increase in concentration of Tb^{3+} ions shows a decrease in the emission intensity of Eu^{2+} followed by monotonous increase in emission intensity of Tb^{3+} ions, indicating resonant energy transfer from Eu^{2+} to Tb^{3+} ions in the host via a dipole-dipole mechanism. This is further supported from the fluorescent decay curve that shows gradual decrease in the lifetime of Eu^{2+} ions from 365.6 to 230.5 nm at excitation wavelength 365 nm. The influence of Tb^{3+} concentrations on the CIE shows that the color can be tuned from blue (0.137, 0.148) to green (0.221, 0.391) by adjusting the Eu^{2+}/Tb^{3+} ratio. The prepared phosphor shows potential applications in n-UV excitable LED-based display systems.

6.7 CONCLUSION

Over the past years, there has been vast growth in rare-earth activated phosphor materials yielding a wide range of emission colors (red, green, blue, yellow, and white) and receiving a prime position in the field of displays and illuminations. LED light sources offer significant benefits such as energy savings, long life, and high performance in almost any application and hence are in a position to replace conventional light sources soon. The rare-earth activation enables a myriad of display technologies and exhibits strong and spectrally narrow emissions in the suitable host. The excitation sources such as cathode beam, electric fields, and photons, among others, lead to device configuration and applications. The continuous optimization of host materials, doping and co-doping, and energy transfer mechanism results in a promising display followed by excellent brightness, resolution, color purity, and low power consumption. The present chapter is brief on the fundamental and electronic structure of rare-earth, along with the impact of doping and co-doping on the luminescence performance of phosphor materials.

REFERENCES

[1] N.C. George, K.A. Denault, R. Seshadri, Phosphors for solid-state white lighting, *Annu. Rev. Mater. Res.* 43 (2013) 481–501. https://doi.org/10.1146/annurev-matsci-073012-125702

[2] J. Meyer, F. Tappe, Photoluminescent materials for solid-state lighting: State of the art and future challenges, *Adv. Opt. Mater.* 3 (2015) 424–430. https://doi.org/10.1002/adom.201400511

[3] S. Chhajed, Y. Xi, Y.L. Li, T. Gessmann, E.F. Schubert, Influence of junction temperature on chromaticity and color-rendering properties of trichromatic white-light sources based on light-emitting diodes, *J. Appl. Phys.* 97 (2005). https://doi.org/10.1063/1.1852073

[4] S. Nakamura, N. Iwasa, M. Senoh, T. Mukai, Hole compensation mechanism of p-type GaN Films, *Jpn. J. Appl. Phys.* 31 (1992) 1258–1266. https://doi.org/10.1143/JJAP.31.1258

[5] S. Nakamura, M. Senoh, N. Iwasa, S. Nagahama, T. Yamada, T. Mukai, Superbright green InGaN single-quantum-well-structure light-emitting diodes, *Jpn. J. Appl. Phys.* 34 (1995) L1332–L1335. https://doi.org/10.1143/JJAP.34.L1332

[6] S. Nakamura, M. Senoh, N. Iwasa, S.I. Nagahama, High-brightness InGaN blue, green and yellow light-emitting diodes with quantum well structures, *Jpn. J. Appl. Phys.* 34 (1995) L797–L799. https://doi.org/10.1143/JJAP.34.L797

[7] G.B. Nair, H.C. Swart, S.J. Dhoble, A review on the advancements in phosphor-converted light emitting diodes (pc-LEDs): Phosphor synthesis, device fabrication and characterization, *Prog. Mater. Sci.* 109 (2020) 100622. https://doi.org/10.1016/j.pmatsci.2019.100622

[8] Y.-D. Huh, J.-H. Shim, Y. Kim, Y.R. Do, Optical properties of three-band white light emitting diodes, *J. Electrochem. Soc.* 150 (2003) H57. https://doi.org/10.1149/1.1535914

[9] R. Yu, H.M. Noh, B.K. Moon, B.C. Choi, J.H. Jeong, K. Jang, S.S. Yi, J.K. Jang, Synthesis and luminescence properties of a novel red-emitting phosphor $Ba_3La(PO_4)_3:Eu^{3+}$ for solid-state lighting, *J. Alloys Compd.* 576 (2013) 236–241. https://doi.org/10.1016/j.jallcom.2013.04.150

[10] K. Knorr, M. Meschke, B. Winkler, Structural and magnetic properties of $Co_2Al_4Si_5O_{18}$ and $Mn_2Al_4Si_5O_{18}$ cordierite, *Phys. Chem. Miner.* 26 (1999) 521–529. https://doi.org/10.1007/s002690050215

[11] G.B. Nair, S.J. Dhoble, A perspective perception on the applications of light-emitting diodes, *Luminescence.* 30 (2015) 1167–1175. https://doi.org/10.1002/bio.2919

[12] T. Ogi, H. Iwasaki, A.B.D. Nandiyanto, F. Iskandar, W.N. Wang, K. Okuyama, Direct white light emission from a rare-earth-free aluminium-boron-carbon- oxynitride phosphor, *J. Mater. Chem. C.* 2 (2014) 4297–4303. https://doi.org/10.1039/c3tc32314e

[13] K.H. Kwon, W. Bin Im, H.S. Jang, H.S. Yoo, D.Y. Jeon, Luminescence properties and energy transfer of site-sensitive $Ca6\text{-}_{x-yMgx-z}(PO_4)_4:Eu_y^{2+},Mn_z^{2+}$ hosphors and their application to near-UV LED-based white LEDs, *Inorg. Chem.* 48 (2009) 11525–11532. https://doi.org/10.1021/ic900809b

[14] R. Gautier, X. Li, Z. Xia, F. Massuyeau, Two-step design of a single-doped white phosphor with high color rendering, *J. Am. Chem. Soc.* 139 (2017) 1436–1439. https://doi.org/10.1021/jacs.6b12597

[15] M. Chen, Z. Xia, M.S. Molokeev, T. Wang, Q. Liu, Tuning of photoluminescence and local structures of substituted cations in $xSr_2Ca(PO_4)_{2\text{-}(1\text{-}x)}Ca_{10}Li(PO_4)_7:Eu^{2+}$ Phosphors, *Chem. Mater.* 29 (2017) 1430–1438. https://doi.org/10.1021/acs.chemmater.7b00006

[16] P.P. Dai, C. Li, X.T. Zhang, J. Xu, X. Chen, X.L. Wang, Y. Jia, X. Wang, Y.C. Liu, A single Eu^{2+}-activated high-color-rendering oxychloride white-light phosphor for white light-emitting diodes, *Light Sci. Appl.* 5 (2016) 1–9. https://doi.org/10.1038/lsa.2016.24

[17] X. Liu, C. Li, Z. Quan, Z. Cheng, J. Lin, Tunable luminescence properties of $CaIn_2O_4:Eu^{3+}$ phosphors, *J. Phys. Chem. C.* 111 (2007) 16601–16607. https://doi.org/10.1021/jp074868o

[18] T. Jüstel, H. Nikol, C. Ronda, New developments in the field of luminescent materials for lighting and displays, *Angew. Chemie - Int. Ed.* 37 (1998) 3084–3103. https://doi.org/10.1002/(SICI)1521-3773(19981204)37:22<3084::AID-ANIE3084>3.0.CO;2-W

[19] S. V. Eliseeva, J.C.G. Bünzli, Rare earths: Jewels for functional materials of the future, *New J. Chem.* 35 (2011) 1165–1176. https://doi.org/10.1039/c0nj00969e

[20] C.K. Gupta, N. Krishnamurthy, Extractive metallurgy of rare earths, *Int. Mater. Rev.* 37, 1992 (2013) 197–248. https://doi.org/10.1179/imr.1992.37.1.197

[21] P.C. De Sousa Filho, J.F. Lima, O.A. Serra, From lighting to photoprotection: Fundamentals and applications of rare earth materials, *J. Braz. Chem. Soc.* 26 (2015) 2471–2495. https://doi.org/10.5935/0103-5053.20150328

[22] J.C.G. Bünzli, Benefiting from the unique properties of lanthanide ions, *Acc. Chem. Res.* 39 (2006) 53–61. https://doi.org/10.1021/ar0400894

[23] K.S. Pitzer, Relativistic Effects on Chemical Properties, *Acc. Chem. Res.* 12 (1979) 271–276. https://doi.org/10.1021/ar50140a001

[24] B.G. Wybourne, L. Smentek, Relativistic effects in lanthanides and actinides, *J. Alloys Compd.* 341 (2002) 71–75. https://doi.org/10.1016/S0925-8388(02)00066-X

[25] M. Seth, P. Schwerdtfeger, M. Dolg, P. Fulde, Lanthanide and actinide contractions: Relativistic and shell structure effects, *J. Am. Chem. Soc.* 117 (1995) 6597–6598. https://doi.org/10.1021/ja00129a026

[26] L.J.F. Broer, C.J. Gorter, J. Hoogschagen, On the intensities and the multipole character in the spectra of the rare earth ions, *Physica.* 11 (1945) 231–250. https://doi.org/10.1016/S0031-8914(45)80009-5
[27] G.S. Opelt, Intensities of crystal spectra of rare-earth ions, *J. Chem. Phys.* 37 (1962) 511–520. https://doi.org/10.1063/1.1701366
[28] B.R. Judd, Optical absorption intensities of rare-earth ions, *Phys. Rev.* 127 (1962) 750–761. https://doi.org/10.1103/PhysRev.127.750
[29] P.A. Tanner, Spectra, energy levels and energy transfer in high symmetry lanthanide compounds, *Transit. Met. Rare Earth Compd.* 241 (2012) 167–278. https://doi.org/10.1007/b96863
[30] D.C. Harris, M.D. Bertolucci, *Symmetry and spectroscopy: An introduction to vibrational and electronical spectroscopy*, Dover Publications, Inc., New York, 1978.
[31] M.J. Weber, A V Dotsenko, L.B. Glebov, V.A. Tsekhomsky, *Handbook ofOptical materials*, CRC Press., New York, 2003.
[32] G.H. Dieke, *Spectra and energy levels of rare earths ions in crystal*, John Wiley & Sons, Inc., New York, 1968.
[33] B.G. Wybourne, *Spectroscopic properties of rare earths*, Wiley Inter-science, New York, 1965.
[34] O. Laporte, W.F. Meggers, Some rules of spectral structure, *J. Opt. Soc. Am.* 11 (1925) 459. https://doi.org/10.1364/JOSA.11.000459
[35] C.R. Ronda, *Luminescence: From theory to applications*, Wiley Online Library, 2008.
[36] L. Zhao, P. Xu, F. Fan, J. Yu, Y. Shang, Y. Li, L. Huang, R. Yu, Synthesis and photoluminescence properties of Sm^{3+} and Dy^{3+} ions activated double perovskite Sr_2MgTeO_6 phosphors, *J. Lumin.* 207 (2019) 520–525. https://doi.org/10.1016/j.jlumin.2018.10.120
[37] M. Shang, C. Li, J. Lin, How to produce white light in a single-phase host?, *Chem. Soc. Rev.* 43 (2014) 1372–1386. https://doi.org/10.1039/c3cs60314h
[38] K. Dev, A. Selot, G.B. Nair, C.M. Mehare, F.Z. Haque, M. Aynyas, S.J. Dhoble, Synthesis and photoluminescence study of Dy^{3+} activated $SrAl_{12}O_{19}$ phosphor, *Optik (Stuttg).* 194 (2019) 163051. https://doi.org/10.1016/j.ijleo.2019.163051
[39] K. Dev, A. Selot, G.B. Nair, V.L. Barai, F.Z. Haque, M. Aynyas, S.J. Dhoble, Energy transfer from $Ce^{3+}$to Dy^{3+} ions for white light emission in $Sr_2MgAl_{22}O_{36}$:Ce^{3+}, Dy^{3+} phosphor, *J. Lumin.* 206 (2019) 380–385. https://doi.org/10.1016/j.jlumin.2018.10.092
[40] W. Zhang, Y. Chen, X. Geng, Y. Yang, L. Xiao, Synthesis and luminescence properties of $LiBaB_9O_{15}$:Eu^{3+} single-component red-light emitting phosphors, *J. Lumin.* 224 (2020) 117324. https://doi.org/10.1016/j.jlumin.2020.117324
[41] W. Geng, X. Zhou, J. Ding, Y. Wang, $NaBaY(BO_3)_2$:Ce^{3+},Tb^{3+}: A novel sharp green-emitting phosphor used for WLED and FEDs, *J. Am. Ceram. Soc.* 101 (2018) 4560–4571. https://doi.org/10.1111/jace.15693
[42] H. Pengde, J. Xiaoping, X. Mengyao, Y. Fangli, J. Baoxiang, G. Rongfeng, Z. Qitu, Three primary colors upconversion phosphors and combined white upconversion luminscence in Y_2O_2S matrix, *Mater. Res. Bull.* 70 (2015) 658–662. https://doi.org/10.1016/j.materresbull.2015.05.036
[43] B. Sundarakannan, M. Kottaisamy, Synthesis and characterization of near UV excitable Y_2O_2S:Eu^{3+} entrapped ZnO for white light emitting diode applications, *J. Solid State Chem.* 293 (2021) 121739. https://doi.org/10.1016/j.jssc.2020.121739
[44] T.W. Chou, S. Mylswamy, R.S. Liu, S.Z. Chuang, Eu substitution and particle size control of Y_2O_2S for the excitation by UV light emitting diodes, *Solid State Commun.* 136 (2005) 205–209. https://doi.org/10.1016/j.ssc.2005.07.032
[45] G. Li, L. Li, M. Li, W. Bao, Y. Song, S. Gan, H. Zou, X. Xu, Hydrothermal synthesis and luminescent properties of $NaLa(MoO_4)_2$:Eu^{3+},Tb^{3+} phosphors, *J. Alloys Compd.* 550 (2013) 1–8. https://doi.org/10.1016/j.jallcom.2012.09.125
[46] L.C. MacHado, M.T.D.O. De Azeredo, H.P.S. Corrêa, J. Do Rosário Matos, Í.O. Mazali, Formation of oxysulfide LnO_2S_2 and oxysulfate LnO_2SO_4 phases in the thermal decomposition process of lanthanide sulfonates (Ln = La, Sm), *J. Therm. Anal. Calorim.* 107 (2012) 305–311. https://doi.org/10.1007/s10973-011-1451-7

[47] Y. Tian, X. Jiao, J. Zhang, N. Sui, D. Chen, G. Hong, Molten salt synthesis of $LaF_3:Eu^{3+}$ nanoplates with tunable size and their luminescence properties, *J. Nanoparticle Res.* 12 (2010) 161–168. https://doi.org/10.1007/s11051-009-9590-5

[48] L. Song, P. Du, Q. Jiang, H. Cao, J. Xiong, Synthesis and luminescence of high-brightness $Gd_2O_2SO_4:Tb^{3+}$ nanopieces and the enhanced luminescence by alkali metal ions co-doping, *J. Lumin.* 150 (2014) 50–54. https://doi.org/10.1016/j.jlumin.2014.01.043

[49] Z. Xia, Q. Liu, Progress in discovery and structural design of color conversion phosphors for LEDs, *Prog. Mater. Sci.* 84 (2016) 59–117. https://doi.org/10.1016/j.pmatsci.2016.09.007

[50] W. Sun, Y. Jia, R. Pang, H. Li, T. Ma, D. Li, J. Fu, S. Zhang, L. Jiang, C. Li, $Sr_9Mg_{1.5}(PO_4)_7:Eu^{2+}$: A novel broadband orange-yellow-emitting phosphor for blue light-excited warm white LEDs, *ACS Appl. Mater. Interfaces.* 7 (2015) 25219–25226. https://doi.org/10.1021/acsami.5b06961

[51] B. Wang, Y. Liu, Z. Huang, M. Fang, Photoluminescence properties of a Ce^{3+} doped $Sr_3MgSi_2O_8$ phosphor with good thermal stability, *RSC Adv.* 8 (2018) 15587–15594. https://doi.org/10.1039/c8ra00526e

[52] X. Bu, Y. Gai Liu, B. Wang, R. Mi, Z. Wang, Z. Huang, Photoluminescent properties of single-phase white-light $Ca_8ZnGd(PO_4)_7:Eu^{2+}$, Mn^{2+} phosphor, *Chem. Phys. Lett.* 743 (2020) 137185. https://doi.org/10.1016/j.cplett.2020.137185

[53] J. Ding, Q. Wu, Y. Li, Q. Long, C. Wang, Y. Wang, $Sr_{7.3}Ca_{2.7}(PO_4)_6F_2:Eu^{2+}$,Mn^{2+} : A novel single-phased white light-emitting phosphor for NUV-LEDs, *Dalt. Trans.* 44 (2015) 9630–9636. https://doi.org/10.1039/C5DT00907C

[54] J. Zhou, Z. Xia, Synthesis, luminescence properties and energy transfer behavior of $Na_2CaMg(PO_4)_2:Eu^{2+}$, Mn^{2+} phosphors, *J. Lumin.* 146 (2014) 22–26. https://doi.org/10.1016/j.jlumin.2013.09.031

[55] M. Jiao, Y. Jia, W. Lü, W. Lv, Q. Zhao, B. Shao, H. You, $Sr_3GdNa(PO_4)_3F:Eu^{2+}$,$Mn^{2+}$: A potential color tunable phosphor for white LEDs, *J. Mater. Chem. C.* 2 (2014) 90–97. https://doi.org/10.1039/c3tc31837k

[56] B. Wang, Y.G. Liu, Z. Huang, M. Fang, X. Wu, Discovery of novel solid solution $Ca_3Si3\text{-}_xO_{3+xN4-2x}$: Eu^{2+} phosphors: Structural evolution and photoluminescence tuning, *Sci. Rep.* 7 (2017) 1–14. https://doi.org/10.1038/s41598-017-18319-5

[57] F.S. Freitas, A.S. Gonçalves, A. De Morais, J.E. Benedetti, A.F. Nogueira, Graphene-like MoS_2 as a low-cost counter electrode material for dye-sensitized solar cells, *J. NanoGe J. Energy Sustain.* (2012) 11002–11003. https://doi.org/10.1039/c0xx00000x

[58] F. Perrin, Théorie quantique des transferts d'activation entre molécules de même espèce. Cas des solutions fluorescentes, Dalt. *Trans.* 17 (1932) 283–314. https://doi.org/10.1051/anphys/193210170283

[59] I. Mayer, R.S. Roth, W.E. Brown, Rare earth substituted fluoride-phosphate apatites, *J. Solid State Chem.* 11 (1974) 33–37. https://doi.org/10.1016/0022-4596(74)90143-1

[60] I. Mayer, S. Cohen, The crystal structure of $Ca_6Eu_2Na_2(PO_4)_6F_2$, *J. Solid State Chem.* 48 (1983) 17–20. https://doi.org/10.1016/0022-4596(83)90054-3

[61] J. Qiao, J. Zhang, X. Zhang, Z. Hao, Y. Liu, Y. Luo, The energy transfer and effect of doped Mg^{2+} in $Ca_3Sc_2Si_3O_{12}:Ce^{3+}$, Pr^{3+} phosphor for white LEDs, *Dalt. Trans.* 43 (2014) 4146–4150. https://doi.org/10.1039/c3dt52902a

[62] D. Wen, J. Shi, A novel narrow-line red emitting $Na_2Y_2B_2O_7:Ce^{3+}$,Tb^{3+},Eu^{3+} phosphor with high efficiency activated by terbium chain for near-UV white LEDs, *Dalt. Trans.* 42 (2013) 16621–16629. https://doi.org/10.1039/c3dt52214h

[63] M. Xie, S. Ou, H. Liang, D. Hou, Y. Huang, Z. Gao, Y. Tao, White-emitting phosphors $Ca_6La_2Na_2(PO_4)_6F_2:Dy^{3+}$ and luminescence enhancement through $Ce^{3+} \rightarrow Dy^{3+}$ energy transfer, *Mater. Chem. Phys.* 142 (2013) 339–344. https://doi.org/10.1016/j.matchemphys.2013.07.026

[64] K. Li, D. Geng, M. Shang, Y. Zhang, Color-tunable luminescence and energy transfer properties of $Ca_9Mg(PO_4)_6F_2:Eu^{2+}$,$Mn^{2+}$ phosphors for UV-LEDs, *Am. Chem. Soc.* 118 (2014) 11026–11034. https://doi.org/10.1021/jp501949m

[65] C. Zhang, S. Huang, D. Yang, X. Kang, M. Shang, C. Peng, J. Lin, Tunable luminescence in Ce^{3+}, Mn^{2+}-codoped calcium fluorapatite through combining emissions and modulation of excitation: A novel strategy to white light emission, *J. Mater. Chem.* 20 (2010) 6674–6680. https://doi.org/10.1039/c0jm01036g

[66] W.R. Liu, C.H. Huang, C.W. Yeh, J.C. Tsai, Y.C. Chiu, Y.T. Yeh, R.S. Liu, A study on the luminescence and energy transfer of single-phase and color-tunable $KCaY(PO_4)_2:Eu^{2+},Mn^{2+}$ phosphor for application in white-light LEDs, *Inorg. Chem.* 51 (2012) 9636–9641. https://doi.org/10.1021/ic3007102

[67] Z. Yang, D. Xu, J. Sun, J. Du, X. Gao, Luminescence properties and energy transfer investigations of $Sr_3Lu(PO_4)_3:Ce^{3+}$, Tb^{3+} phosphors, *Mater. Sci. Eng. B Solid-State Mater. Adv. Technol.* 211 (2016) 13–19. https://doi.org/10.1016/j.mseb.2016.05.015

[68] B. Cui, Z. Chen, Q. Zhang, H. Wang, Y. Li, A single-composition $CaSi_2O_2N_2$:RE (RE=Ce^{3+}/Tb^{3+}, Eu^{2+}, Mn^{2+}) phosphor nanofiber mat: Energy transfer, luminescence and tunable color properties, *J. Solid State Chem.* 253 (2017) 263–269. https://doi.org/10.1016/j.jssc.2017.04.012

[69] L. Lin, T. Wanjun, Effects of Tb^{3+} doping on luminescence properties of $NaMg_4(PO_4)_3:Eu^{2+}$, *J. Lumin.* 198 (2018) 405–409. https://doi.org/10.1016/j.jlumin.2018.03.001

[70] Z. Yang, C. Ji, G. Zhang, G. Han, H. Wang, H. Bu, X. Tan, D. Xu, J. Sun, Tunable blue-green color emission and energy transfer of $Sr_3NaSc(PO_4)_3F:Eu^{2+},Tb^{3+}$ phosphors with near-UV broad band excited for white LEDs, *J. Lumin.* 206 (2019) 585–592. https://doi.org/10.1016/j.jlumin.2018.10.078

CHAPTER 7

Effect of Singly, Doubly and Triply Ionized Ions on Photoluminescent Energy Materials

Abhijeet R. Kadam

R. T. M. Nagpur University, Nagpur, India

Fuad Ameen

King Saud University, Riyadh, Saudi Arabia

Sanjay J. Dhoble

R. T. M. Nagpur University, Nagpur, India

7.1 INTRODUCTION

Phosphor-converted white light-emitting diodes (pc-WLEDs) are becoming more and more popular and are widely regarded as the most promising next generation solid-state lighting (SSL) sources due to their exceptional luminous efficiency and luminosity [1], low power consumption [2, 3], dependability [4], extended operational life [5, 6], and non-toxicity [7] when compared to other lighting technologies, such as incandescent, halogen, xenon, fluorescent lamps, and so forth. The pc-WLEDs have many benefits over the aforementioned alternative light sources, particularly in terms of energy conservation [8–12]. No other type of lighting system could save energy as much as this one. As a result, they play a critical role in lowering the demand for power globally and reducing the consumption of fossil fuels. Furthermore, pc-WLEDs are great backlighting options for usage in display spaces like liquid crystal panels (LCDs) [13–15]. Many factors need be taken into account and improved in order to create highly efficient WLED devices, comprising luminaires, phosphor-conversion components, semiconductor chips (often InGaN semiconductor chips), and packaging techniques [16, 17]. Phosphors are essential parts of pc-WLED devices, so discovering and developing pc-WLED phosphors to serve as lighting and

DOI: 10.1201/9781003315261-9

display backlight sources is among the most significant and urgent tasks that may be solved by cutting-edge science and technology. Using pc-WLEDs, strong blue or near-ultraviolet light (n-UV) LEDs are typically connected with downconverting phosphors to produce white light. Two primary tactics are employed in this process: the first involves combining highly efficient (In, Ga)N blue LEDs (420–480 nm) with yellow-emitting phosphors [18–20], as in a device with a schematic structure. The most popular and straightforward method, for instance, is still in use today and consists of combining a 460 nm blue InGaN LED chip with a yellow phosphor made of a yttrium aluminum garnet (YAG:Ce^{3+}) doped with cerium(III) [21, 22]. Nevertheless, due to the absence of a red component, such a device has a low color rendering index (R_a = 70–80) and a significant correlated color temperature (CCT = 4000–7500 K). The improved methods for producing white light with a high R_a and an acceptable CCT include mixing green and red multi-phased phosphors on 460 nm blue LEDs or coating yellow-emitting phosphors with an additional red component [12].

Systems usually result in increased manufacturing costs and low luminous efficiency. The second technique entails combining n-UV LED chips with red, green, and blue phosphor emission in a system with diagrammatic structure [23–27]. It is like the previous technique wherein less efficiency is unavoidable which is attributable to re-absorption, problems with deposition, and challenges with the homogenous distribution of phosphors in silicon resin. In exchange, several advantages can be obtained, such as a very high color rendering index, broad color spectrum distribution, and steady light color output at various driving currents [28–30]. Phosphors have undergone substantial research and development in order to lessen the re-absorption effect between phosphors and improve their luminosity and color rendering index. Phosphors play a major role in the growth of the WLED lighting and backlight display industries [25]. Extremely effective and emission-tunable phosphors not only increase the energy efficiency of WLEDs but also have the ability to improve their color rendering index, associated color temperature, and color spectrum [31–33], expanding the range of applications for them in areas such as vehicle illumination, architectural ornamentation [34], streetlights [35], guiding lights [36], and clinical medical lighting [37]. Consequently, it is necessary to produce new and palatable fluorescent materials to enhance the performance of solid-state lighting. Up to this point, various researchers around the world have all provided helpful studies describing the pc-WLED phosphors for solid-state illumination or backlight displays. The synthesis, optical characteristics, and use of pc-WLED phosphors, with an attention to how luminescence attributes rely on composition, have been well reviewed in recent studies. For example, Smet and co-authors [38] presented, identified, and discussed six main criteria dealing with the shape and position of the emission and excitation spectra, the thermal quenching behavior, the quantum efficiency, the chemical and thermal stability, and finally, the occurrence of saturation effects.

7.2 SPECTRAL TUNING IN PHOTOLUMINESCENCE (PL)

The excitation and emission spectra of luminescence, also referred as luminescence spectra, are typically used to record the colors of light from pc-WLED lighting and display systems based on rare-earth ion triggered phosphors. The luminescence spectra are tightly connected with various crucial luminescence performance metrics in pc-WLED devices, including: CCT [39, 40], Commission Internationale de lÉclairage (CIE) color coordinates [41–44], color rendering index (CRI or R_a) [45, 46], luminous efficiency [47], quantum yield (QY) [48, 49], and so on. These variables, including CCT, CIE color coordinates, and R_a, can be perfectly managed by modifying the position and full-width at half maxima (FWHM) of photoluminescence (PL) emission spectra of rare-earth doped phosphors. This enhances the luminescence outcomes of WLED devices. For instance, by adjusting the phosphors' emission positions, a single pc-WLED device based on multivariate phosphors might spread its color coordinates across the whole white light region, with varying CCTs and R_a. Moreover, altering the excitation spectra of pc-WLED phosphors may boost both their and WLED devices' illumination. To increase the luminescence performances of WLEDs, it is generally possible to adequately optimize the performance characteristics of rare-earth activated phosphors, which can be summed up by the various aspects. The luminous efficiency of WLEDs can be increased by altering the spectral position to get closer to the eye sensitivity curve. The color rendering properties of WLEDs can also be optimized by synchronizing the spectral position with conventional illuminators. By modifying the spectral position, the required color temperature can be obtained. As a result, it is crucial and essential to properly tune the luminescence spectra of rare-earth ion triggered phosphors for the rapid evolution of WLED lighting and backlighting display projects. Focusing on the alteration of phosphors' structures and constituent parts to tune their luminescence spectra requires a lot of work from researchers working in the phosphor field.

7.3 FUNDAMENTAL ASPECTS OF RARE-EARTH ACTIVATED MATERIALS

Because to their exceptional luminescent properties, lanthanide ions are essential to modern science and technology. Ce^{3+}, Sm^{3+}, Eu^{2+}/Eu^{3+}, Tb^{3+}, Dy^{3+} and other downconverting luminous lanthanide ions are the most often employed ones [50–56]. Due to the wide range of allowable energy levels, they can transfer energy between distinct lanthanide ions and emit light from the ultraviolet region across the visible spectrum. Broad band emitting phosphors that emit light in the 5d-4f range and narrow band emitting phosphors that emit light in the 4f range are two types of lanthanide-activated phosphors, respectively. The overall luminous properties of the 5d-4f and 4f-4f transitions are discussed in the next section to serve as a foundation for the tuning of lanthanide luminescence in phosphors.

7.3.1 5d-4f Emission

Because of the electric dipole f-d transition's parity-allowed nature, Ce^{3+} and Eu^{2+}, which exhibit the 5d-4f transition, frequently exhibit broad emission bands, brief lifespan, and strong oscillators. The host lattice has a stronger influence on the spectroscopic features of Ce^{3+}/Eu^{2+} because the 5d electrons are no longer protected from their surroundings by the $5s^2$ and $5p^6$ shells when they are excited. The impacts of the host matrix on the PL characteristics of Ce^{3+} and Eu^{2+} are depicted in Figure 7.1. Large energy gaps exist between the

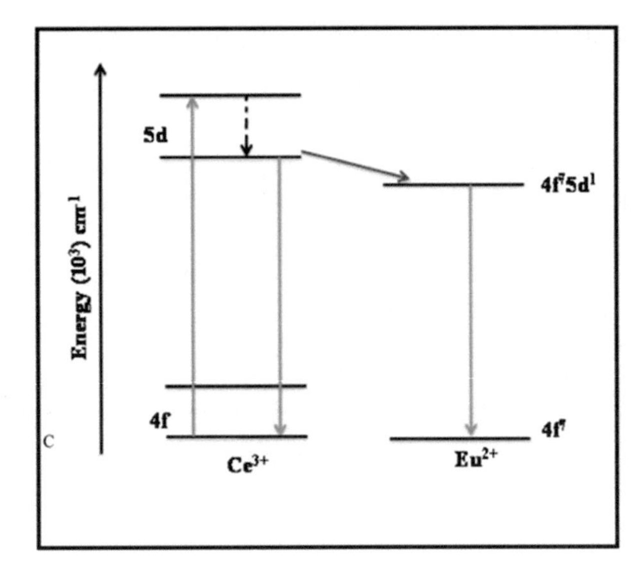

FIGURE 7.1 Energy level diagram of Ce^{3+} to Eu^{2+}. (Reproduced with permission from the Ref. [57], © 2020 Elsevier B.V.)

TABLE 7.1 Overview of Trivalent Rare-Earth Transitions

Rare-Earth Ion	Ground State	Excited State	Emission Wavelength (nm)	Emission Color
Dy^{3+}	$^4F_{9/2}$	$^6H_{J\,15/2,\,13/2}$	573	Blue/yellow
Tb^{3+}	$^5D_4$	$^7F_{J\,6-0}$	545	Green
Sm^{3+}	$^4G_{5/2}$	$^6H_{J5/2-15/2}$	600	Orange/red
Eu^{3+}	$^5D_0$	$^7F_{J\,(0\text{-}6)}$	614	Red

ground state ($4f^n$ energy level) and the lowest excited state ($4f^{n-1}5d^1$ energy level) for free Ce^{3+} and Eu^{2+} ions. These gaps are around 6.2 eV for Ce^{3+} and 4.2 eV for Eu^{2+}.

7.3.2 4f-4f Emission

Due to the 4f electrons' strong shielding from the functionality offered by the filled $5s^2 5p^6$ subshells, the 4f-4f transitions have atomic-like characteristics, sharp emission lines, prolonged lifespan, and outstanding coherence features. The 4f orbitals are less susceptible to the crystal field, exhibit fewer exchange disturbances, and have weaker electron-phonon couplings than the outer 5s and 5p orbitals because of their lower radial extensions. As a result, the color of the emission is generally independent of the lanthanide ion's surroundings and is reliant on the species of rare-earth ions [58, 59]. Eu^{3+}, Tb^{3+}, Sm^{3+}, and Dy^{3+}, for instance, each emit a different color of light [60–62]. Table 7.1 lists the major luminous transitions. Because the intrinsic 4f-4f transitions are parity-forbidden, they exhibit low molar absorption coefficients in the n-UV and blue bands. With the exception of particular hosts' highly effective and potent red Eu^{3+} emission, lanthanides with 4f-4f transitions are not appropriate for solid-state illumination due to these characteristics.

7.4 EFFECT OF SINGLY, DOUBLY, AND TRIPLY IONIZED IONS

The host lattice has a significant impact on the rare-earths' luminous characteristics. The excitation and emission wavelengths of the activator can be adjusted by altering the chemical makeup of the host. The crystal lattice environment around doubly/triply ionized rare-earth ions will typically change when a constituent element of the host is replaced with an element with a different ionic radius, leading to a spectrum shift in the rare-earth activated luminescence. Three categories of substitutions can be distinguished in the host: cationic, anionic, and cationic-anionic co-substitution.

A host lattice contains two different types of cations: host cations (C_h) and cations in ionic complexes (C_{ic}). Alkaline earth metal ions (Mg^{2+}, Ca^{2+}, Sr^{2+}, and Ba^{2+}), rare-earth metal ions (Y^{3+}, Sc^{3+}, La^{3+}, Gd^{3+}, and Lu^{3+}), as well as alkali metal ions (Li^+, Na^+, and K^+), can all cationically substitute at C_h sites. The B^{3+} in $(BO_3)^{3-}$ or $(BO_4)^{5-}$, the Al^{3+} in $(AlO_4)^{3-}$ or $(AlO_6)^{9-}$, the Si^{4+} in $(SiO_4)^{4-}$, and the P^{5+} in $(PO_4)^{3-}$ are a few examples of other locations where it can happen.

Because of the immediate impact of the activator-anion couplings on the 5d levels and ground states of Eu^{2+}/Ce^{3+}, specifically the nephelauxetic effect, color point tuning via anionic replacement is yet another basic method for photoluminescence tuning. A lesser electronegativity, as per crystal field theory, leads to a greater nephelauxetic effect (covalency), and a severe negative charge on an anion causes a larger crystal field separation of the 5d orbitals of Ce^{3+} or Eu^{2+}. Selenides > Sulfides > Nitrides > Oxides > Fluorides, or $Se^{2-} > S^{2-} > N^{3-} > O^{2-} > F^-$, are the compounds that have the greatest nephelauxetic effect (centroid shift) on an anion.

It is widely recognized in the scientific community that cationic-anionic co-substitution is a viable technique for adjusting the luminous characteristics of Ce^{3+} or Eu^{2+} triggered systems [63]. The crystal field and nephelauxetic effect surrounding Ce^{3+} or Eu^{2+} can be tuned controllably by simultaneous cationic-anionic substitutions, which may produce a controllable spectrum shift in the emission of the phosphors. According to the charge balancing concept in compounds, the following forms of cationic-anionic co-substitutions are most common:

$$(1)\, C^{3+} + Y^{2-} \leftrightarrow D^{4+} + Z^{3-} \left(Al^{3+} + O^{2-} \leftrightarrow Si^{4+} + N^{3-}\right)$$

$$(2)\, D^{4+} + Y^{2-} \leftrightarrow C^{3+} + X^{-} \left(Si^{4+} + O^{2-} \leftrightarrow Al^{3+} + F^{-}\right)$$

$$(3)\, C^{3+} + U^{4-} \leftrightarrow B^{2+} + C^{3-} \left(Y^{3+} + C^{4-} \leftrightarrow Sr^{2+} + N^{3-}\right)$$

$$(4)\, D^{4+} + U^{4-} \leftrightarrow C^{3+} + Z^{3-} \left(Si^{4+} + C^{4-} \leftrightarrow Al^{3+} + N^{3-}\right)$$

$$(5)\, D^{4+} + Y^{2-} \leftrightarrow E^{5+} + Z^{3-} \leftrightarrow C^{3+} + X^{-} \left(Si^{4+} + O^{2-} \leftrightarrow P^{5+} + N^{3-} \leftrightarrow Al^{3+} + F^{-}\right)$$

$$(6)\, E^{5+} + Y^{2-} \leftrightarrow D^{4+} + X^{-} \left(P^{5+} + N^{3-} \leftrightarrow Si^{4+} + F^{-}\right)$$

The periodic table's alkali metals, IIIA and IIIB metals, IVA metals, and VA metals are each represented by B, C, D, and E in this example. As an illustration, group B consists of Mg, Ca, Sr, and Ba; group C consists of Sc, Y, La, Gd, and Lu (IIIB metals) or Al, Ga, and

In (IIIA metals); group D consists of Si and Ge; and group E is P. While the elements F, O, N, and C, respectively, are represented by X, Y, Z, and U. Exemplary cationic-anionic co-substitution cases are the formulations included in brackets.

7.4.1 Eu^{3+} Doped $Na_2Sr_2Al_2PO_4Cl_9$ Phosphor

According to Kadam et al. [64], the doping of singly, doubly, and triply ionized ions altered the red PL of the Eu^{3+} doped $Na_2Sr_2Al_2PO_4Cl_9$ phosphor. The artificial phosphors exhibited excellent crystalline nature. The three peaks at 317, 395, and 467 nm were seen in the phosphor's PL excitation spectrum. When excited at 395 and 467 nm, the Eu^{3+} doped $Na_2Sr_2Al_2PO_4Cl_9$ phosphor displayed a bright red hue. However. the phosphor's PL intensity was higher for 395 nm excitation. The PL intensity of the doped Eu^{3+} $Na_2Sr_2Al_2PO_4Cl_9$ phosphor was dramatically reduced when the singly, doubly, and triply ionized ions (F^-, WO_4^{2-}, MoO_4^{2-}, VO_4^{3-}, La^{3+}, and Y^{3+}) are co-doped. The variation in local crystal structure brought about by these ions is what caused the PL intensity to drop. It's noteworthy that when the (Y^{3+}) ion-incorporated phosphor was excited at 317 nm wavelength, the PL intensity of the phosphor increased significantly.

Figure 7.2 illustrates the PLE spectra of the $Na_2Sr_2Al2\text{-}_cLa_cVO_4Cl_6F_3$:1 mol% Eu^{3+} phosphor at emission wavelengths of 593 and 618 nm. The spectra of the Eu^{3+} ion's $^7F_0 \rightarrow {}^5L_6$ and $^7F_0 \rightarrow {}^5D_2$ transitions have two prominent peaks at 395 and 467 nm, respectively. The charge transfer band (CTB) of the $O^{2-} \rightarrow Eu^{3+}$ ion, the 4f-4f transition, and a harmonic of the monitoring wavelength overlap in the spectra, causing an additional strong and broad peak to be seen at a wavelength of 317 nm. Together with the CTB of $O^{2-} \rightarrow Eu^{3+}$ ion, the CTB of WO_4^{2-}, MoO_4^{2-}, and VO_4^{3-} ions are likewise located in the UV area. The $^7F_0 \rightarrow {}^5L_6$ transition of the Eu^{3+} ion is attributed as the cause of the 4f-4f transition at 317 nm in wavelength.

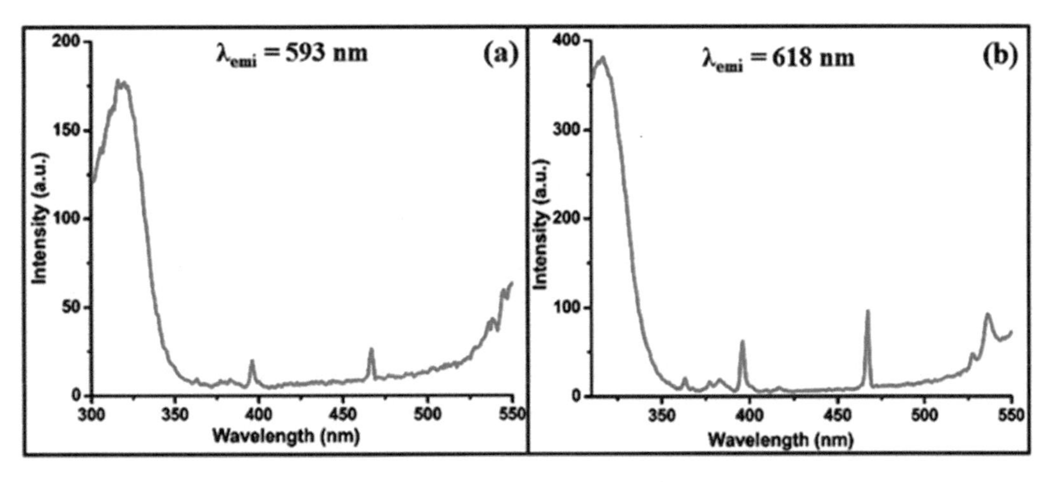

FIGURE 7.2 PL excitation spectra of $Na_2Sr_2Al2\text{-}_cLa_cVO_4Cl_6F_3$:1 mol% Eu^{3+} phosphor at the emission wavelengths of (a) 593 nm and (b) 618 nm. (Reproduced with permission from Ref. [64], © 2019 Elsevier B.V.)

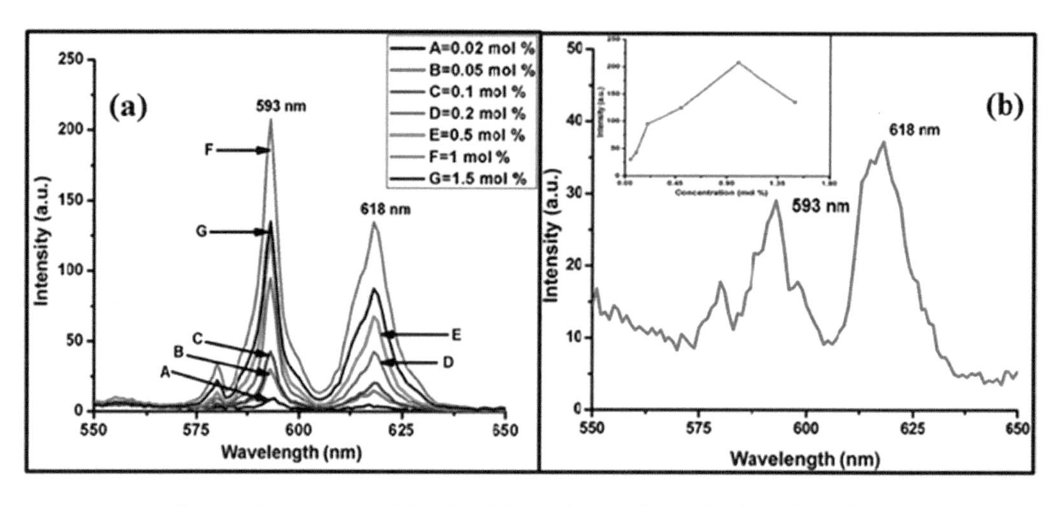

FIGURE 7.3 PL emission spectra of $Na_2Sr_2Al_2PO_4Cl_9$:x mol% Eu^{3+} phosphors (i.e. x = 0.02, 0.05, 0.1, 0.2, 0.5, 1 and 1.5 mol%) on excitations with (a) 395 nm and (b) 467 nm (only one emission spectrum was obtained for this excitation). The inset in (b) shows photoluminescence intensity versus concentration of $Na_2Sr_2Al_2PO_4Cl_9$:x mol% Eu^{3+} phosphors (i.e. x = 0.02, 0.05, 0.1, 0.2, 0.5, 1 and 1.5 mol%) for the 593 nm emission band under 395 nm excitation. (Reproduced with permission from Ref. [64], © 2019 Elsevier B.V.)

The PL emission spectra of $Na_2Sr_2Al_2PO_4Cl_9$:x mol% Eu^{3+} (x = 0.02, 0.05, 0.1, 0.2, 0.5, 1 and 1.5 mol%) phosphors were observed between 550 and 650 nm on excitation with 395 and 467 nm light, as shown in Figure 7.3. Although the concentration of Eu^{3+} ions varied, the emission peaks always matched one another, and their peak positions were unaltered. Two emission peaks from the phosphor sample were seen at 593 and 618 nm, respectively. These peaks are attributed to the transitions of the Eu^{3+} ion from $^5D_0$ to $^7F_1$ and $^5D_0$ to $^7F_2$, respectively. The 618 nm emission band's PL intensity was lower than the 593 nm emission band (Figure 7.3(a)). This was caused by the excitation wavelength matching the energy level of Eu^{3+}. The magnetic dipole type is clearly attributed to the $^5D_0$ to $^7F_1$ transition, and the selection criteria $J = 1$ is then used. However, the $^5D_0 \rightarrow {}^7F_2$ transition, which is characterized by $J = 2$ as an electric dipole type transition, is extremely sensitive to the environment surrounding the Eu^{3+} ion. Hypersensitive transition is another name for this transition. The proportion of electric to magnetic dipole transitions provides insight into the asymmetry of the phosphor sample.

The amount of dopant ions present affects the photoluminescence intensity the phosphor samples can exhibit. The concentration of Eu^{3+} ions used to prepare the phosphor samples ranged from 0.02, 0.05, 0.1, 0.2, 0.5, 1 and 1.5 mol%. The PL intensity initially rose with Eu^{3+} ion concentration and then decreased. The PL intensity was observed to improve synchronously as the ion concentration was raised from 0.02 to 1.0 mol%. The intensity decreased with additional increments in the Eu^{3+} ion concentration. This was brought on by the concentration quenching effect. A technique to explain a reduction in PL intensity is concentration quenching. In this mechanism, at greater concentrations, the space between Eu^{3+} ions is reduced below the normal distance. The unexcited ions are subsequently given the excitation energy, and this is accompanied by a non-radiative relaxation.

This is what caused the PL intensity of Eu^{3+} ions to diminish. Researchers found that 1.0 mol% of Eu^{3+} ions produced the strongest signal. The red photoluminescence was produced by the $Na_2Sr_2Al_2PO_4Cl_9$:x mol% Eu^{3+} phosphors (i.e., $x = 0.02, 0.05, 0.1, 0.2, 0.5$, 1 and 1.5 mol%) excited with 467 nm (see Figure 7.3(b)). Here, only one emission peak with a Eu^{3+} ion concentration of 1.0 mol% was seen by researchers. For the 593 nm emission band generated at 395 nm, the inset in Figure 7.3(b) also depicts a fluctuation in the PL intensity with varied concentrations of Eu^{3+} ions; the PL intensity was optimal at 1.0 mol% concentration.

In this study, Kadam et al. [64] reported the PL intensity of several phosphors in the presence and absence of singly, doubly, and triply ionized ions for the 593 and 618 nm emission bands upon 395 and 467 nm excitations (Figure 7.4). It has been found that the doping ions have a substantial impact on how bright the phosphor materials' PLs are. Due to their unique ionic radii, the F^-, WO_4^{2-}, MoO_4^{2-}, VO_4^{3-}, La^{3+}, and Y^{3+} doping ions are expected to alter the crystallinity of phosphor samples, which in turn affects how intense the phosphors' PL is. The XRD peak position shifts when smaller or bigger ionic sized ions

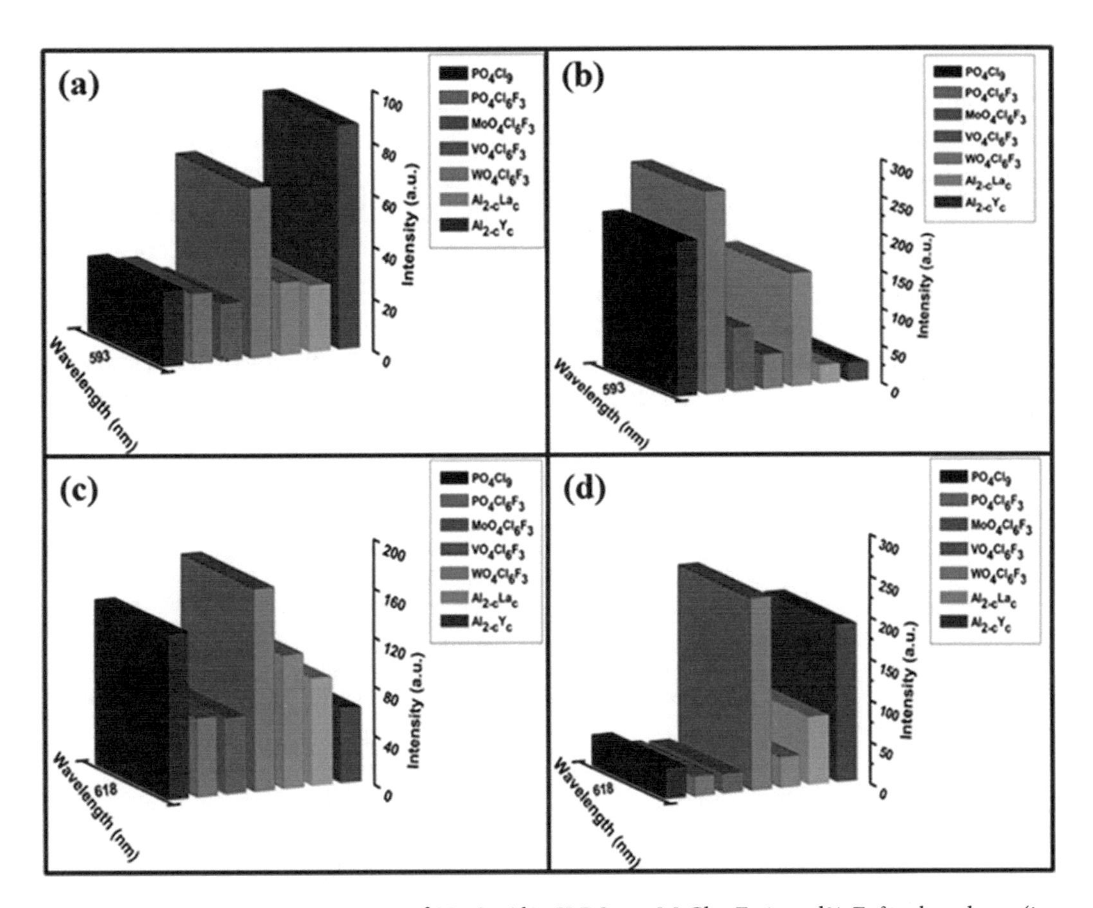

FIGURE 7.4 PL emission spectra of Na$_2$Sr$_2$Al2-$_c$X$_c$PO$_{4(1-a)}$M$_a$Cl$_{9-b}$F$_b$:1 mol% Eu^{3+} phosphors (i.e. X = La and Y), (M = MoO_4, WO_4 and VO_4) for the 593 nm emission band on excitations with (a) 395 nm and (b) 467 nm; and for the 618 nm emission band on excitations with (c) 395 nm and (d) 467 nm. (Reproduced with permission from the Ref. [64], Copyright © 2019 Elsevier B.V.)

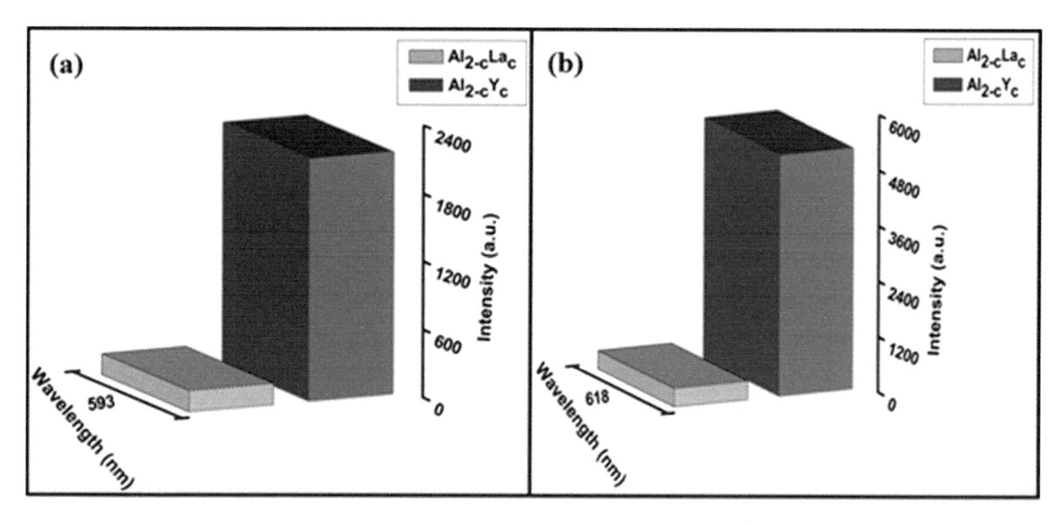

FIGURE 7.5 PL emission spectra of $Na_2Sr_2Al2\text{-}_cX_cVO_4Cl_6F_3$:1 mol% Eu^{3+} phosphors (i.e. X = La, Y) on excitation with 317 nm for the emission bands (a) 593 and (b) 618 nm. (Reproduced with permission from Ref. [64], © 2019 Elsevier B.V.)

are doped in the host lattice, which affects the crystallinity of the phosphor. With F^- and Y^{3+} doping under 395 and 467 nm excitations, respectively, the PL intensity of the 593 nm emission band is increased as a result. Nevertheless, when VO_4^{3-}, La^{3+}, and Y^{3+} are doped with identical excitations, the PL intensity of the 618 nm emission band is greater.

Conversely, the PL intensity is double when the La^{3+} and Y^{3+} ions are co-doped at the Al^{3+} sites of the $Na_2Sr_2Al2\text{-}_cX_cVO_4Cl_6F_3$:1 mol% Eu^{3+} phosphor. Comparison of the PL spectra of $Na_2Sr_2Al2\text{-}_cX_cVO_4Cl_6F_3$:1 mol% Eu^{3+} phosphors stimulated at 317 nm is shown in Figure 7.5. It's noteworthy to observe that for Y^{3+} doping in the 593 and 618 nm emission bands, the PL intensity is significantly higher. At 317 nm excitation, the 618 nm emission band's PL intensity is more than twice as strong as the 593 nm emission band.

TABLE 7.2 Comparison of PL Intensity of the Phosphors Upon 395, 467 and 317 nm Excitations

	Excitation Wavelength 395 nm		Excitation Wavelength 467 nm		Excitation Wavelength 317 nm	
Phosphor	**Emission Wavelength 593 nm**	**Emission Wavelength 618 nm**	**Emission Wavelength 593 nm**	**Emission Wavelength 618 nm**	**Emission Wavelength 593 nm**	**Emission Wavelength 618 nm**
$Na_2Sr_2Al_2PO_4Cl_9$	1	1	1	1		
$Na_2Sr_2Al_2PO_4Cl_6F_3$	1.308	0.478	0.691	0.713		
$Na_2Sr_2Al_2(PO_4)_{1-x}(MoO_4)_xCl_6F_3$	0.405	0.459	0.699	0.861		
$Na_2Sr_2Al_2(PO_4)_{1-x}(WO_4)_xCl_6F_3$	0.728	0.804	0.834	1.021		
$Na_2Sr_2Al_2(PO_4)_{1-x}(VO_4)_xCl_6F_3$	0.222	1.22	1.293	6.442		
$Na_2Sr_2Al_{2-c}La_cVO_4Cl_6F_3$	0.118	0.644	0.871	2.266	188.683	404.908
$Na_2Sr_2Al_{2-c}Y_cVO_4Cl_6F_3$	0.117	0.450	2.948	5.250	2139.69	5220.18

(Reproduced with permission from the Ref. [64], Copyright © 2019 Elsevier B.V.)

Table 7.2 provides a summary of quantitative comparison of various phosphors' PL intensities. Table 7.2 shows that there is a clear relationship between the doping ions and the PL intensity. The three excitation wavelengths, 317, 395, and 467 nm, have differing PL intensities for the 593 and 618 nm emission bands. It is fascinating to learn from Table 7.2 that the introduction of the Y^{3+} ion causes a greater PL intensity for the 593 and 618 nm emission bands upon 317 nm excitation, and they are determined to be, correspondingly, 2140 and 5220 times higher than the pure phosphor.

7.4.2 *x* mol% Eu(III)-Doped $Ca_{3(1-x-z)}M_z(PO_4)_2A_x$

Kadam et al. [65] reported the Eu^{3+}-activated double and triple phosphates that were synthesized via a high-temperature solid-state reaction. The luminescence properties of trivalent europium ion activated phosphors strongly depend on its valence states. Typically, Eu^{3+} ions can give luminescence from orange to red originating from 4f-4f transitions. The PL emission spectrum for the above-mentioned excitation peaks consisted of two identical peaks which is a wonder of this work. The first emission peak was located at 591 nm (orange) corresponding to $^5D_0 \rightarrow {}^7F_1$ transition of Eu^{3+} ion, and another intense peak was located at 614 nm (red)) that can be associated with the $^5D_0 \rightarrow {}^7F_2$ transition of Eu^{3+} ion. The detailed statistical analysis regarding emission intensity ratio for Eu^{3+} characteristics emission (591 and 614 nm) in $Ca_3(PO_4)_2$, $Ca_3(PO_4)_{2(x-1)}(WO_4)_x$, $Ca_3(PO_4)_{2(x-1)}(SO_4)_x$, $Ca_3(PO_4)_{2(x-1)}(MoO_4)_x$, $Ca_{1-x-y}Sr_yMoO_4$ and $Ca_{1-x-y}Ba_yMoO_4$ phosphors in this study is tabulated in Table 7.3. The detailed graphical analysis is presented in Figure 7.6.

By co-doping the trivalent rare-earth ions into the $Ca_{3(1-x-z)}M_z(PO_4)_2A_x{:}x$ host and adjusting their relative concentration ratio, multicolor tunable emissions are acquired by changing the ratio of Eu^{3+}. The energy transfer from rare-earth to host has been investigated by photoluminescence. The influence of W substitution by Mo on the luminescence properties of $Ca_{3(1-x-z)}M_z(PO_4)_2A_x{:}x$ has also been investigated. The emission intensities of Eu^{3+} ions were found to reach a maximum when the molar ratio of Mo/W-O tetrahedron was varied. During the replacement, two lattice sites of Ca^{2+} are substituted by two Eu^{3+} ions with the formation of a vacancy site at Ca^{2+} site for the prerequisite of charge compensation (i.e., $3Ca^{2+} \rightarrow 2Eu^{3+}$ (vacancy)). This further adds to more asymmetric atmosphere

TABLE 7.3 Statistical Analysis of PL Emission Intensity Ratio

Emission Wavelength →	Excitation 258 nm		Excitation 395 nm		Excitation 466 nm		Excitation 536 nm	
Phosphor ↓	591 nm	614 nm	591 nm	614 nm	591 nm	614 nm	591 nm	614 nm
$Ca_3(PO_4)_2$	427.196	717.565	630.299	938.064	338.19	504.704	343.84	621.075
$Ca_3(PO_4)_{2(x-1)}(WO_4)_x$	49.772	216.857	78.976	367.047	271.102	1015.82	196.593	855.144
$Ca_3(PO_4)_{2(x-1)}(SO_4)_x$	184.114	373.376	188.197	504.765	114.656	331.656	182.029	306.086
$Ca_3(PO_4)_{2(x-1)}(MoO_4)_x$	511.6	2966.16	179.18	1195.8	181.12	1081.84	405.2	2375.6
$Ca_{1-x-y}Sr_yMoO_4$	533.08	2922.8	110.36	713.04	225.47	1396.12	248.8	1604.2
$Ca_{1-x-y}Ba_yMoO_4$	6501.72	4751.8	113.16	900.36	444.6	2957.04	638.92	4093.44

(Reproduced with permission from Ref. [65], © 2021 Royal Society of Chemistry).

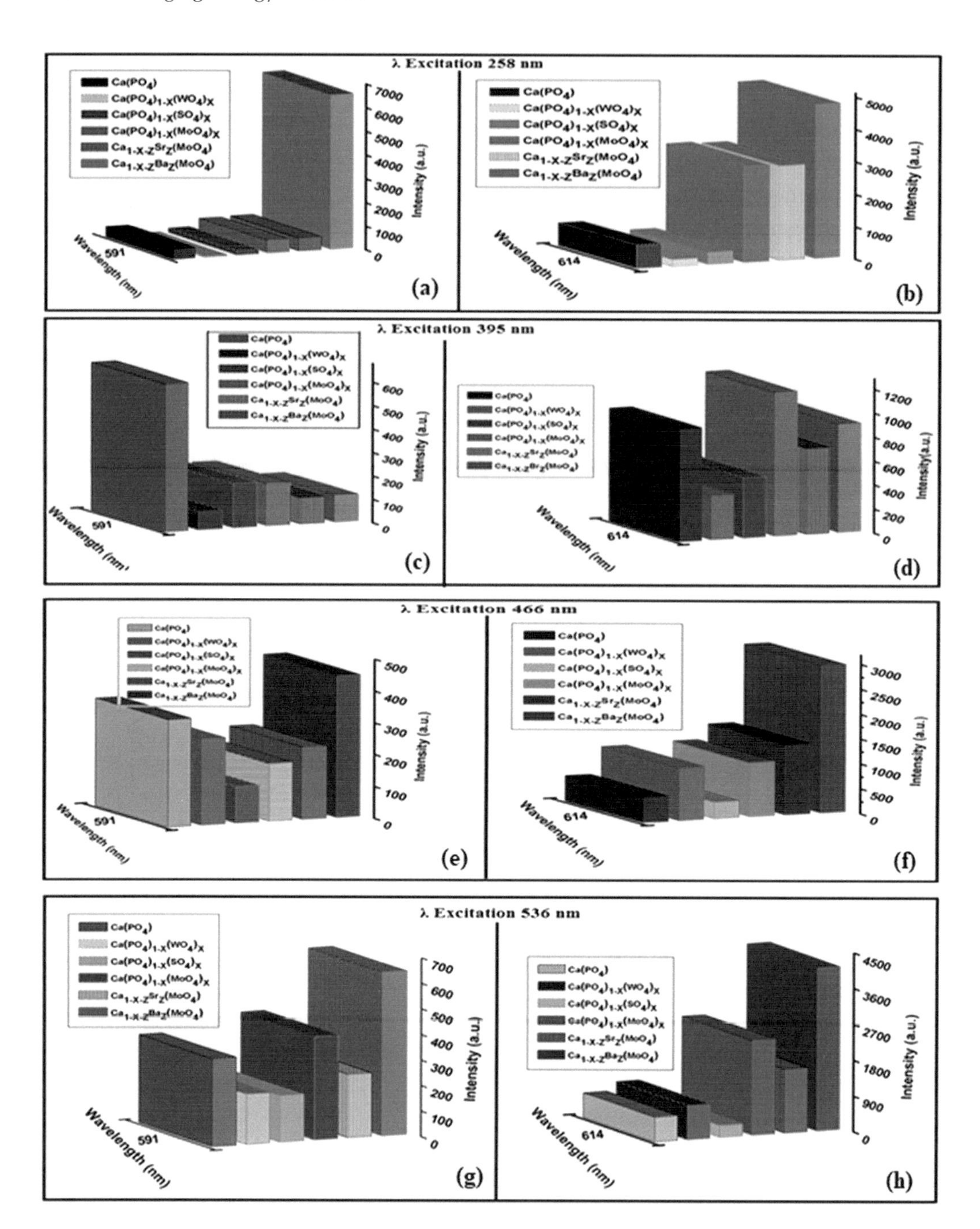

FIGURE 7.6 Variation in PL emission of $Ca_{3(1-x-z0}M_z(PO_4)_{2(1-x)}A_x$:$x$ mol% of Eu^{3+} phosphor (M = Sr, Ba), (A = SO_4, MoO_4, VO_4,WO_4) for 614 nm at excitation of 258, 395, 466 and 536 nm (x, y, z = 0.05,0.1,0.2,0.5,1), (a) λ emission 591 nm, (b) λ emission 614 nm. (Reproduced with permission from Ref. [65], © 2021Royal Society of Chemistry.)

near the Eu^{3+} environment. Thus, it might have again directed to more deviation from actual S4 site symmetry.

7.5 CONCLUDING REMARKS

In view of the current global energy crisis, the innovation of pc-WLEDs as appropriate green lighting sources for the future generation lighting sector was motivated by their exceptional benefits of energy conservation, excellent optical efficiencies, and high switching rates, among others. Lighting accounts for around 20% of the world's energy production, which, considering the current energy shortage, stimulates the production of pc-WLEDs. The demanding technical and practical requirements that have emerged due to the quick advancement of current technological and scientific knowledge have led to stringent guidelines for pc-WLEDs with respect to both their luminous efficiency and tunable emission color. In accordance with the intrinsic luminous characteristics of the activators, the two ways that have been most frequently used for PL tuning of lanthanide-activated phosphors were extensively examined in this chapter: (1) Since the energy levels of Ce^{3+} and Eu^{2+} outermost 5d orbitals are sensitive to the electronic configuration, changing the host composition through cation/anion replacement is a viable technique for adjusting the emission color in Ce^{3+} and Eu^{2+} singly doped phosphors; and (2) Pertaining to 4f-4f transition brightness through energy transfer between various 'ionized ion' couples, including phosphate, sulphate, vanadate, tungstate, lanthanum, and other co-doping species, lanthanide-activated phosphors, co-doping with singly, doubly, and triply ionized ions, may not only increase the efficiency of absorption, but also generate an extensive variety of tunable emission colors. With the aid of these techniques, novel luminous materials can be created and discovered for high-performance pc-WLED devices.

There have been significant and amazing advancements made in the PL tuning of lanthanide-activated phosphors for pc-WLED over the past years, yet there continue to be a number of problems that must be solved: (1) it is challenging to achieve comprehensive emission in a single-phase host that is comparable to sunlight, which would be intended to resolve a number of technological issues, notably color balancing, color re-absorption, and complex device design. Warm white lighting will undoubtedly become more common over time due to needs for human comfort and health; (2) to meet the constantly rising need for high-resolution and high-saturation color displays, novel phosphors with narrow-band emissions and monodisperse spherical forms must be discovered; and (3) it is difficult to theoretically anticipate the emission position, complete width at half maxima, and thermal quenching of the lanthanide-activated luminescence mechanism. Researchers have demonstrated that theoretical computation is a strong and effective method for screening fresh, highly effective luminous phosphors and that it also offers workable PL tuning assistance. Theoretical simulations can also offer a thorough understanding of the thermal quenching and spectrum blue-/red-shifts, which is helpful in developing deliberate, well-thought-out, and effective techniques for phosphor screening and PL tuning.

REFERENCES

[1] J. Huo, A. Yu, Q. Ni, D. Guo, M. Zeng, J. Gao, Y. Zhang, Q. Wang, Efficient energy transfer from trap levels to Eu^{3+} leads to antithermal quenching effect in high-power white light-emitting diodes, *Inorg. Chem.* 59 (2020) 15514–15525. https://doi.org/10.1021/acs.inorgchem.0c02541

[2] Y.N. Ahn, K. Do Kim, G. Anoop, G.S. Kim, J.S. Yoo, Design of highly efficient phosphor-converted white light-emitting diodes with color rendering indices (R1 – R15) ≥ 95 for artificial lighting, *Sci. Rep.* 9 (2019) 16848. https://doi.org/10.1038/s41598-019-53269-0

[3] P. Dang, D. Liu, Y. Wei, G. Li, H. Lian, M. Shang, J. Lin, Highly efficient cyan-green emission in self-activated $Rb_3RV_2O_8$ (R = Y, Lu) vanadate phosphors for full-spectrum white light-emitting diodes (LEDs), *Inorg. Chem.* 59 (2020) 6026–6038. https://doi.org/10.1021/acs.inorgchem.0c00015

[4] X. Zhang, D. Zhang, Z. Zheng, B. Zheng, Y. Song, K. Zheng, Y. Sheng, Z. Shi, H. Zou, A single-phase full-visible-spectrum phosphor for white light-emitting diodes with ultra-high color rendering, *Dalt. Trans.* 49 (2020) 17796–17805. https://doi.org/10.1039/d0dt03797d

[5] D. Huang, P. Dang, H. Lian, Q. Zeng, J. Lin, Luminescence and energy-transfer properties in Bi^{3+}/Mn^{4+}-codoped Ba_2GdNbO_6 double-perovskite phosphors for white-light-emitting diodes, *Inorg. Chem.* 58 (2019) 15507–15519. https://doi.org/10.1021/acs.inorgchem.9b02558

[6] G.B. Nair, S.J. Dhoble, A perspective perception on the applications of light-emitting diodes, *Luminescence.* 30 (2015) 1167–1175. https://doi.org/10.1002/bio.2919

[7] D. Chen, X. Chen, Luminescent perovskite quantum dots: Synthesis, microstructures, optical properties and applications, *J. Mater. Chem. C.* 7 (2019) 1413–1446. https://doi.org/10.1039/C8TC05545A

[8] G. Anoop, J.R. Rani, J. Lim, M.S. Jang, D.W. Suh, S. Kang, S.C. Jun, J.S. Yoo, Reduced graphene oxide enwrapped phosphors for long- term thermally stable phosphor converted white light emitting diodes, *Nat. Publ. Gr.* (2016) 1–9. https://doi.org/10.1038/srep33993

[9] T.S. Sreena, P.P. Rao, A.K.V. Raj, T.R.A. Thara, Narrow-band red-emitting phosphor, $Gd_3Zn_2Nb_3O_{14}$:Eu^{3+} with high color purity for phosphor-converted white light emitting diodes, *J. Alloys Compd.* 751 (2018) 148–158. https://doi.org/10.1016/j.jallcom.2018.04.135

[10] B. Han, Y. Dai, J. Zhang, B. Liu, H. Shi, Development of near-ultraviolet-excitable single-phase white-light-emitting phosphor $KBaY(BO_3)_2$:Ce^{3+} ,Dy^{3+} for phosphor-converted white light-emitting-diodes, *Ceram. Int.* 44 (2018) 14803–14810. https://doi.org/10.1016/j.ceramint.2018.05.111

[11] X. Huang, J. Liang, S. Rtimi, B. Devakumar, Z. Zhang, Ultra-high color rendering warm-white light-emitting diodes based on an efficient green-emitting garnet phosphor for solid-state lighting, *Chem. Eng. J.* 405 (2021) 126950. https://doi.org/10.1016/j.cej.2020.126950

[12] G.B. Nair, H.C. Swart, S.J. Dhoble, A review on the advancements in phosphor-converted light emitting diodes (pc-LEDs): Phosphor synthesis, device fabrication and characterization, *Prog. Mater. Sci.* 109 (2020) 100622. https://doi.org/10.1016/j.pmatsci.2019.100622

[13] J. Cho, Y.K. Jung, J.K. Lee, H.S. Jung, Surface coating of gradient alloy quantum dots with oxide layer in white-light-emitting diodes for display backlights, *Langmuir.* 33 (2017) 13040–13050. https://doi.org/10.1021/acs.langmuir.7b03335

[14] Y. Won, H.S. Shin, M. Jo, Y.J. Lim, R. Manda, S.H. Lee, An electrically switchable dye-doped liquid crystal polarizer for organic light emitting-diode displays, *J. Mol. Liq.* 333 (2021) 115922. https://doi.org/10.1016/j.molliq.2021.115922

[15] M. Khoury, H. Li, P. Li, Y.C. Chow, B. Bonef, H. Zhang, M.S. Wong, S. Pinna, J. Song, J. Choi, J.S. Speck, S. Nakamura, S.P. DenBaars, Polarized monolithic white semipolar (20–21) InGaN light-emitting diodes grown on high quality (20–21) GaN/sapphire templates and its application to visible light communication, *Nano Energy.* 67 (2020) 104236. https://doi.org/10.1016/j.nanoen.2019.104236

[16] S. Jeong, S.K. Oh, J.H. Ryou, K.S. Ahn, K.M. Song, H. Kim, Monolithic inorganic ZnO/GaN semiconductors heterojunction white light-emitting diodes, *ACS Appl. Mater. Interfaces.* 10 (2018) 3761–3768. https://doi.org/10.1021/acsami.7b15946

[17] J. Jeong, D.K. Jin, J. Choi, J. Jang, B.K. Kang, Q. Wang, W. Il Park, M.S. Jeong, B.S. Bae, W.S. Yang, M.J. Kim, Y.J. Hong, Transferable, flexible white light-emitting diodes of GaN p–n junction microcrystals fabricated by remote epitaxy, *Nano Energy.* 86 (2021) 106075. https://doi.org/10.1016/j.nanoen.2021.106075

[18] Q. Sun, S. Wang, L. Sun, J. Liang, B. Devakumar, X. Huang, Achieving full-visible-spectrum LED lighting via employing an efficient Ce^{3+}-activated cyan phosphor, *Mater. Today Energy.* 17 (2020) 100448. https://doi.org/10.1016/j.mtener.2020.100448

[19] T.H. Kim, A.R. White, J.P. Sirdaarta, W. Ji, I.E. Cock, J. St John, S.E. Boyd, C.L. Brown, Q. Li, Yellow-emitting carbon nanodots and their flexible and transparent films for white LEDs, *ACS Appl. Mater. Interfaces.* 8 (2016) 33102–33111. https://doi.org/10.1021/acsami.6b12113

[20] L. Sun, B. Devakumar, J. Liang, S. Wang, Q. Sun, X. Huang, A broadband cyan-emitting $Ca_2LuZr_2(AlO_4)_3:Ce^{3+}$ garnet phosphor for near-ultraviolet-pumped warm-white light-emitting diodes with an improved color rendering index, *J. Mater. Chem. C.* 8 (2020) 1095–1103. https://doi.org/10.1039/c9tc04952e

[21] S. Liu, M. He, X. Di, P. Li, W. Xiang, X. Liang, $CsPbX_3$ nanocrystals films coated on YAG:Ce^{3+} PiG for warm white lighting source, *Chem. Eng. J.* 330 (2017) 823–830. https://doi.org/10.1016/j.cej.2017.08.032

[22] Q. Yao, P. Hu, P. Sun, M. Liu, R. Dong, K. Chao, Y. Liu, J. Jiang, H. Jiang, YAG:Ce^{3+} transparent ceramic phosphors brighten the next-generation laser-driven lighting, *Adv. Mater.* 32 (2020) 1–7. https://doi.org/10.1002/adma.201907888

[23] Q. Dong, P. Xiong, J. Yang, Y. Fu, W. Chen, F. Yang, Z. Ma, M. Peng, Bismuth activated blue phosphor with high absorption efficiency for white LEDs, *J. Alloys Compd.* 885 (2021) 160960. https://doi.org/10.1016/j.jallcom.2021.160960

[24] K. Li, J. Xu, X. Cai, J. Fan, Y. Zhang, M. Shang, H. Lian, J. Lin, An efficient green-emitting α-Ca $_{1.65}Sr_{0.35}SiO_4:Eu^{2+}$ phosphor for UV/n-UV w-LEDs: Synthesis, luminescence and thermal properties, *J. Mater. Chem.* C. 3 (2015) 6341–6349. https://doi.org/10.1039/c5tc00796h

[25] H. Li, Y. Liang, S. Liu, W. Zhang, Y. Bi, Y. Gong, Y. Chen, W. Lei, Highly efficient green-emitting phosphor $Sr_4Al_{14}O_{25}$:Ce, Tb with low thermal quenching and wide color gamut upon UV-light excitation for backlighting display applications, *J. Mater. Chem. C.* 9 (2021) 2569–2581. https://doi.org/10.1039/d0tc04618c

[26] J. Zhao, C. Guo, T. Li, X. Su, N. Zhang, J. Chen, Synthesis, electronic structure and photoluminescence properties of $Ba_2BiV_3O_{11}$: Eu^{3+} red phosphor, *Dye. Pigment.* 132 (2016) 159–166. https://doi.org/10.1016/j.dyepig.2016.04.052

[27] B. Fan, J. Liu, W. Zhou, L. Han, Luminescence properties of new red-emitting phosphor $Li_2Al_2Si_3O_{10}:Eu^{3+}$ for near UV-based white LED, *Opt. Mater. (Amst).* 98 (2019) 109499. https://doi.org/10.1016/j.optmat.2019.109499

[28] P. Dang, Q. Zhang, D. Liu, G. Li, H. Lian, M. Shang, J. Lin, Hetero-valent substitution strategy toward orange-red luminescence in Bi^{3+} doped layered perovskite oxide phosphors for high color rendering index white light-emitting diodes, *Chem. Eng. J.* 420 (2021) 127640. https://doi.org/10.1016/j.cej.2020.127640

[29] M. Qu, X. Zhang, X. Mi, H. Sun, Q. Liu, Z. Bai, Luminescence color tuning of Ce^{3+} and Tb^{3+} co-doped $Ca_2YZr_2Al_3O_{12}$ phosphors with high color rendering index via energy transfer, *J. Lumin.* 228 (2020) 117557. https://doi.org/10.1016/j.jlumin.2020.117557

[30] Z. Zhang, B. Devakumar, S. Wang, L. Sun, N. Ma, W. Li, X. Huang, Using an excellent near-UV-excited cyan-emitting phosphor for enhancing the color rendering index of warm-white LEDs by filling the cyan gap, *Mater. Today Chem.* 20 (2021) 100471. https://doi.org/10.1016/j.mtchem.2021.100471

[31] W.L. Wu, M.H. Fang, W. Zhou, T. Lesniewski, S. Mahlik, M. Grinberg, M.G. Brik, H.S. Sheu, B.M. Cheng, J. Wang, R.S. Liu, High color rendering index of $Rb_2GeF_6:Mn^{4+}$ for light-emitting diodes, *Chem. Mater.* 29 (2017) 935–939. https://doi.org/10.1021/acs.chemmater.6b05244

[32] X. Li, B. Milićević, M.D. Dramićanin, X. Jing, Q. Tang, J. Shi, M. Wu, Eu^{3+}-activated $Sr_3ZnTa_2O_9$ single-component white light phosphors: Emission intensity enhancement and color rendering improvement, *J. Mater. Chem. C.* 7 (2019) 2596–2603. https://doi.org/10.1039/c9tc00159j

[33] Y. Cui, T. Song, J. Yu, Y. Yang, Z. Wang, G. Qian, Dye encapsulated metal-organic framework for warm-white LED with high color-rendering index, *Adv. Funct. Mater.* 25 (2015) 4796–4802. https://doi.org/10.1002/adfm.201501756

[34] K.M. Zielinska-Dabkowska, J. Hartmann, C. Sigillo, LED light sources and their complex set-up for visually and biologically effective illumination for ornamental indoor plants, *Sustain.* 11 (2019). https://doi.org/10.3390/su11092642

[35] S. Park, B. Kang, M. I. Choi, S. Jeon, S. Park, A micro-distributed ESS-based smart LED street-light system for intelligent demand management of the micro grid, *Sustain. Cities Soc.* 39 (2018) 801–813. https://doi.org/10.1016/j.scs.2017.10.023

[36] S. Hou, M.K. Gangishetty, Q. Quan, D.N. Congreve, Efficient blue and white perovskite light-emitting diodes via manganese doping, *Joule.* 2 (2018) 2421–2433. https://doi.org/10.1016/j.joule.2018.08.005

[37] V.H. Panhoca, R. de Fatima Zanirato Lizarelli, S.C. Nunez, R.C.A. de Pizzo, C. Grecco, F.R. Paolillo, V.S. Bagnato, Comparative clinical study of light analgesic effect on temporomandibular disorder (TMD) using red and infrared led therapy, *Lasers Med. Sci.* 30 (2015) 815–822. https://doi.org/10.1007/s10103-013-1444-9

[38] P.F. Smet, A.B. Parmentier, D. Poelman, Selecting conversion phosphors for white light-emitting diodes, *J. Electrochem. Soc.* 158 (2011) R37. https://doi.org/10.1149/1.3568524

[39] A.R. Bansod, A.R. Kadam, P.S. Bokare, S.J. Dhoble, Luminescence study of Sm^{3+},Eu^{3+}-doped $Y_2Zr_2O_7$ host: Optical investigation of greenish yellow to red color tunable pyrochlore phosphor, *Luminescence.* (2022) 1–9. https://doi.org/10.1002/bio.4305

[40] R.G. Deshmukh, A.R. Kadam, S.J. Dhoble, Energy transfer mechanism in $K_2Ba(PO_4)F:Dy^{3+}$, Eu^{3+} co-activated phosphor: Spectral tuning phosphor for photovoltaic efficiency enhancement, *J. Mol. Struct.* 1257 (2022) 132603. https://doi.org/10.1016/j.molstruc.2022.132603

[41] A.R. Kadam, R.B. Kamble, M. Joshi, A.D. Deshmukh, S.J. Dhoble, Eu(iii), Tb(iii) activated/co-activated K_2NaAlF_6 host array: Simultaneous approach for improving photovoltaic efficiency and tricolor emission, *New J. Chem.* 46 (2022) 334–344. https://doi.org/10.1039/d1nj04836h

[42] K. Dubey, A.R. Kadam, N. Baig, N.S. Dhoble, S.J. Dhoble, Wet chemically prepared terbium activated sodium calcium chlorosulfate phosphor for solid state lighting industry, *Radiat. Eff. Defects Solids.* 2 (2020). https://doi.org/10.1080/10420150.2020.1855179

[43] P. Tadge, A. Kadam, S. Sapra, S. Dhoble, S. Ray, Efficient near-infrared quantum cutting in Y_2O_3 codoped with Ho^{3+} ,Yb^{3+} phosphor synthesized by solution route, *ECS Trans.* 107 (2022) 17717–17730. https://doi.org/10.1149/10701.17717ecst

[44] A.R. Kadam, S.B. Dhoble, G.C. Mishra, A.D. Deshmukh, S.J. Dhoble, Combustion assisted spectroscopic investigation of Dy^{3+} activated $SrYAl_3O_7$ phosphor for LED and TLD applications, *J. Mol. Struct.* 1233 (2021) 130150. https://doi.org/10.1016/j.molstruc.2021.130150

[45] X. Yang, Y. Zhang, X. Zhang, J. Chen, H. Huang, D. Wang, X. Chai, G. Xie, M.S. Molokeev, H. Zhang, Y. Liu, B. Lei, Facile synthesis of the desired red phosphor $Li_2Ca_2Mg_2Si_2N_6:Eu^{2+}$ for high CRI white LEDs and plant growth LED device, *J. Am. Ceram. Soc.* 103 (2020) 1773–1781. https://doi.org/10.1111/jace.16858

[46] X. Huang, Q. Sun, B. Devakumar, Preparation, crystal structure, and photoluminescence properties of high-brightness red-emitting $Ca_2LuNbO_6:Eu^{3+}$ double-perovskite phosphors for high-CRI warm-white LEDs, *J. Lumin.* 225 (2020) 117373. https://doi.org/10.1016/j.jlumin.2020.117373

[47] H. Hua, S. Feng, Z. Ouyang, H. Shao, H. Qin, H. Ding, Q. Du, Z. Zhang, J. Jiang, H. Jiang, YAG:Ce transparent ceramics with high luminous efficiency for solid-state lighting application, *J. Adv. Ceram.* 8 (2019) 389–398. https://doi.org/10.1007/s40145-019-0321-9

[48] P. Ma, X. Sun, W. Pan, G. Yu, J. Wang, Green and orange emissive carbon dots with high quantum yields dispersed in matrices for phosphor-based white LEDs, *ACS Sustain. Chem. Eng.* 8 (2020) 3151–3161. https://doi.org/10.1021/acssuschemeng.9b06008

[49] M. Liu, N. Jiang, H. Huang, J. Lin, F. Huang, Y. Zheng, D. Chen, Ni^{2+}-doped $CsPbI_3$ perovskite nanocrystals with near-unity photoluminescence quantum yield and superior structure stability for red light-emitting devices, *Chem. Eng. J.* 413 (2021) 127547. https://doi.org/10.1016/j.cej.2020.127547

[50] T.S. Dhapodkar, A.R. Kadam, N. Brahme, S.J. Dhoble, Efficient white light-emitting $Mg_{21}Ca_4Na_4(PO_4)_{18}$:$Dy^{3+}$,$Tb^{3+}$,$Eu^{3+}$ triple-doped glasses: A multipurpose glasses for WLEDs, solar cell efficiency enhancement, and smart windows applications, *Mater. Today Chem.* 24 (2022) 100938. https://doi.org/10.1016/j.mtchem.2022.100938

[51] T.S. Dhapodkar, A.R. Kadam, A. Duragkar, N.S. Dhoble, S.J. Dhoble, Recent progress in phosphate based luminescent materials: A case study, *J. Phys. Conf. Ser.* 1913 (2021). https://doi.org/10.1088/1742-6596/1913/1/012024

[52] S.J. Helode, A.R. Kadam, S.J. Dhoble, Luminescence investigation of Sm^{3+} activated $CaAl_2(SiO_4)_2Cl_2$ chlorapatite phosphor for solid state lighting applications, *Chem. Data Collect.* 40 (2022) 100881. https://doi.org/10.1016/j.cdc.2022.100881

[53] N. Baig, A.R. Kadam, K. Dubey, N.S. Dhoble, S.J. Dhoble, Wet chemically synthesized $Na_3Ca_2(SO_4)_3Cl$:RE^{3+} (RE = Ce, Dy, Eu) phosphors for solid-state lighting, *Radiat. Eff. Defects Solids.* 176 (2021) 493–507. https://doi.org/10.1080/10420150.2021.1871735

[54] A.R. Kadam, G.C. Mishra, M. Michalska-Domanska, S.J. Dhoble, Theoretical analysis of electron vibrational interaction (EVI) parameters in 5d states of Eu^{2+} activated $BaSiF_6$ downconversion phosphor, *J. Mol. Struct.* 1229 (2021) 129505. https://doi.org/10.1016/j.molstruc.2020.129505

[55] G.B. Nair, S.J. Dhoble, Highly enterprising calcium zirconium phosphate [$CaZr_4(PO_4)_6$:Dy^{3+}, Ce^{3+}] phosphor for white light emission, *RSC Adv.* 5 (2015) 49235–49247. https://doi.org/10.1039/C5RA07306E

[56] G.B. Nair, S.J. Dhoble, Orange light-emitting $Ca_3Mg_3(PO_4)_4$:Sm^{3+} phosphors, *Luminescence.* 32 (2017) 125–128. https://doi.org/10.1002/bio.3194

[57] A.R. Kadam, G.C. Mishra, A.D. Deshmukh, S.J. Dhoble, Enhancement of blue emission in Ce^{3+}, Eu^{2+} activated $BaSiF_6$ downconversion phosphor by energy transfer mechanism: A photochromic phosphor, *J. Lumin.* 229 (2021) 117676. https://doi.org/10.1016/j.jlumin.2020.117676

[58] Q. Wang, M. Liao, Q. Lin, M. Xiong, Z. Mu, F. Wu, A review on fluorescence intensity ratio thermometer based on rare-earth and transition metal ions doped inorganic luminescent materials, *J. Alloys Compd.* 850 (2021) 156744. https://doi.org/10.1016/j.jallcom.2020.156744

[59] I. Gupta, S. Singh, S. Bhagwan, D. Singh, Rare earth (RE) doped phosphors and their emerging applications: A review, *Ceram. Int.* 47 (2021) 19282–19303. https://doi.org/10.1016/j.ceramint.2021.03.308

[60] G.B. Nair, S.J. Dhoble, Photoluminescence properties of Eu^{3+}/Sm^{3+} activated $CaZr_4(PO_4)_6$ phosphors, *J. Fluoresc.* 26 (2016) 1865–1873. https://doi.org/10.1007/s10895-016-1880-6

[61] Z. Jianghui, C. Qijin, W.U. Shunqing, C. Rong, Electronic structure and luminescence properties of Tb^{3+}-activated $NaBaBO_3$ green-emitting phosphor, *J. Rare Earths.* 33 (2015) 933–938. https://doi.org/10.1016/S1002-0721(14)60508-1

[62] G.B. Nair, S.J. Dhoble, White light emitting $MZr_4(PO_4)_6$:Dy^{3+} (M = Ca, Sr, Ba) phosphors for WLEDs, *J. Fluoresc.* 27 (2017) 575–585. https://doi.org/10.1007/s10895-016-1985-y

[63] S. Tamboli, G.B. Nair, R.E. Kroon, S.J. Dhoble, H.C. Swart, Color tuning of the $Ba_{1.9}6Mg(PO_4)_2$:$0.04Eu^{2+}$ phosphor induced by the chemical unit co-substitution of the $(BO3)^{3-}$anion group, *J. Alloys Compd.* 864 (2021) 158124. https://doi.org/10.1016/j.jallcom.2020.158124

[64] A.R. Kadam, R.S. Yadav, G.C. Mishra, S.J. Dhoble, Effect of singly, doubly and triply ionized ions on downconversion photoluminescence in Eu^{3+} doped $Na_2Sr_2Al_2PO_4Cl_9$ phosphor: A comparative study, *Ceram. Int.* 46 (2020) 3264–3274. https://doi.org/10.1016/j.ceramint.2019.10.032
[65] A.R. Kadam, R.L. Kohale, G.C. Mishra, S.J. Dhoble, Eu(iii)-doped tri-calcium $Ca_{3(1-X-Z)}MZ(PO_4)_2A_X$:X host array: Optical investigations of down-conversion red phosphor for boosting display intensity and high color purity, *New J. Chem.* 45 (2021) 7285–7307. https://doi.org/10.1039/d0nj05930g

III

Photovoltaics and Energy-Harvesting Materials

CHAPTER 8

Highly Stable Inorganic Hole Transport Materials in Perovskite Solar Cells

Sarojini Jeeva Panchu and Hendrik C. Swart

University of the Free State, Bloemfontein, South Africa

8.1 INTRODUCTION

In recent times, energy consumption has gradually increased with the increase in population. It is important to reduce fossil fuel consumption for the production of energy to avoid environmental issues. Solar energy is among the cleanest and most abundant sources of renewable energy. There have been many successful applications of solar energy in various fields, including solar architecture, solar heating, photosynthesis, photocatalysis, and photovoltaics. Amongst these, photovoltaics, which converts solar energy into electricity via the photovoltaic effect, has gained much attention.

Solar cells can be divided into three categories: wafers, thin films, and organic-inorganic hybrids. Perovskite solar cells (PSCs) have gained increased attention since the introduction of third-generation solar cells, primarily because they are easy to fabricate using solution processing methods and have a high power conversion efficiency (PCE) of 25.5%, approaching silicon solar cell records of 26.7% [1, 2]. But the Si-based solar cells require high processing costs for the fabrication of the device. In 2009, dye-sensitized solar cells became the first to employ perovskite as a light-harvesting material, and the PCE was improved from 3.8% to 6.5%. In the presence of iodide/triiodide-based electrolytes, perovskite sensitized on the titania surface was highly problematic, and it affected the stability of the devices. Instead of an electrolyte, a solid-state hole-transporting material (HTM), spiro-MeOTAD (2,2′7,7′-tetrakis (N,N-di-p-methoxyphenyl amine)-9,9′-spirobifluorene) was used. A solid-state PSC exhibit with a PCE value of 10.9% was reported in 2012 [3, 4]. By using dual-source thermal evaporation of $MAPbI_{3-x}Cl_x$, the efficiency of PSCs was improved (15.4%) in 2013 [5]. Subsequently, in 2014, Yang's group achieved a PCE of 19.3% in planar PSCs using interfacial engineering [6]. A study completed by Seok and co-workers in 2015 showed that $FAPbI_3$ films exhibit 20.2% PCE [7]. According to Grätzel's group, they were able to create high-quality perovskite films by using polymethyl

DOI: 10.1201/9781003315261-11

methacrylate (PMMA) as a matrix, which resulted in a PCE of 21.6% (certified PCE: 21.02%) in 2016 [8]. With additional iodide ions introduced, the formamidine-based PSCs achieved a PCE (certified) of 22.1% in 2017 [9]. An increase up to 23.32% in PCE was reported by You's group in 2019 [10]. Within a decade, PSCs have surpassed the PCE of 25.5% (Figure 8.1) in the photovoltaics market which is fairly close to the 26.7% for Si photovoltaic cells [11]. Moreover, solid-state PSC was stable over a period of 500 h, which was demonstrated in a stability study executed on non-encapsulated devices exposed to the ambient environment. This indicates that perovskite is rapidly evolving as a photovoltaic material with a low processing cost.

In general, the name 'perovskite' was derived from the mineral name calcium titanate ($CaTiO_3$), whose crystal structure is related to the chemical formula ABX_3 where A is a symbol for an organic cation (Cs^+), methylene ammonium (MA^+, $CH_3NH_3^+$) or formamidine (FA^+, $CH_3CH_2NH_3^+$), B represents the inorganic cations like Sn^{2+}, Pb^{2+}, and X stands for halide ions such as I, Cl, and Br. Perovskites have a cubic symmetry, and their backbones are corner-shared octahedra, along with A cations in cuboctahedral spaces. The development of PSCs has been facilitated by their unique properties including wide absorption coefficients, varying compositions in terms of absorbance edges, longer diffusion lengths of ions, and efficient ambipolar charge transport. Perovskites with A-site cations larger than B-site cations can form organic-inorganic hybrid materials, and $MAPbI_3$ and $FAPbI_3$ have been studied most extensively for PSCs. The PSCs consist of several layers,

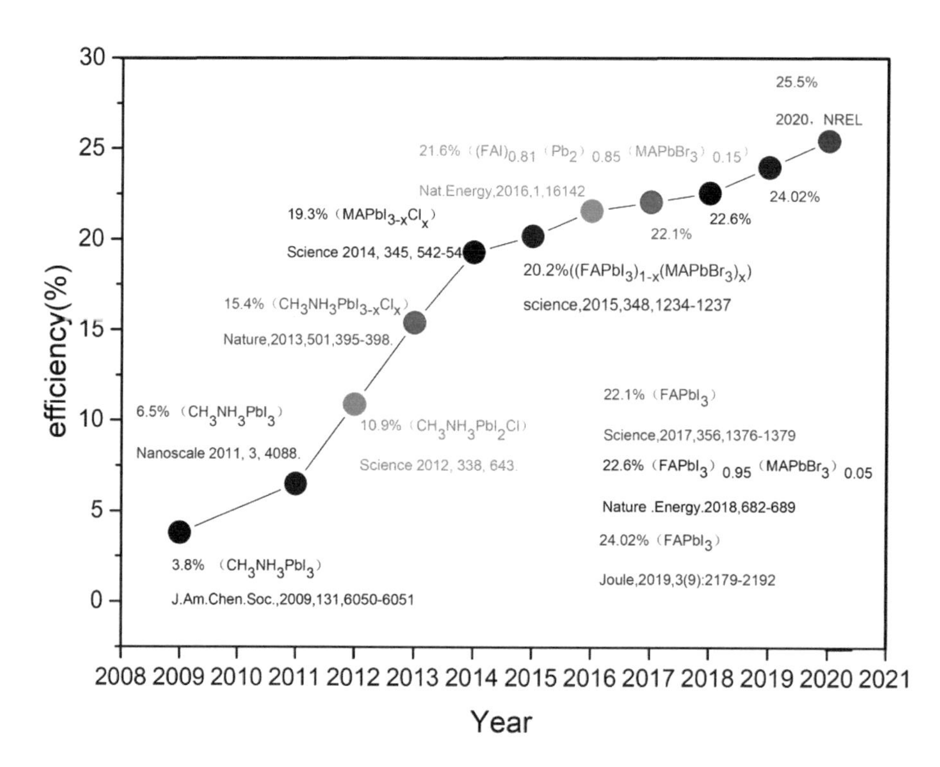

FIGURE 8.1 Graph of PCE improvement of PSCs. (Reproduced under the Creative Commons license from Ref. [12], © 2021, T. Dai et al.)

where perovskites act as the active layer, in addition to a charge collection layer, and an electrode layer that collects the current. Moreover, the organic-inorganic perovskite active layer possesses exceptional properties such as a large Bohr radius, low binding energy, high dielectric constant, rapid diffusion of carriers and diffusion length, and outstanding light absorption.

Even though, PSCs exhibit high PCE values, long-term stability, and cost-effective materials and methods are preferable for large-scale and commercialization processes. For large-scale processes, solution-processable materials are advantageous in terms of low-cost fabrication. The Pb-based perovskite has a limitation due to the toxic nature of Pb that can lead to environmental issues. By careful encapsulation techniques and adopting Pb-free perovskites, it is possible to tackle these environmental issues. However, the stability of the PSCs is important when it comes to the commercialization process. The stability of the PSC devices is determined by the photo-absorbers as well as the HTMs. The best performances of PSCs are observed by employing $MAPbI_3$ and $FAPbI_3$. When these MA^+ and FA^+ organic cations are substituted with Cs^+, the stability of the PSCs can be considerably improved from the theoretical point of view. As an inorganic perovskite, $CsPbI_3$-based PSCs have shown a theoretical PCE value of 28% with a narrow band gap of 1.73 eV. Still, the $CsPbI_3$-based PSCs showed degradation, posing doubts over their commercialization [13]. A thorough investigation of the degradation pathways involved in all-inorganic perovskites is necessary due to their complete absence of organic components.

Furthermore, another important material in PSCs that determines the device performance and stability of the device is the hole-transporting layer (HTL), which consists of hole-transporting materials (HTMs). The HTL generally consists of organic, polymers, or inorganic counterparts. Most HTLs are made from organic materials, which are easily degradable in the presence of moisture moieties. In this chapter, we have focused on inorganic HTMs and their synthesis.

8.2 DEVICE ARCHITECTURE AND WORKING PRINCIPLES

The PSC device architecture consists of a perovskite absorber layer sandwiched between the electron-transporting layer (ETL) and the HTL (Figure 8.2). Upon irradiation, exciton formation occurs in the active layer and the electrons and holes are extricated by the ETL and HTL and finally collected at their corresponding electrodes.

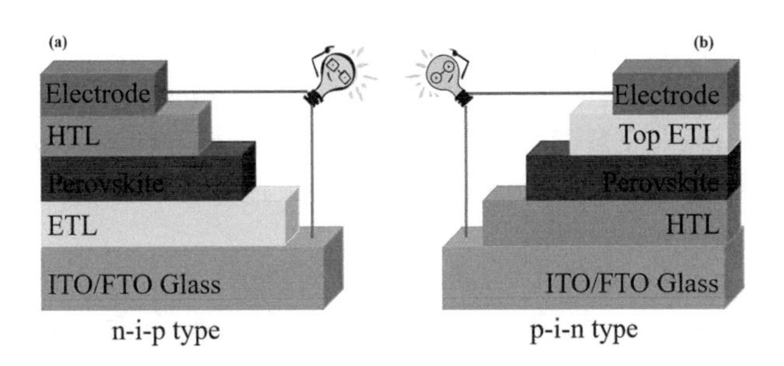

FIGURE 8.2 General device architecture of n-i-p and p-i-n PSCs. (Reproduced under the Creative Commons license from Ref. [14], © 2021, T. Dai et al.)

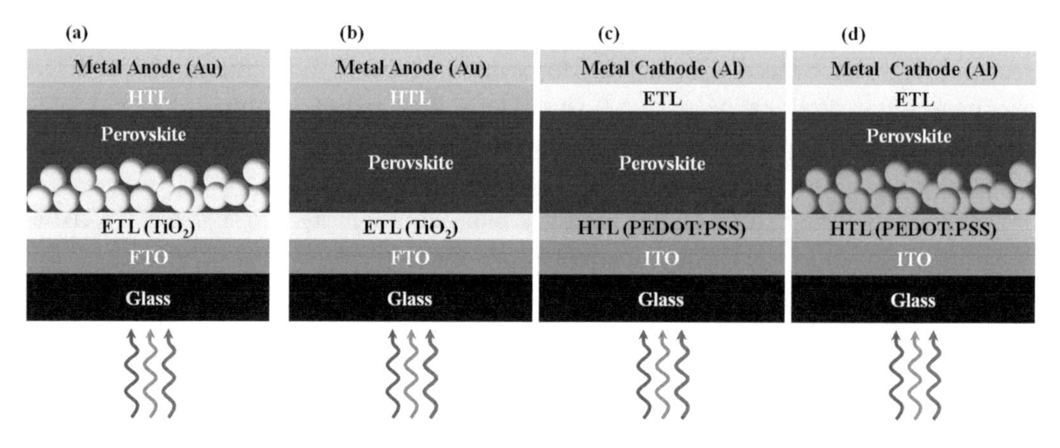

FIGURE 8.3 (a)–(d) Examples of mesoscopic and planar structured perovskite devices with (n-i-p and p-i-n) architectures. (Reproduced with permission from Ref. [15], © 2021 Elsevier Ltd.)

As planar and mesoscopic architectures, PSC architecture can be classified as n-i-p or p-i-n. The mesoscopic structure contains a mesoporous layer as a scaffold for perovskite material and the absence of a mesoporous layer is the planar structure. Figure 8.3(a) and (b) exhibit an n-i-p structure with the TCO (Transparent Conductive Oxide)/ETM (Electron Transport Material/Perovskite/HTM/Au architecture of PSC. The p-i-n structure of PSC with TCO/HTM/perovskite/ETM/Au architecture is shown in Figure 8.3(c) and (d). Depending on the device configuration materials such as TCO, ETM, HTM, perovskite, and metal contacts may vary. As of today, spiro-OMeTAD has proven to be an efficient HTM for the n-i-p as well as the p-i-n configurations due to its energy alignment with perovskites to facilitate efficient electron blocking and hole transfer. The fabrication of good quality perovskites will go on to achieve high PCE and the performance of the overall device is being improved by modifying HTMs.

8.3 HOLE-TRANSPORTING MATERIALS

In PSCs, HTMs extract holes from the exciton and transmit them to the cathode layer for collection. So, HTMs should maintain a proper energy alignment with the active layer to separate the holes and prevent the electrons from passing through them (Figure 8.4). Another criterion for HTMs is high mobility, along with high optical, thermal, and chemical stability. To prevent the degradation of perovskites, HTMs are deposited on the active layer of PSCs. In contrast, the p-i-n architecture active layer is coated on the HTM. It is, therefore, important that the HTMs are insoluble in high-polarity organic solvents such as dimethyl formamide (DMF) and dimethyl sulfoxide (DMSO) to prevent damage during the deposition of perovskite layers.

Generally, organic spiro-OMeTAD has been utilized as HTM in PSCs. The spiro-OMeTAD is a p-type semiconductor (2.98 eV) and it has low hole mobility, as well as low conductivity. To improve the mobility and conductivity, p-type dopants as additives have been added to the spiro-OMeTAD. The p-type dopants are 4-tert-butylpyridine (t-BP),

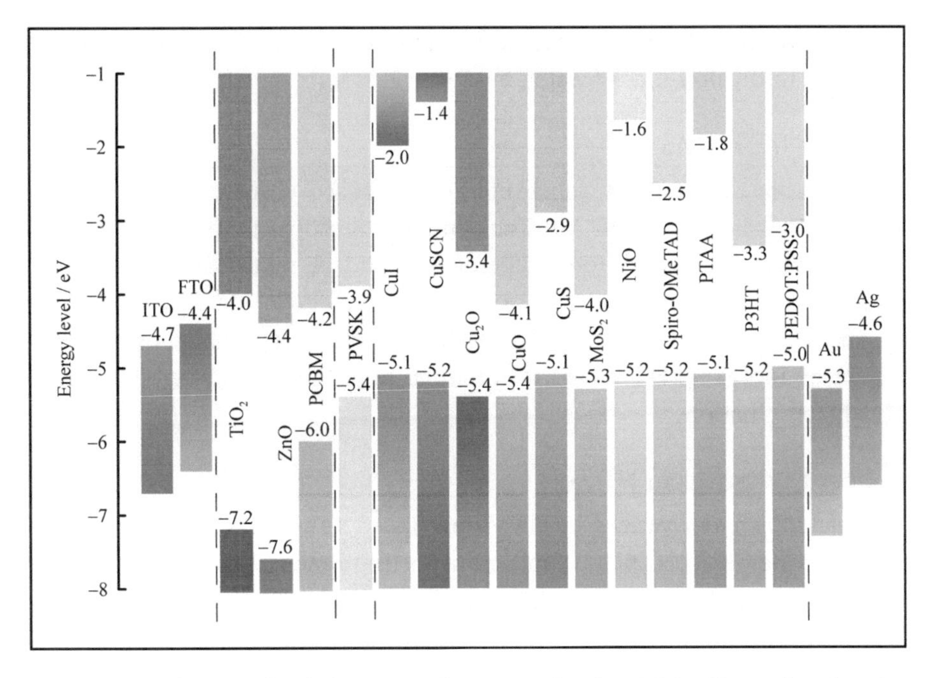

FIGURF 8.4 General energy level diagrams of HTMs utilized in PSCs. (Reproduced with permission from Ref. [16], © 2021, Youke Publishing Co. Ltd.)

lithium bis (trifluoromethane-sulfonyl) imide (LiTFSI). Despite the fact that t-BP and LiTFSI enhance hole mobility in PSCs, the compounds tend to corrode the perovskite and have a hygroscopic nature, which negatively impacts the devices' performance. Apart from these limitations, the cost of spiro-OMeTAD limits the practical application in terms of large-scale production. Poly[bis(4-phenyl)(2,4,6-trimethylphenyl) amine] (PTAA) was also used as a substitute for Spiro in both n-i-p and p-i-n architectures of PSCs, but PTAA was unsuitable for mass production due to its low mobility and higher cost.

A recent advance in PSCs has focused research on large-area production. The industrialization of PSCs requires them to be dopant-free, and thermally, chemically, and optically stable, as well as using low-cost materials. In PSCs, HTMs are primarily responsible for extracting holes from the perovskite layer upon illumination.

In an ideal HTM, the valence band maximum (VBM) or the highest occupied molecular orbital (HOMO) must align with the valence band of the active layer or the absorber material to prevent recombination. Furthermore, the HTMs should possess higher mobility to efficiently transport the collected holes to back contact electrodes. The solubility of HTMs plays an important role, along with thin film formation which should be inert for the perovskite active material. The organic solvents used to form film on the active materials do not affect the perovskite layer. Finally, high photo and chemical stability are important, along with good transparency, which is important when the inverted architecture is being fabricated with HTMs.

To date, numerous organic HTMs have been developed, such as spiro-OMeTAD, PEDOT:PSS, PTAA, and P3HT, that exhibit higher V_{oc} and higher PCE. But, in large-scale development with organic HTMs there are limitations due to the higher cost and their unstable nature in the presence of water, heat as well as light.

8.4 INORGANIC HTMS

To control the above limitations, substitute organic-based HTMs must be replaced with lower-cost and higher-stability HTMs. Thus, numerous inorganic-based HTMs have been identified and developed. In addition to being economical, widely available, toxicity-free, and energy efficient, transition metal oxides also have good stability and higher hole mobility, and are low cost. As well as increasing the efficiency of hole extraction from a perovskite layer, inorganic HTMs assist in maintaining the device's stability and PCE by preventing recombination.

8.4.1 Copper Derivatives in HTMs

The inorganic copper derivatives in HTMs are listed as CuSCN, CuI, Cu_2ZnSnS_4, and CuS are one of the potential p-type sole transporting materials in PSCs. These copper-based HTMs are already utilized in quantum dots (QDs)-sensitized dye-sensitized solar cells (DSSCs) as a hole transporting layer (HTL). Cu-based HTMs have the advantage of being easily solution-processed. They are wide bandgap semiconductors with higher conductivity.

8.4.1.1 CuI-HTM

CuI is among the best copper-based p-type hole conductors for PSCs. Jeffrey et al. utilized CuI as an HTL by solution processing method in $MAPbI_3$ ($CH_3NH_3PbI_3$) perovskite-sensitized solar cell utilizing. Owing to the nature of suitable VBM position, p-type conductivity, and solution-processable deposition that is compatible with the perovskite layer are utilized. The CuI-based PSCs exhibit 6.0% PCE and higher short-circuit current density compared to the spiro-OMeTAD which results in higher fill factor (FF) values. Even though CuI shows higher J_{sc}, the V_{oc} is still low compared to spiro-OMeTAD which makes a lower PCE than the control device. Apart from lower PCE values, the CuI devices show better photostability upon continuous 180 min illumination in normal PSCs. On the other hand, CuI was utilized in PSCs for inverted structures as an alternative to polymer-based HTMs as a result of its high optical transparency and high hole mobility. Under the same experimental conditions, CuI as the HTL achieved a PCE value of 13.58%, slightly exceeding the PCE of 13.28% achieved by the PEDOT:PSS-based device. Additionally, the CuI-based devices were more stable in the air than those based on PEDOT:PSS. This suggests that sol-processed CuI could be a viable substitute for PEDOT:PSS HTL used in inverted phase heterojunction (PHJ) perovskites.

Stability is an important characteristic for large-scale production. A PEDOT:PSS layer acts as an HTL layer in an inverted PSC. However, PEDOT:PSS has low stability in PSCs due to its acidic and hygroscopic nature. The CuI is a hydrophobic material in an ambient condition which is an excellent property for solar cells to maintain their stability. CuI-based devices were exposed to air continuously for 14 days and maintained 90% of their initial PCE values, while PEDOT:PSS-based devices demonstrated 27%.

8.4.1.2 Copper Oxide

Numerous techniques have been applied to form an HTL layer based on copper oxide HTMs. Yet solution-processable methods are preferable to reduce the cost and from the commercialization point of view. There are two ways to produce the CuO films, the first is solution-processed CuI thin film immersed in NaOH solution to form Cu_2O followed by drying at 100°C. CuO is then formed by annealing Cu_2O films at 250°C in the presence of oxygen. Alternatively, CuO is prepared by using CuI films that can be annealed in the presence of air at higher than 300°C resulting in the formation of CuO films. The prepared Cu_2O and CuO films are utilized as HTL in PSCs and compared against PEDOT:PSS-based PSCs where the $MAPbI_3$ is used as a light harvester. PSCs based on Cu_2O and CuO show higher V_{oc} and J_{sc} values than the control device (PEDOT:PSS). The PCE value of PEDOT:PSS, Cu_2O, and CuO are 11.04%, 13.35%, and 12.16%. The devices have been stored for 70 days and analyzed the performance. The Cu_2O-based PSCs shower better performance along and its PCE value shows 90% of its initial value. As a result of their lower VBMs PEDOT:PSS, Cu_2O, and CuO-based devices have higher V_{oc}.

8.4.1.3 Copper Sulphide

Copper sulphide is one of the important metal chalcogenides research communities due to its unique properties and high electrocatalytic properties. It has been widely utilized as a counter electrode (CE) component for dye- or QDs-sensitized solar cells. It is a binary inorganic compound denoted as Cu_xS_y, and categorized as CuS and Cu_2S. Copper sulphide is a p-type material with unique physical and optical properties which is beneficial for the inverted architecture of PSCs. A CuS-based HTM has two important characteristics: Due to the proper alignment of its energy levels (Figure 8.5(a)), it improves hole transport rates. In addition, it has high intrinsic hole mobility, which enhances the charge transfer characteristics and reduces the recombination of electrons and holes. Lei et al. synthesized $Cu_{1.75}S$ (anilite) by the vapor deposition method in a normal PSCs device without any thermal annealing. Thus, the perovskite layer is protected from unwanted solution methods to form an HTL. In order to improve the efficiency and stability of the device, $Cu_{1.75}S$ was placed between the spiro-OMeTAD and the Au metal contact. Thus, the hole transport rate increased and the underlying perovskite layer was protected by depositing in the spiro layer. Comparing the $Cu_{1.75}S$ device with the control device, the $Cu_{1.75}S$ device exhibits more stable performance in air over 1,000 hours: the PCE is 18.58% (Figure 8.5(b)) [17].

8.4.1.4 Copper Thiocyanate (CuSCN)

Copper cyanate (CuSCN) is an important HTM in solar cells. It has optical transparency as well as high hole mobility (10^{-2}–10^{-1} $cm^2 V^{-1} S^{-1}$) with a bandgap value of 3.7–3.9 eV [18]. Due to excess SCN-, CuSCN exhibits p-type conductivity due to copper vacancies in its crystal lattice.

For the solution process method, CuSCN works very well with a proper solvent with appropriate concentration at room temperature. So, CuSCN is an excellent solution for processable HTMs that works as a cost-effective, large-scale process and has flexible substrates. Mostly, inorganic HTMs with low-temperature processing are most preferable for

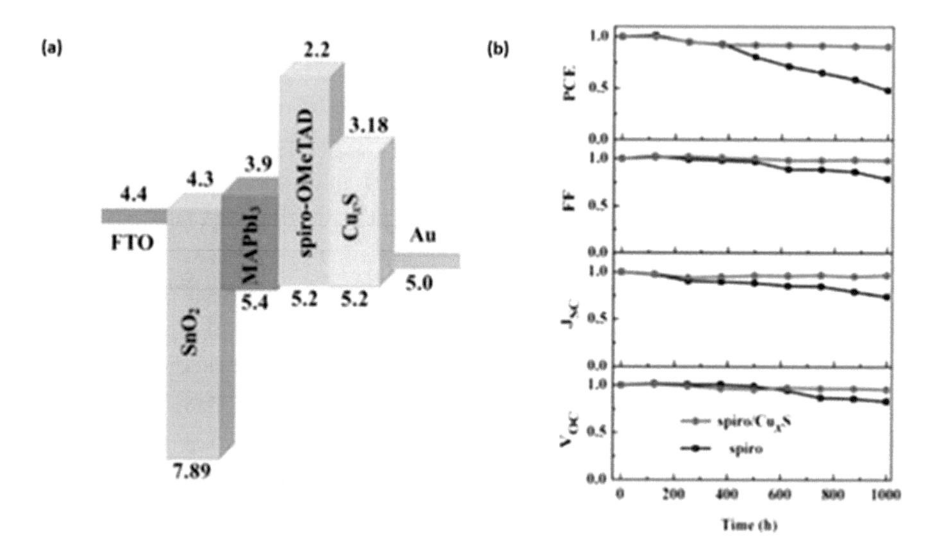

FIGURE 8.5 (a) Energy band diagram of the PSC. (b) PCE, FF, J_{sc}, and V_{oc} as a function of ambient storage time of devices with different HTLs. (Reproduced with permission from Ref. [17], © 2017 WILEY-VCH Verlag GmbH & Co. KGaA, Weinheim.)

PSCs. It has been widely utilized in DSSCs [19–21], organic photovoltaics (OPVs) [22, 23], and PSCs [24–26]. Nilushi et al. have replaced the organic solvent of diethylsulfide with aqueous solvents (Figure 8.6(a)) for OPV that achieved a PCE value of 10.7% higher than the PEDOT: PSS device (9.2%). The same method was utilized for an inverted PSC as an HTL and MAPbI3 perovskite that showed a PCE of 17.2% higher than the PEDOT:PSS device (13.6%) (Figure 8.6(b)) [27].

Various methods have been used to deposit CuSCN, such as the doctor blade technique, electrodeposition, spin casting, and spray pyrolysis. But solution-based techniques are

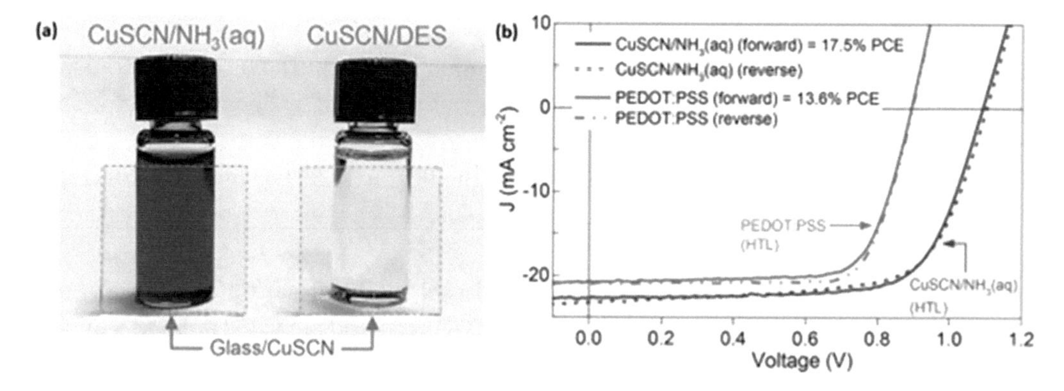

FIGURE 8.6 (a) Photographs of the vials containing the $CuSCN/NH_3(aq)$ and CuSCN/DES solutions placed behind 2×2 cm^2 glass substrates coated with thin layers of CuSCN. (b) J–V characteristics measured under simulated 1-sun solar illumination for two $CH_3NH_3PbI_3$ solar cells based on $CuSCN/NH_3(aq)$ and PEDOT:PSS HTLs. (Reproduced with permission from Ref. [27]. © 2017 WILEY-VCH Verlag GmbH & Co. KGaA, Weinheim.)

always preferable and uncomplicated techniques. Usually, a thin layer of CuSCN (dissolved in DES) is deposited on triple cationic ($CsFAMAPBI_{3-x}Br_x$) perovskite by a spin coating technique without affecting the perovskite layer. In general, the CuSCN exists as two polymorphs, such as alpha-CuSCN and beta-CuSCN. A PMMA polymer coating was applied to the devices to evaluate their thermal stability at 85 °C and the power conversion efficiency was recorded as 20.2%. After 1,000 h of operation, the devices exhibited about 85% of their initial PCE value [28]. The uniform morphology of the CuSCN film blocked the metal diffusion reducing the degradation rate of the device.

8.4.2 Nickel Oxide Hole Transporting Materials

Nickel oxide (NiO_x) is a cubic structure and belongs to the *Fm3m* space group and is shown in Figure 8.8.

The normal stoichiometric formula of NiO has an insulating nature, and the non-stoichiometric state of NiO_x having an excess of oxygen leads to nickel vacancies which result in a *p*-type semiconducting material. The E_g value of NiO_x varies between 3.6 and 4 eV. The wide bandgap nature of NiO_x enables extreme transparency at near UV and visible wavelengths. Thus, the NiO_x is suitable for n-i-p or p-i-n as mesoscopic as well as planar device architecture of PSCs.

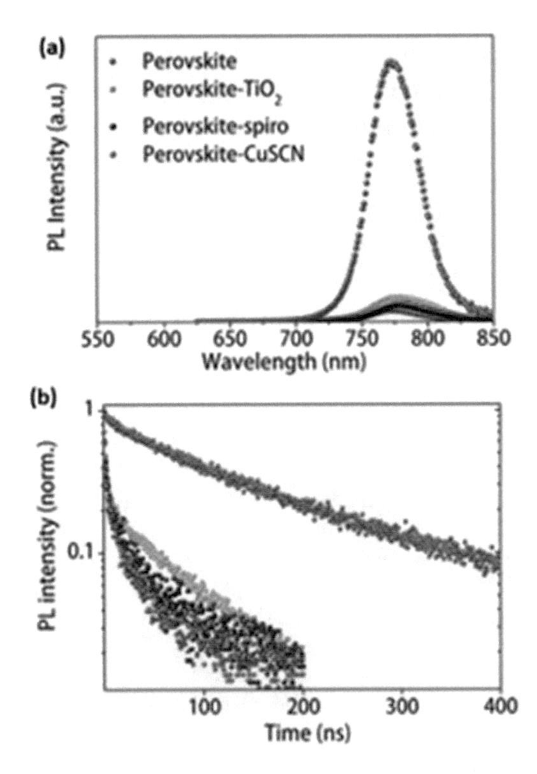

FIGURE 8.7 (a) A steady-state PL spectrum shows that the pristine perovskite film strongly quenches intense PL. (b) In the time-correlated single photon counting (TCSPC), charge carriers stay in the perovskite film for a very long time and are rapidly injected into the electron and hole extraction layers. (Reproduced under the Creative Commons license from Ref. [28]. © 2017 N. Arora et al.)

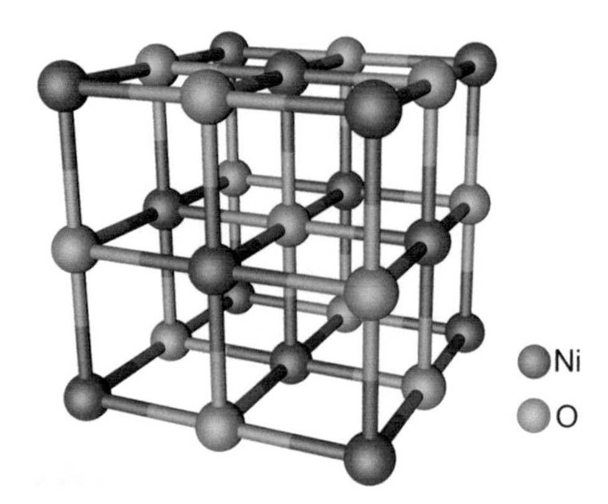

FIGURE 8.8 Crystal structure of NiO_x. (Reproduced with permission from Ref. [29], © 2019 WILEY-VCH Verlag GmbH & Co. KGaA, Weinheim.)

NiO_x has been widely utilized in numerous solar cells, in particular, organic photovoltaics, and DSSCs, as well as optoelectronic systems, as a hole transporting material owing to high hole mobility with excellent optical properties and thermally stable material. In picoseconds, NiO_x and perovskite extract and transfer holes efficiently. The recombination of electrons and holes takes place from 100 ps to ns due to low conductivity at the surface of NiO_x that can be rectified by the efficient appropriate doping to improve the conductivity in NiO_x and reduce the recombination rate. By appropriate doping of NiO_x like standard spiro OmATED, the mobility of holes will be increased, and the hole extraction and transfer rate will be improved which results in reduced recombination rate [30]. The band positioning concerning the perovskite and the NiO_x is shown in Figure 8.9.

In order for holes to be extracted from the interface between NiOx and perovskites, band alignment is crucial (Figures 8.9 and 8.10(a)). The holes that are generated from excitons are transferred to the NiO_x via the proper band alignment at the VBM, but the electron transfer is blocked due to higher CBM. Based on the previous report, the NiO_x

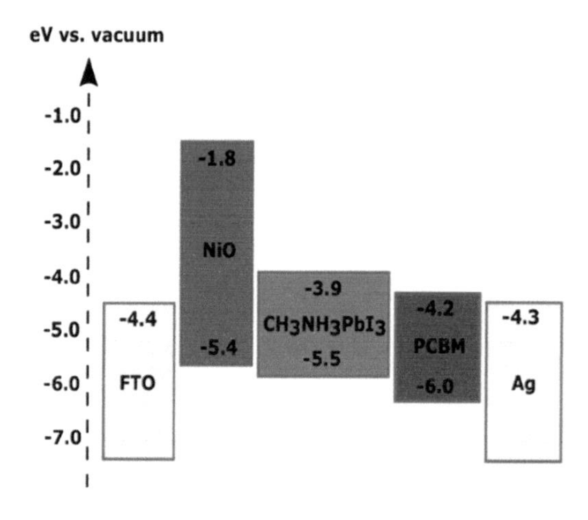

FIGURE 8.9 The band position of NiO_x film and $CH_3NH_3PbI_3$. (Reproduced with permission from Ref. [31], © 2019 Mustafa Aboulsaad et al.)

photoanode *p*-type DSSCs thus act as a scaffold in mesoscopic PSCs for the absorber layer as well as acting as an HTL for hole transport. In planar PSCs, the absorber layer was sandwiched between the ETL and HTL. So the quality of the films is crucial to obtaining high PCE values. Numerous technologies have been widely applied for the fabrication of NiOx films, including physical vapor deposition (PVD), chemical vapor deposition (CVD), electrodeposition, and solution process methods. The solution process techniques include spin coating and spray pyrolysis. In a sol-gel technique, the NiO_x is easily prepared and utilized as an HTL layer by using spray or spin coating techniques. In the planar PSCs, the deposited NiO_x films on transparent conducting oxides are annealed at higher than 300°C to obtain the crystalline NiO_x films. In 2014 Zhu et al. reported inverted planar PSCs based on NiO_x as an HTL layer where the NiO_x nanocrystals (10–20 nm) were deposited on fluorine-doped tin oxide (FTO). The thickness of the NiO_x (20–70 nm) will provide sufficient transparency which paves the way for sunlight to reach the absorber layer. The NiO_x were deposited at 20, 40, and 70 nm thickness. From 20 to 40 nm, V_{oc} and J_{sc} improved because of improved electron blocking, improved charge recombination, and reduced leakage of current. In contrast, as the film thickness is raised to 70 nm, photocurrent and voltage decrease, most likely as a result of increased series resistance and photo-absorption. As shown in Figure 8.4, PEDOT:PSS performs well as an HTM in PSCs. Also, 40-nm NiO NC layers display the highest V_{oc} and J_{sc} values and the highest PCE values (9.11%) (Figure 8.10(b)) [32]. The band positioning of the device architectures (relative to the vacuum level) for perovskite solar cell with NiO_x HTL is shown in Figure 8.10(a) [32].

The thickness of NiOx films impacts the functionality of the device. In 2015 Yin et al. utilized solution-based NiO_x as HTL in inverted PSCs, and PCE values of 14.42% were achieved. In this case, the perovskite layer that was directly deposited on the surface of TCO showed poor performance (PCE = 1.15%). Later a thin layer (5 nm) of NiO_x was

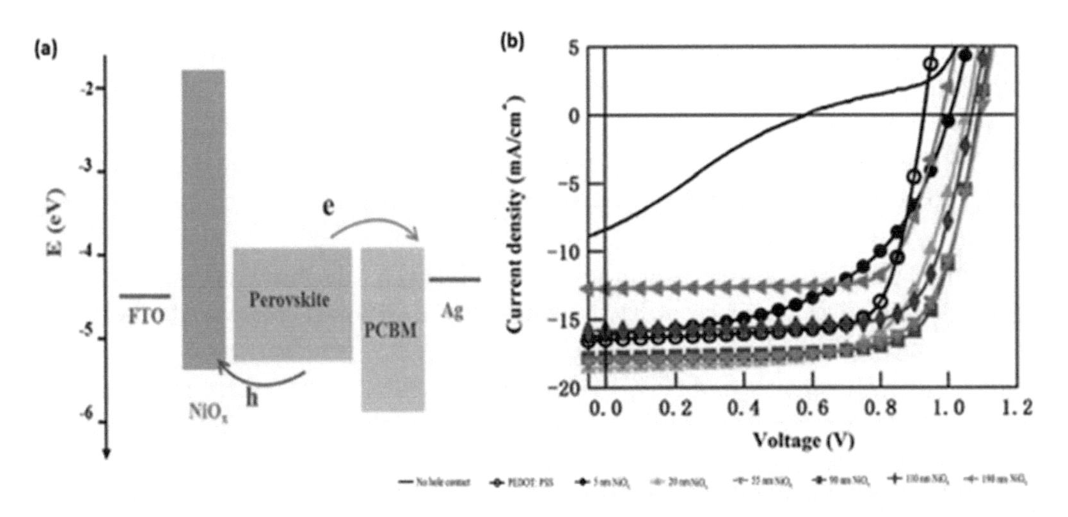

FIGURE 8.10 (a) Illustration of band diagram of PSC with NiO_x HTL. (b) J–V curves of devices with PEDOT:PSS and NiO_x HTLs. (Reproduced with permission from Ref. [33], © 2015 The Royal Society of Chemistry.)

introduced between the TCO and absorber layer which resulted in an improvement in J_{sc} and V_{oc} values and the PCE value was 8.39%. As the thickness increases, rising J_{sc} values result, while the V_{oc} values are maintained at ~1.09 V. The voltage-current properties of NiOx decrease when the film thickness is doubled (190 nm), affecting device performance. Thus, the 90 nm thick NiOx shows better performance (14.42%) as shown in Figure 8.10 [33]. In 2016, Zhu et al. deposited a NiO layer on FTO by spin casting and annealing at 400°C in 1 h under air, and the PSCs device with NiO_x showed an efficiency of 18.8%, and initial PCE values remained at 90% after 30 days storage in ambient atmosphere [34]. Even though NiO_x was successfully employed as an HTL in planar PSCs, the low hole conductivity of the NiOx affects device performance and the fill factor values of the devices. This can be rectified either by surface modification or doping of NiO_x. NiOx-based devices experience low J_{sc} and FF due to their low conductivity, poor crystallinity on the perovskite layer, poor contact, and poorly aligned band alignment. For these reasons, doping and surface modification methods are proposed. For further improving perovskite solar cell performance, interfacial engineering has also been considered an effective method. NiO_x films were introduced by the diethnolamine, 4-bromobenzoic acid, ferrocenedicarboxylic acid and poly (trial amine) (PTAA) by self-assembled monolayers technique. These functional groups are introduced between the NiO_x and perovskite layer. In diethanolamine-based surface modification, the amine groups form a chemical bonding with NiO_x via and N and Ni, and the crystalline quality of the perovskite layer also improves. Consequently, hole extraction and transport are enhanced by forming a molecular dipole layer. The hole kinetics and light absorption properties of amine-functionalized NiO_x was studied by using optical absorption and photoluminescence spectroscopy.

A current-voltage measurement was conducted on NiO_x/amine (cell 1) and NiO_x (cell 2) to determine their photovoltaic performance. in comparison, NiO_x/amine showed higher photovoltaic performance and higher J_{sc} values with a PCE value of ~16%. The charge transfer resistance was measured under full sun illumination for cells 1 and 2. Cell 1 shows lower charge transfer resistance (414.3 ohm values compared to 512 ohm for cell 2). The increased J_{sc} values increase the fill factor for cell 2 resulting in high PCE. A significant improvement in solar cell stability was also achieved by the DEA modification strategy, with PCE retention of more than 85% after storing in a glove box for 60 days unencapsulated, apparently as a result of improved interface contact and a compact and morphology of the film without pinholes [35].

Zhang et al. utilized ferrocenedicarboxylic acid (FDA) for the modification of NiO_x. The ferrocene derivatives are widely utilized in catalysis and OPVs are attributed to their ability to withstand heat and chemical corrosion.

In Figure 8.11, the surface of NiO_x/FDA shows a smooth surface rather than a bare NiO_x surface and the contact angle of FDA modified NiO_x surface is increased to 57.5° which results in lower surface roughness. Due to the reduced surface roughness of NiO_x, the perovskite nucleation density will be reduced and resulting in a larger grain size. In PSCs, the NiO_x/FDA exhibit with better ultraviolet resistance properties. The photovoltaic performance was measured and the J-V properties are improved compared to bare NiO_x The FDA modification improves the hole extraction properties at the NiO_x/perovskite interface

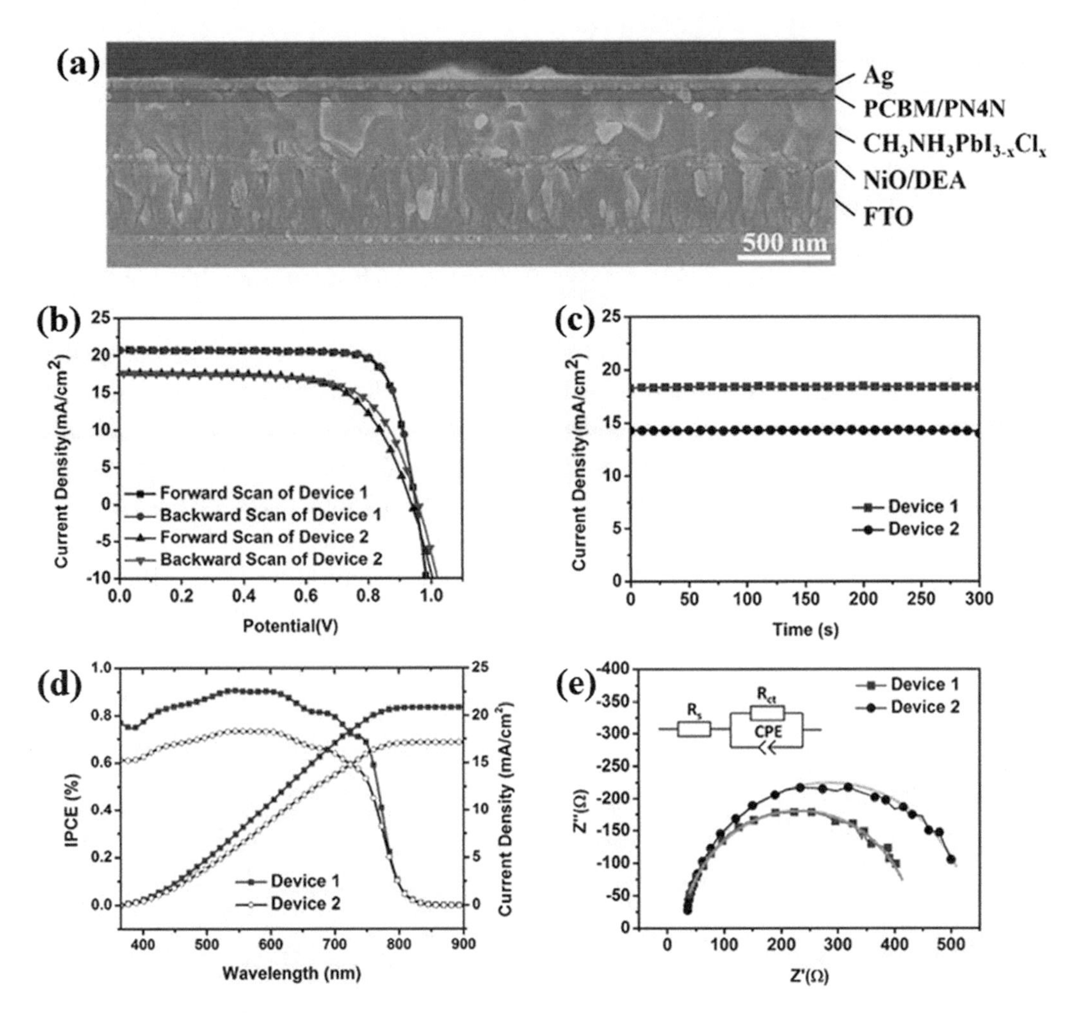

FIGURE 8.11 (a) Cross-sectional imaging of the device with NiO/DEA HTL. (b) J–V curves of PSC with NiO and NiO/DEA. (c) Time profiles of photocurrent density at 0.8V. (d) IPCE Spectra. (e) EIS Nyquist plots of the devices. (Reproduced with permission from Ref. [35], © 2016 WILEY-VCH Verlag GmbH & Co. KGaA, Weinheim.)

and the PCE values are obtained as 18.20% which is higher than pure NiO_x and PCBM hole-transport-based devices [36]. Furthermore, surface modification of NiO_x by poly triaryl amine (PTAA) was employed in PSCs by Du et al. The PTAA improves the connection between the NiOx/perovskite layer. One of the main advantages of PTAA modification on NiOx improves the V_{oc} of the device due to the well-matched band alignment. Planar PSCs with an optimal concentration of PTAA of 0.5 mg/mL yielded the highest PCE of 17.1% combined with almost non-existent hysteresis [37].

The hole conductivity of NiOx can also be improved by doping. The p-type conductivity of the NiOx arises due to excess Ni atoms that are presents in NiO. The doping can improve the optoelectronic properties of the NiOx. The low conductivity of NiOx is the main reason for poor FF and J_{sc} values. When elements have similar ionic radii to Ni2+, mismatches will be minimized and lattice stability will be enhanced. In addition, the dopants should increase the relative value. Several effective dopants, including Cu, Li, Mg, Ag, and Cs, have shown that the relative Ni^{3+} acceptor content must be increased by the dopants.

NiO_x-HTMs are commonly doped with copper, among other elements. In the crystal structure, it often substitutes the Ni site for this metal with an ionic radius similar to nickel's (0.69). A planar PSC device incorporating a $MAPbI_3$ perovskite-based device was achieved by using copper-doped NiOx in sol-gel processing at 500C annealing temperature, as described by Kim et al. [38]. A 15.4% PCE value can be seen from the photovoltaic parameters.

The unique electronic and structural properties of NiO_x-based solution are further enhanced by doping it with Cu atoms, which have been characterized using conductive atomic force microscopy (c-AFM), as shown in Figure 8.12. From the IV-curves of NiO_x/Cu-NiO_x, Cu-NiO_x displays increased conductivity 8.4×10^{-4} S cm^{-1} when compared with pristine NiO_x (2.2×10^{-6} S cm^{-1}). Moreover for NiO_x, there are high nickel vacant sites. This indicates that the strong correlation between carrier concentration and Cu content is linked to Ni vacancy generation, at least in a certain composition range. The weight

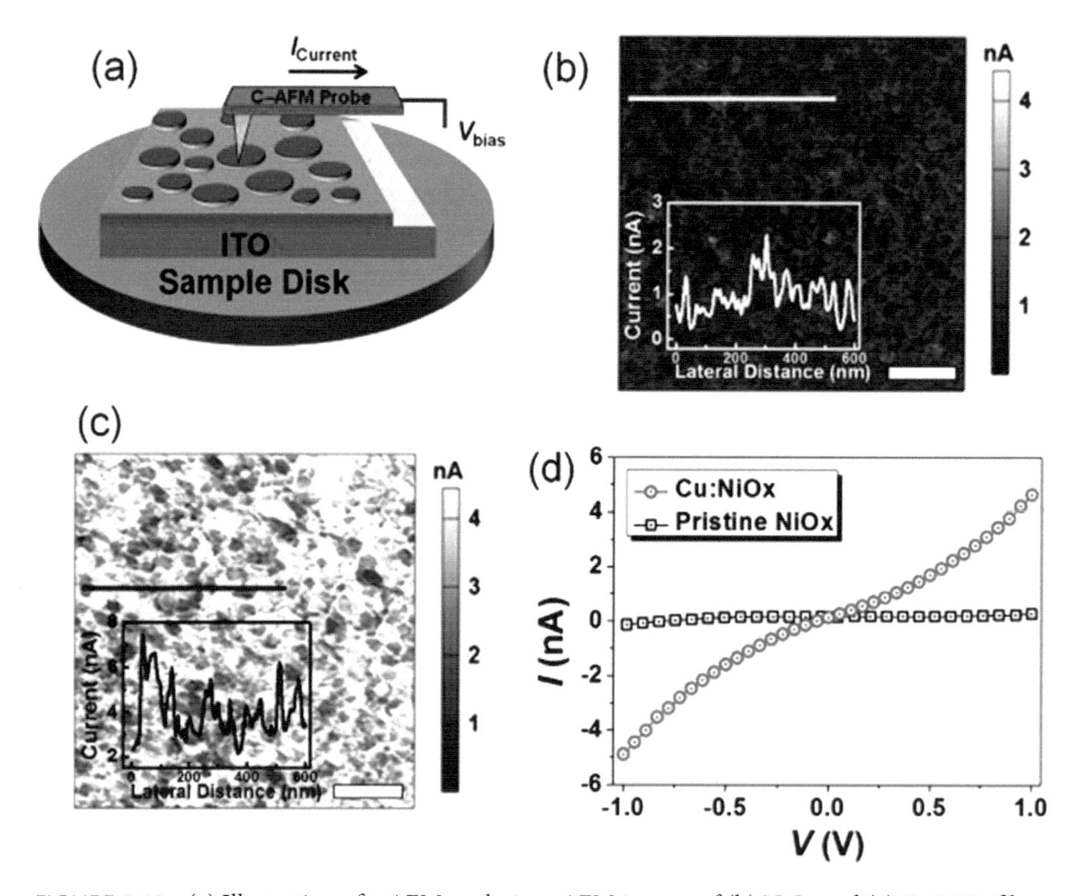

FIGURE 8.12 (a) Illustration of c-AFM analysis. c-AFM images of (b) NiO_x and (c) Cu:NiO_x films. (d) I–V curves of pristine NiO_x and Cu:NiO_x measured by c-AFM. J–V curves (Reproduced with permission from Ref. [38], © 2014 WILEY-VCH Verlag GmbH & Co. KGaA, Weinheim.)

percentage of Cu doping plays a crucial role because upon increasing the Cu dopant, the carrier concentration and also the transmittance of the NiO_x gradually decrease. These occur due to large NiO_x grains due to Cu doping. A high PCE of 15.40% is achieved with decent environmental stability in the NiO_x-based pristine devices with minimal J_{SC} and FF loss. In addition, Cu:NiO_x has also been shown to apply to larger perovskite solar cells [e.g. $CH_3NH_3Pb(I_{1-x}\ Br_x)_3$]. NiO_x exhibits a deeper VB than PEDOT: PSS and therefore fewer potential losses and a higher V_{oc} of 1.11–1.16 V, outperforming the PEDOT:PSS device. In perovskite tandem solar cells, Cu:NiO_x is capable of providing promising results [38]. Consequently, Cu:NiO_x films result in better perovskite films and enhanced PL, leading to relatively weak hysteresis and high PCE. To boost the PCE values of PSCs, lithium, magnesium, cobalt, cesium, yttrium, and silver were doped into NiO_x to align the band energy diagram. Compared to the doping method, co-doped NiO_x also showed a high-performance level for PSCs. For instance, the Li-Mg co-doped NiO_x where Li (ionic radius = 0.76 Å) contributes to a relatively lower VB energy level in NiO_x and increases h^+ mobility significantly whereas Mg ion serves to counteract the adverse positive shifting of the VB energy level induced by the insertion of Li-ion. Increased carrier concentration from 2.66×10^{17} cm^{-3} for the undoped NiO_x film to 6.46×10^{18} cm^{-3} for the Li-Mg-NiO_x film.

8.5 CONCLUSION

In summary, the purpose of this chapter is to provide an overview of the modifications and critical issues that arise when synthesizing different HTMs, as well as how to improve the efficiency and stability of PSC devices. The advantages and deficiencies of various HTMs in PSCs arere also discussed which are compared with low-cost inorganic HTMs. PSCs are in the process of being commercialized, so it is hoped that this chapter will facilitate future innovations in inorganic HTMs and PSCs.

ACKNOWLEDGMENT

This work is supported by the South African Research Chairs Initiative of the Department of Science and Technology and National Research Foundation of South Africa (Grant 84415). The financial assistance from the University of the Free State is highly recognized.

REFERENCES

[1] Z. Xie, Y. Do, S.J. Choi, H.-Y. Park, H. Kim, J. Kim, D. Song, T. Gokulnath, H.-B. Kim, I.W. Choi, Y. Jo, D.S. Kim, S.-Y. Yoon, Y.-R. Cho, S.-H. Jin, Perovskite solar cells approaching 25% PCE using side chain terminated hole transport materials with low concentration in a non-halogenated solvent process, *J. Mater. Chem. A.* 11 (2023) 9608–9615. https://doi.org/10.1039/D2TA09964K

[2] H. Liu, L. Xiang, P. Gao, D. Wang, J. Yang, X. Chen, S. Li, Y. Shi, F. Gao, Y. Zhang, Improvement strategies for stability and efficiency of perovskite solar cells, *Nanomaterials.* 12 (2022) 3295. https://doi.org/10.3390/nano12193295

[3] H.-S. Kim, C.-R. Lee, J.-H. Im, K.-B. Lee, T. Moehl, A. Marchioro, S.-J. Moon, R. Humphry-Baker, J.-H. Yum, J.E. Moser, M. Grätzel, N.-G. Park, Lead iodide perovskite sensitized all-solid-state submicron thin film mesoscopic solar cell with efficiency exceeding 9%, *Sci. Rep.* 2 (2012) 591. https://doi.org/10.1038/srep00591

[4] M.M. Lee, J. Teuscher, T. Miyasaka, T.N. Murakami, H.J. Snaith, Efficient hybrid solar cells based on meso-superstructured organometal halide perovskites, *Science.* 338 (2012) 643–647. https://doi.org/10.1126/science.1228604

[5] M. Liu, M.B. Johnston, H.J. Snaith, Efficient planar heterojunction perovskite solar cells by vapour deposition, *Nature.* 501 (2013) 395–398. https://doi.org/10.1038/nature12509

[6] H. Zhou, Q. Chen, G. Li, S. Luo, T. -B. Song, H.-S. Duan, Z. Hong, J. You, Y. Liu, Y. Yang, Interface engineering of highly efficient perovskite solar cells, *Science.* 345 (2014) 542–546. https://doi.org/10.1126/science.1254050

[7] W.S. Yang, J.H. Noh, N.J. Jeon, Y.C. Kim, S. Ryu, J. Seo, S.I. Seok, High-performance photovoltaic perovskite layers fabricated through intramolecular exchange, *Science.* 348 (2015) 1234–1237. https://doi.org/10.1126/science.aaa9272

[8] D. Bi, C. Yi, J. Luo, J.D. Décoppet, F. Zhang, S.M. Zakeeruddin, X. Li, A. Hagfeldt, M. Grätzel, Polymer-templated nucleation and crystal growth of perovskite films for solar cells with efficiency greater than 21%, *Nat. Energy.* 1 (2016) 1–5. https://doi.org/10.1038/nenergy.2016.142

[9] W.S. Yang, B.-W. Park, E.H. Jung, N.J. Jeon, Y.C. Kim, D.U. Lee, S.S. Shin, J. Seo, E.K. Kim, J.H. Noh, S. Il Seok, Iodide management in formamidinium-lead-halide–based perovskite layers for efficient solar cells, *Science.* 356 (2017) 1376–1379. https://doi.org/10.1126/science.aan2301

[10] Q. Jiang, Y. Zhao, X. Zhang, X. Yang, Y. Chen, Z. Chu, Q. Ye, X. Li, Z. Yin, J. You, Surface passivation of perovskite film for efficient solar cells, *Nat. Photonics.* 13 (2019) 460–466. https://doi.org/10.1038/s41566-019-0398-2

[11] X. Yin, Z. Song, Z. Li, W. Tang, Toward ideal hole transport materials: A review on recent progress in dopant-free hole transport materials for fabricating efficient and stable perovskite solar cells, *Energy Environ. Sci.* 13 (2020) 4057–4086. https://doi.org/10.1039/D0EE02337J

[12] T. Dai, Q. Cao, L. Yang, M. Aldamasy, M. Li, Q. Liang, H. Lu, Y. Dong, Y. Yang, Strategies for high-performance large-area perovskite solar cells toward commercialization, *Crystals.* 11 (2021) 295. https://doi.org/10.3390/cryst11030295

[13] S. Lim, S. Han, D. Kim, J. Min, J. Choi, T. Park, key factors affecting the stability of $CsPbI_3$ perovskite quantum dot solar cells: A comprehensive review, *Adv. Mater.* 35 (2023) 2203430. https://doi.org/10.1002/adma.202203430

[14] B. Yang, S. Peng, W.C.H. Choy, Inorganic top electron transport layer for high performance inverted perovskite solar cells, *EcoMat.* 3 (2021). https://doi.org/10.1002/eom2.12127

[15] A. Mahapatra, S. Kumar, P. Kumar, B. Pradhan, Recent progress in perovskite solar cells: Challenges from efficiency to stability, *Mater. Today Chem.* 23 (2022) 100686. https://doi.org/10.1016/j.mtchem.2021.100686

[16] S. Li, Y.-L.L. Cao, W.-H.H. Li, Z.-S.S. Bo, A brief review of hole transporting materials commonly used in perovskite solar cells, *Rare Met.* 40 (2021) 2712–2729. https://doi.org/10.1007/s12598-020-01691-z

[17] H. Lei, G. Yang, X. Zheng, Z.-G. Zhang, C. Chen, J. Ma, Y. Guo, Z. Chen, P. Qin, Y. Li, G. Fang, Incorporation of high-mobility and room-temperature-deposited Cu_xS as a hole transport layer for efficient and stable organo-lead halide perovskite solar cells, *Sol. RRL.* 1 (2017) 1700038. https://doi.org/10.1002/solr.201700038

[18] P. Pattanasattayavong, N. Yaacobi-Gross, K. Zhao, G.O.N. Ndjawa, J. Li, F. Yan, B.C. O'Regan, A. Amassian, T.D. Anthopoulos, Hole-transporting transistors and circuits based on the transparent inorganic semiconductor copper(I) thiocyanate (CuSCN) processed from solution at room temperature, *Adv. Mater.* 25 (2013) 1504–1509. https://doi.org/10.1002/adma.201202758

[19] G.R.R. Kumara, A. Konno, G.K. Senadeera, P.V. Jayaweera, D.B.R.. De Silva, K. Tennakone, Dye-sensitized solar cell with the hole collector p-CuSCN deposited from a solution in n-propyl sulphide, *Sol. Energy Mater. Sol. Cells.* 69 (2001) 195–199. https://doi.org/10.1016/S0927-0248(01)00027-7

[20] K.J. Chen, A.D. Laurent, F. Boucher, F. Odobel, D. Jacquemin, Determining the most promising anchors for CuSCN: Ab initio insights towards p-type DSSCs, *J. Mater. Chem. A.* 4 (2016) 2217–2227. https://doi.org/10.1039/C5TA10421A

[21] B. O'Regan, F. Lenzmann, R. Muis, J. Wienke, A solid-state dye-sensitized solar cell fabricated with pressure-treated P25–TiO_2 and CuSCN: Analysis of pore filling and IV characteristics, *Chem. Mater.* 14 (2002) 5023–5029. https://doi.org/10.1021/cm020572d
[22] N. Yaacobi-Gross, N.D. Treat, P. Pattanasattayavong, H. Faber, A.K. Perumal, N. Stingelin, D.D.C. Bradley, P.N. Stavrinou, M. Heeney, T.D. Anthopoulos, High-efficiency organic photovoltaic cells based on the solution-processable hole transporting interlayer copper thiocyanate (CuSCN) as a replacement for PEDOT:PSS, *Adv. Energy Mater.* 5 (2015) 1–7. https://doi.org/10.1002/aenm.201401529
[23] A. Mishra, T. Rana, A. Looser, M. Stolte, F. Würthner, P. Bäuerle, G.D. Sharma, High performance A–D–A oligothiophene-based organic solar cells employing two-step annealing and solution-processable copper thiocyanate (CuSCN) as an interfacial hole transporting layer, *J. Mater. Chem. A.* 4 (2016) 17344–17353. https://doi.org/10.1039/C6TA07640H
[24] P. Xu, J. Liu, J. Huang, F. Yu, C.-H. Li, Y.-X. Zheng, Interfacial engineering of CuSCN-based perovskite solar cells via PMMA interlayer toward enhanced efficiency and stability, *New J. Chem.* 45 (2021) 13168–13174. https://doi.org/10.1039/D1NJ02454J
[25] A. Wijesekara, S. Varagnolo, G.D.M.R. Dabera, K.P. Marshall, H.J. Pereira, R.A. Hatton, Assessing the suitability of copper thiocyanate as a hole-transport layer in inverted $CsSnI_3$ perovskite photovoltaics, *Sci. Rep.* 8 (2018) 15722. https://doi.org/10.1038/s41598-018-33987-7
[26] V.E. Madhavan, I. Zimmermann, A.A.B. Baloch, A. Manekkathodi, A. Belaidi, N. Tabet, M.K. Nazeeruddin, CuSCN as hole transport material with 3D/2D perovskite solar cells, *ACS Appl. Energy Mater.* 3 (2020) 114–121. https://doi.org/10.1021/acsaem.9b01692
[27] N. Wijeyasinghe, A. Regoutz, F. Eisner, T. Du, L. Tsetseris, Y.-H. Lin, H. Faber, P. Pattanasattayavong, J. Li, F. Yan, M.A. McLachlan, D.J. Payne, M. Heeney, T.D. Anthopoulos, Copper(I) thiocyanate (CuSCN) hole-transport layers processed from aqueous precursor solutions and their application in thin-film transistors and highly efficient organic and organometal halide perovskite solar cells, *Adv. Funct. Mater.* 27 (2017) 1701818. https://doi.org/10.1002/adfm.201701818
[28] N. Arora, M.I. Dar, A. Hinderhofer, N. Pellet, F. Schreiber, S.M. Zakeeruddin, M. Grätzel, Perovskite solar cells with CuSCN hole extraction layers yield stabilized efficiencies greater than 20%, *Science.* 358 (2017) 768–771. https://doi.org/10.1126/science.aam5655
[29] X. Yin, Y. Guo, H. Xie, W. Que, L.B. Kong, Nickel oxide as efficient hole transport materials for perovskite solar cells, *Sol. RRL.* 3 (2019) 1900001. https://doi.org/10.1002/solr.201900001
[30] A. Corani, M.-H. Li, P.-S. Shen, P. Chen, T.-F. Guo, A. El Nahhas, K. Zheng, A. Yartsev, V. Sundström, C.S. Ponseca, Ultrafast dynamics of hole injection and recombination in organometal halide perovskite using nickel oxide as p-Type contact electrode, *J. Phys. Chem. Lett.* 7 (2016) 1096–1101. https://doi.org/10.1021/acs.jpclett.6b00238
[31] M. Aboulsaad, A. El Tahan, M. Soliman, S. El-Sheikh, S. Ebrahim, Thermal oxidation of sputtered nickel nano-film as hole transport layer for high performance perovskite solar cells, *J. Mater. Sci. Mater. Electron.* 30 (2019) 19792–19803. https://doi.org/10.1007/s10854-019-02345-2
[32] Z. Zhu, Y. Bai, T. Zhang, Z. Liu, X. Long, Z. Wei, Z. Wang, L. Zhang, J. Wang, F. Yan, S. Yang, High-performance hole-extraction layer of sol-gel-processed NiO nanocrystals for inverted planar perovskite solar cells, *Angew. Chemie Int. Ed.* 53 (2014) 12571–12575. https://doi.org/10.1002/anie.201405176
[33] X. Yin, M. Que, Y. Xing, W. Que, High efficiency hysteresis-less inverted planar heterojunction perovskite solar cells with a solution-derived NiO_x hole contact layer, *J. Mater. Chem. A.* 3 (2015) 24495–24503. https://doi.org/10.1039/C5TA08193A
[34] Z. Zhu, Y. Bai, X. Liu, C.-C. Chueh, S. Yang, A.K.Y. Jen, Enhanced efficiency and stability of inverted perovskite solar cells using highly crystalline SnO_2 nanocrystals as the robust electron-transporting layer, *Adv. Mater.* 28 (2016) 6478–6484. https://doi.org/10.1002/adma.201600619
[35] Y. Bai, H. Chen, S. Xiao, Q. Xue, T. Zhang, Z. Zhu, Q. Li, C. Hu, Y. Yang, Z. Hu, F. Huang, K.S. Wong, H.-L. Yip, S. Yang, Effects of a molecular monolayer modification of NiO nanocrystal layer surfaces on perovskite crystallization and interface contact toward faster hole extraction

and higher photovoltaic performance, *Adv. Funct. Mater.* 26 (2016) 2950–2958. https://doi.org/10.1002/adfm.201505215

[36] J. Zhang, H. Luo, W. Xie, X. Lin, X. Hou, J. Zhou, S. Huang, W. Ou-Yang, Z. Sun, X. Chen, Efficient and ultraviolet durable planar perovskite solar cells: Via a ferrocenecarboxylic acid modified nickel oxide hole transport layer, *Nanoscale.* 10 (2018) 5617–5625. https://doi.org/10.1039/c7nr08750k

[37] Y. Du, C. Xin, W. Huang, B. Shi, Y. Ding, C. Wei, Y. Zhao, Y. Li, X. Zhang, Polymeric surface modification of NiO_x-based inverted planar perovskite solar cells with enhanced performance, *ACS Sustain. Chem. Eng.* 6 (2018) 16806–16812. https://doi.org/10.1021/acssuschemeng.8b04078

[38] J.H. Kim, P.-W. Liang, S.T. Williams, N. Cho, C.-C. Chueh, M.S. Glaz, D.S. Ginger, A.K.Y. Jen, High-performance and environmentally stable planar heterojunction perovskite solar cells based on a solution-processed copper-doped nickel oxide hole-transporting layer, *Adv. Mater.* 27 (2015) 695–701. https://doi.org/10.1002/adma.201404189

CHAPTER 9

Metal-Halide Perovskites

Opportunities and Challenges

Sumedha Tamboli and Govind B. Nair
University of the Free State, Bloemfontein, South Africa

Sanjay J. Dhoble
R.T.M. Nagpur University, Nagpur, India

Hendrik C. Swart
University of the Free State, Bloemfontein, South Africa

9.1 INTRODUCTION

Metal-halide perovskites (MHPs) have emerged as promising photovoltaic material [1, 2]. The last ten years have witnessed significant and vigorous research on perovskite materials for developing high-efficiency solar cells, and in recent studies, there has been an increasing focus on perovskite LEDs (PeLEDs) [3, 4]. The power conversion efficiency (PCE) of perovskite solar cells has reached up to 25.5%, which has been achieved for $FAPbI_3$ (FA = formamidinium-CH_3 $(NH_2)_2^+$) perovskite [5]. Exploration of the properties of bulk perovskite materials dates back to the nineteenth and twentieth centuries. Since then, researchers have designed perovskite materials with good optical, magnetic, and electronic properties. However, in the last ten years, perovskite materials have been extensively researched, as MHP nanocrystals (NCs) have been found to have higher photoluminescence quantum yield (PLQY) than the bulk perovskite. Again, color tunable PL emission was attained by quantum confinement in the MHP nanocrystals. Miyaska and co-worker in 2009, reported the use of organic-inorganic lead halide (hybrid perovskite) as a sensitizing material in dye sensitized solar cells [6], and in 2012, Park and co-workers reported a perovskite sensitized solar cell with enhanced efficiency of 9% [7]. Although perovskites were initially used as sensitizers, their exceptional charge transport properties have been discovered and single cell perovskite-based solar cells have been devised [8]. Kojima et al. demonstrated MHP's exceptional luminescence behavior [9]. At this time, MHP NCs have gained immense interest in the material research field. They possess tunable bandgaps, broader absorption

DOI: 10.1201/9781003315261-12

spectra, strong light absorption, long charge-carrier diffusion length, low defect resistance, and high charge carrier mobility, which makes them suitable for photovoltaic applications [10]. Along with these properties, MHPs possess narrow band emission and high PLQY and, hence, can be used to fabricate LEDs. The low-cost synthesis and facile fabrication process of MHP NCs have enabled researchers to tailor their chemical composition and structure to obtain the desired properties [11]. Again, MHPs have high crystallinity and they can be solution-processed at lower temperatures, which is an advantage over traditional Si and GaAs solar cells. Hence, this opens up the possibility of processing the MHPs ranging from three-dimensional (3D) structures, two-dimensional (2D) layered structures to zero-dimensional (0D) quantum dots on a variety of substrates. In addition, MHPs have found applications in photodetectors [12, 13], lasers [14], field-effect transistors [15–17], resistive-switching memory devices [18–20], supercapacitors [21, 22], fuel cells [23] and batteries [24].

9.1.1 Crystal Structure of Perovskite Materials

Figure 9.1 shows an ideal perovskite structure ABX_3 ($CaTiO_3$). Calcium titanate ($CaTiO_3$), a mineral, was discovered by German mineralogist Gustav Rose. Russian scientist Lev

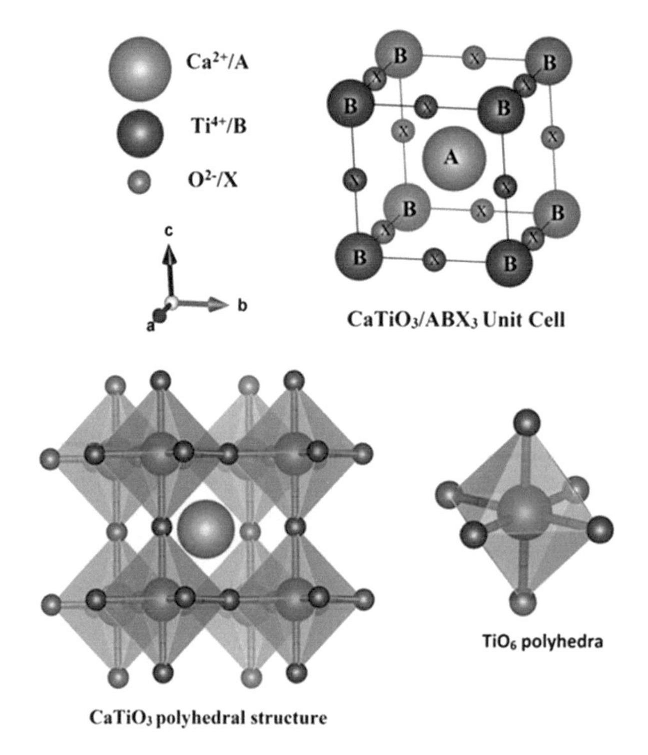

FIGURE 9.1 Unit cell and polyhedral perovskite structured $CaTiO_3$ compound, that is, ABX_3, (here, $A = Ca^{2+}$, $B = Ti^{4+}$ and $X = O^{2-}$) perovskite structure. Cation A is situated at the body center of the cube and BX_6 octahedra situated at each corner of the cube.

(Source: Authors.)

Perovski further researched this material and, hence, its crystal structure takes Perovski's name. Further, compounds having similar crystal structure to that of $CaTiO_3$ are classified as 'perovskites' [25]. Perovskites have a general formula ABX_3, where A and B stand for the cations, while X stands for the anion. It is a three-dimensional structure. Perovskites possess a cubic structure (space group *Pm3m*) with corner sharing BX_6 octahedra and an A-cation situated at the center. For an ideal cubic perovskite structure, the B–X distance should be equal to $a/2$ and the A–X distance should be $\sqrt{2}(a/2)$, where a is the cube unit cell length. The ionic radii should satisfy the relation, $R_A + R_X = \sqrt{2}(R_A + R_B)$. But Goldschmidt revealed that perovskite structure can be maintained even if the above relation is not satisfied, and introduced a tolerance factor (t) given as:

$$t = \frac{R_A + R_X}{\sqrt{2}\left(R_A + R_B\right)}$$

where, the value of t is unity for ideal perovskite structure. However, for a real perovskite structure, the value of t can be given as $0.75 < t \leq 1.0$. When the value of t is lower than 1, the structure distorts to tetragonal, rhombohedral, or orthorhombic symmetry [26].

As stated above, perovskites are represented by a general formula ABX_3, with an A-cation larger than the B-cation. The A-cation is 12-fold coordinated with X-anion and the B-cation is 6-fold coordinated with X-anion. Previous studies on perovskite materials were mainly focused on oxides with bivalent ions such as (Ca^{2+}, Sr^{2+}, Ba^{2+}, etc.) and tetravalent ions such as (Ge^{4+}, Ti^{4+}, Zr^{4+}, etc.). $SrTiO_3$ [27], $CaTiO_3$ [28], $LaGaO_3$ [29], $SrSnO_3$ [30], $BaSnO_3$ and $CaSnO_3$ [31] were some of the perovskite oxides, mostly investigated for their photocatalytic activity.

9.1.2 All Inorganic and Organic–Inorganic Hybrid MHPs

Nowadays, perovskite studies are mainly focused on nanocrystals of lead halides (Cl^-, Br^-, I^- etc.) of monovalent ions (Cs^+, Rb^+, etc.) [32–34]. $CsPbBr_3$ is the most popularly studied MHP in recent years for its exceptional PL properties and the ease of preparation. It was reported to have 97% PLQY. Its emission color can be tuned from blue to red by replacing Br^- with Cl^- and I^- [35], as shown in Figure 9.2. In general, MHPs can be classified into all-inorganic MHP and organic–inorganic hybrid MHP. If the central A-site in the ABX_3 matrix is replaced by an organic cation such as methylammonium (MA^+-$CH_3NH_3^+$) ion or formamidinium (FA^+-$CH_3(NH_2)_2^+$) ion, the perovskite is called as organic–inorganic hybrid MHP [36]. In this type of MHP, the B-site is replaced by bivalent Pb^{2+}, Sn^{2+} or Ge^{2+} ions and the X-anion is replaced by halides (X = Cl, Br, I). $MAPb(Br/I)_3$ is the most researched material among the hybrid MHPs. The bandgap of MHPs can be tuned by varying its elemental composition. For example, in $(MA/FA/Cs)PbI_3$, MA^+, FA^+ or Cs^+ occupies the central A-cation position in ABX_3 matrix, which changes the 'B–X' bond length, thereby, contracting or expanding the unit cell. The effective ionic radii of these ions follow the trend as $R_{Cs} < R_{MA} < R_{FA}$. Expansion of the unit cell of a material leads to decreased bandgap [37]. All the ions in the perovskite materials can be doped with different elements

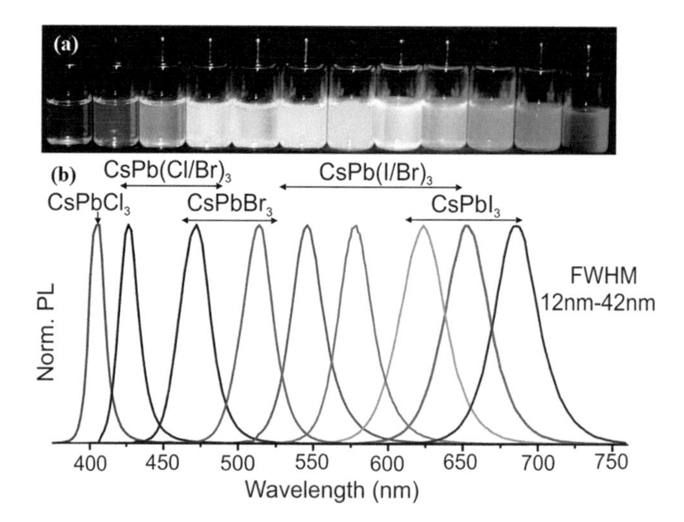

FIGURE 9.2 Colloidal perovskite $CsPbX_3$ NCs (X = Cl, Br, I) (a) light emitted by colloidal solution of $CsPbX_3$ NCs in toluene under UV lamp (λ = 365 nm) (b) PL spectra (λ_{exc} = 400 nm for all but 350 nm for $CsPbX_3$, X = Cl, Br and I). (Reproduced with permission from Ref. [35], © 2015, American Chemical Society.)

resulting in a large family of compounds, with the condition that the Goldschmidt tolerance factor (*t*) should have a value in between 0.75 and 1.

9.1.3 Perovskite-Related Structures

In addition to primary perovskites, some other perovskite phases have also been discovered and categorized as perovskite-related structures. Most of the perovskite research is based on nanocrystals of 3D $APbX_3$ phase, but the instability and internal toxicity of Pb-based MHPs has led to the study of perovskite-related structures *viz.* 0D, 2D, 3D and double perovskites as shown in Figure 9.3. A 3D $APbX_3$ structure consist of corner sharing $[PbX_6]^{4-}$ polyhedra placed at four corners of a cube and A^+ cation placed at the center of this cube. This type of arrangement gives cubic or pseudocubic perovskite structure. Layered or 2D perovskite structures have also received much attention for their stable structures with exceptional photoluminescent (PL) properties. $CsPb_2X_5$ is a layered perovskite which contains alternate Cs^+ and $[Pb_2X_5]^-$ layers. A_2PbX_4 is also a 2D perovskite structure having alternate layers of $[PbX_6]^{4-}$ octahedral and large-sized cations. Cs_4PbX_6 phase is a 0D phase of perovskite, which shows intense green PL emission [38]. In Cs_4PbX_6 phase, $[PbX_6]^{4-}$ octahedra are no longer corner-shared and they are completely decoupled in all directions. Here, Cs^+ ions do not occupy the identical crystallographic sites, but two different sites. The presence of decoupled $[PbX_6]^{4-}$ octahedra reduces the bandgap of the NCs towards the value of bandgap of free $[PbX_6]^{4-}$ clusters in solutions. This results in excitonic band absorption. Hence, photoexcited carriers experience stronger quantum confinement than $CsPbX_3$ perovskite. Therefore, this phase is called the 0D phase.

To remove the toxicity of lead halide perovskite, two strategies are adopted: one is to replace the Pb^{2+} ions by group IV ions such as Sn or Ge, and the other is to replace two Pb^{2+} ions by one monovalent B^+ ion and one trivalent B^{3+} ion. This gives quaternary system

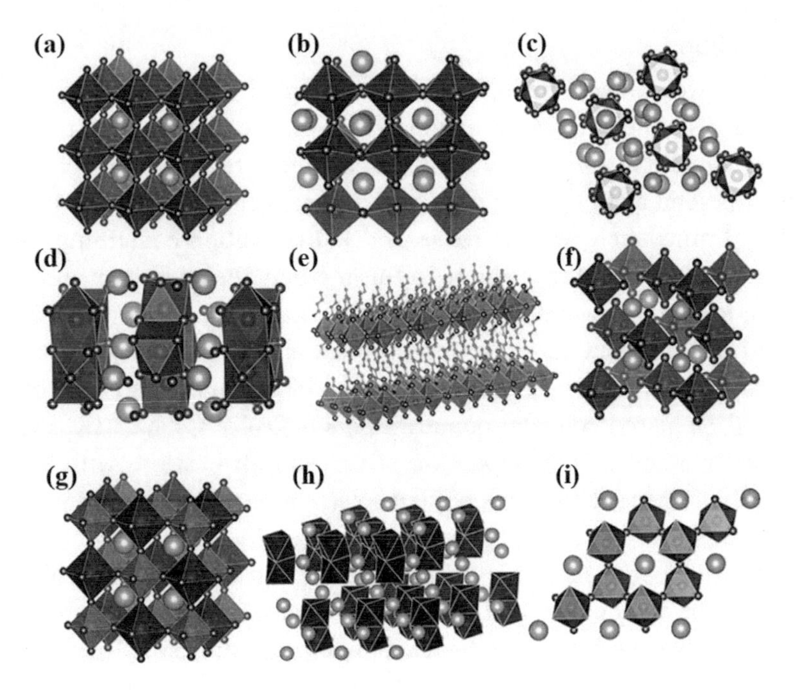

FIGURE 9.3 Types of perovskite materials based on dimensions. (a) ABX_3 (3D) cubic, (b) ABX_3 (3D) pseudocubic, (c) A_4BX_6 (0D), (d) AB_2X_5 (2D), (e) A_2BX_4 (2D), (f) A_2BX_6 (0D), (g) $A_2B^+B^{3+}X_6$ (3D), and (h) and (i) $A_3B_2X_9$ (2D). (Reproduced with permission from Ref. [39], © 2019, American Chemical Society.)

'double perovskite', also called elpasolites. Double perovskites have the general formula $A_2B^+B^{3+}X_6$. Cs_2AgBiX_6 (X = Cl/Br) is one example of a double perovskite investigated for its photovoltaic application [40].

9.2 SYNTHESIS

The synthesis of MHP NCs was a daunting task in the initial years, as the shape, size and quality of MHP NCs play important roles in obtaining the desired optical properties. Lots of efforts were made to design reliable and easy synthesis methods that would produce high quality NCs without compromising its performance. Generally, materials are synthesized by two approaches: 'Top down' or 'Bottom up'. In the top-down approach, macroscopic particles are fragmented and given the desired size and shape by mechanical or chemical routes, whereas in the bottom-up approach, molecules or atoms or ions combine to form a particle through gas- or liquid-phase chemical reactions. Among all the synthesis routes, the bottom-up approach through liquid phase is one of the best strategies to synthesize colloidal MHP NCs. For the synthesis of colloidal MHP NCs, two types of liquid-phase synthesis methods were developed, *viz.* hot injection and ligand-assisted reprecipitation (LARP) method. The hot injection method is a very costly affair since it requires a high temperature and inert atmosphere to initiate the desired reaction. On the other hand, the LARP method is cost-effective and can be used for mass production. The LARP method can be employed at room temperature with ambient atmosphere. Some of the synthesis strategies developed for perovskite MHPs are described below.

9.2.1 Hot Injection

Hot injection is a well-known method for synthesizing CDs NCs and considered as a reliable approach for synthesis of MHP nanoparticles. In the hot injection method, precursors are added rapidly to the hot solution ligands and a high boiling solvent, as shown in Figure 9.4. As soon as the precursor is added, there occurs a burst of the small nuclei formation [39]. Immediately (within five seconds), this solution-containing flask is shifted to an ice bath for quenching the reaction. In this method, the nucleation and growth stage of the particle formation is inhibited and, thereby, narrow size distribution is attained. The shape and size distribution of particles depend on the temperature of the precursor added to the solution, reaction time, concentration of precursor, and ratio of precursors. Protesescu et al. prepared colloidal quantum dots (QDs) of $CsPbX_3$, (X = Cl, Br and I) by hot injection method [35]. Precipitation of Cs^+ and Pb^{2+} cations with X^- anions was arrested by adding their solution to a boiling solvent and cooling them immediately. Cs-oleate solution was added to PbX_2 solution in octadecene boiling at 140–200°C. This solution was cooled suddenly by keeping the solution flask in ice-water bath. Nucleation and growth process occurred quickly, and the growth resulted in 1–3 sec, therefore, tuning the particle size is feasible by controlling the reaction temperature [35]. Mixed halides such as $CsPb(Cl/Br)_3$ and $CsPb(Br/I)_3$ were also prepared by this method by mixing appropriate amounts of PbX_2. Mixed halides of Cl^- and I^- cannot be prepared as they have large difference in their ionic sizes. The hot injection approach was extended to synthesize $FAPbBr_3$ (FA = formamidinium = $CH(NH_2)_2^+$) hybrid perovskite [41]. For the preparation of colloidal nanocrystals of $FAPbBr_3$, FA acetates and Pb were reacted in oleic acid and octadecene, and oleyl ammonium bromide (OAmBr) was injected to this solution, which was boiling at 130°C. After ten seconds, this solution was cooled to room temperature. Purification of the prepared NCs was done by toluene and acetonitrile. These NCs exhibited crystal size around 12 nm and produced bright green luminescence with PL emission situated around

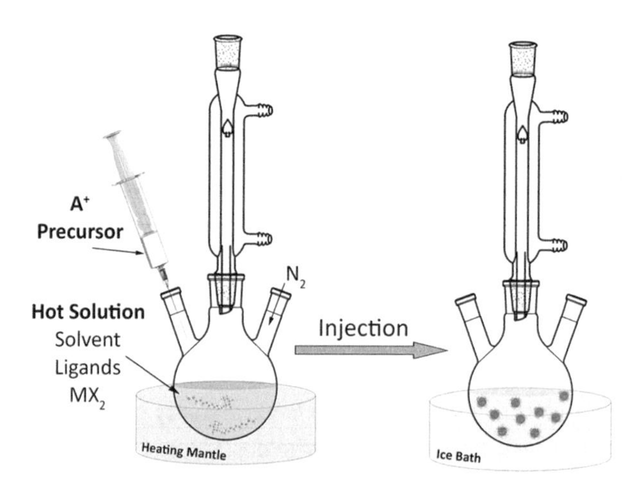

FIGURE 9.4 Schematic diagram for hot injection method. (Reproduced with permission from Ref. [39], © 2019, American Chemical Society.)

530 nm. Particle size can be tuned by varying amount of OAmBr precursor or by controlling the injecting temperature from 70 to 165°C. The hot injection method was developed for synthesizing perovskite-related structures also. One of the drawbacks of the hot injection method is that large scale production is not possible.

9.2.2 Ligand-Assisted Reprecipitation (LARP)

The 'ligand-assisted reprecipitation' (LARP) approach is a simple method for synthesizing metal-halide nanoparticles. This method was initially used for the preparation of organic nanoparticles and polymer dots. In this method, ions are dissolved in the solvent till the equilibrium stage is reached and then the solution is moved to a non-equilibrium state of supersaturation. The state of non-equilibrium can be achieved by adding miscible co-solvent having low solubility of ions, cooling the solution at lower temperature, or by evaporating the solvent. After reaching the non-equilibrium state, the particles start precipitating and their crystallization occurs [39]. This method needs simple apparatus, and the product can be scaled up to achieve large-scale production. For synthesizing the MHPs by the LARP approach, AX and MX_2 salts (where A = Cs, MA, FA; M = Pb, Sn and X = Cl, Br and I) are used as the precursors. In addition, two different types of solvents are used: one is called a good solvent and the other a poor solvent. Solvents such dimethylsulfoxide (DMSO) and dimethylformamide (DMF) in which the solubility of the salt precursor is good are called good solvents, whereas solvents such as toluene, hexane, or octadecene in which the solubility of the precursor salts is poor are called poor solvents. Supersaturation state is achieved by the mixing of these solvents. $CH_3NH_3PbX_3$ ($MAPbBr_3$) was prepared by Zang et al. by the LARP method [42]. They mixed $PbBr_2$ and CH_3NH_3Br to alkyl amines and carboxylic acid. Then, DMF was also added to this solution. This solution was added dropwise to toluene under continuous stirring to form precipitates. It was observed that amine controls the size of nanoparticles and organic acid controls their agglomeration. All-inorganic $CsPbX_3$ (X = Cl, Br and I) colloidal nanoparticles of different shapes and sizes were synthesized by Deng and co-workers at room temperature using the LARP technique [43]. The prepared MHPs were found to have PL emission from blue to red regions with more than 80% PLQY. They achieved different types of morphologies for MHPs such as quantum dot, nanocubes, nanorods, and nanosheets by varying the type of amines and organic acids used. Polar solvent (good solvent) used to dissolve the metal ions was DMF, DMSO, and tetrahydrofuran (THF). Toluene and hexane were used as non-polar solvent (bad solvent). Octylamine and dodecylamine were used as ligands, whereas oleic acid and hexanoic acid were used for suppressing the agglomeration. For synthesizing spherical QDs of $CsPbBr_3$, the $PbBr_2$ salts were mixed with DMF in flask. To this solution, hexanoic acid and octylamine were added. Cs-oleate was added after $PbBr_2$ was completely dissolved. Some quantity of this mixed solution was then added to another flask and toluene was added dropwise leading to the precipitation of spherical QDs of $CsPbBr_3$. A similar process was adopted for synthesizing other MHPs also. Perovskite-related structures such as $CsPb_2Br_5$ [44] and lead-free perovskites such as $MABi_2X_9$ (X = Cl, Br, or I) were also synthesized by the LARP method [45]. Though the LARP method enables large-scale synthesis with the ease of simple experimental setup, it has some disadvantages, too. For example, MHPs are easily affected by polar solvents resulting in the degradation or complete solubility of MHPs in the solvent. This affects the overall yield of the MHPs emerging from the LARP method.

9.2.3 Emulsion LARP

Emulsion LARP is a modified version of the LARP synthesis method. In a conventional LARP method, the metal salts are dissolved in a polar solvent along with the capping ligands, and then, their aliquots are dropped into a vigorously stirring non-polar solvent to precipitate the perovskite phase. Generally, a large quantity of the polar solvent is required to dissolve the precursors, and this may pose a threat to the stability of the yielded perovskite phase. On the other hand, emulsion LARP requires a minimum quantity of the polar solvents, and the metal precursors are dissolved separately in two different solutions of the same or different polar solvents. The organic capping agents are dissolved in the non-polar solvent, and then the precursors solutions are added one after the other into the non-polar solvent to form an emulsion. The emulsion is demulsified and allowed to mix through vigorous stirring for the reaction to take place. The demulsifier plays an important role in this method, as it is miscible with both the solvents used. A supersaturated phase is achieved immediately after adding the demulsifier. It gives rise to the precipitation of MHP NCs. In most cases, acetone or tert-butanol is added as the demulsifier. Huang et al. were the first to employ the emulsion LARP method to synthesize $CH_3NH_3PbBr_3$ ($MAPbBr_3$) QDs. $MAPbBr_3$ is unstable in many polar solvents and therefore, it is difficult to find a pair of polar and non-polar solvents. They selected DMF (aqueous phase) and n-hexane (oil phase) as two immiscible liquids. Oleic acid was used as the surfactant and tert-butanol was used as the demulsifier. They dissolved CH_3NH_3Br and $PbBr_2$ in DMF separately. These solutions were added to hexane, and then the capping agent oleic acid and octylamine were added to this solution to form the emulsion. Acetone was added dropwise to this solution under stirring to achieve demulsification. This resulted in the formation of a colloidal solution of MHPs. Nair et al. synthesized $CsPbBr_3$ by emulsion-based LARP synthesis and observed color-tunable PL emission by varying the acetone volumes [11]. For the synthesis of $CsPbBr_3$ by emulsion LARP, hexane was used as the non-polar solvent, acetone as demulsifier, and DMSO and ethanol as the polar solvents. Oleylamine and oleic acid were used as capping agents. The protocol for the synthesis is shown in the schematic diagram in Figure 9.5. $CsPbBr_3$ MHP NCs prepared by the emulsion LARP method were found to have tunable emission from blue to green with high quantum yield.

9.2.4 Reverse Microemulsion

Ligands in the polar-nonpolar solvent framework can organize themselves in an ordered structure called 'micelles'. These micelles act as nanoreactors, where the precursor diffuses and reacts by suppressing the agglomeration. The process is referred to as 'microemulsion' if the micelles form within the polar solvent, and the oil phase is stabilized by inwardly turned hydrophobic carbon chains. If the micelles form in the non-polar solvent with the hydrophilic head group turned inward and stabilizing the aqueous phase, then the process is called 'reverse microemulsion'. The precursors diffuse and react on the micelles

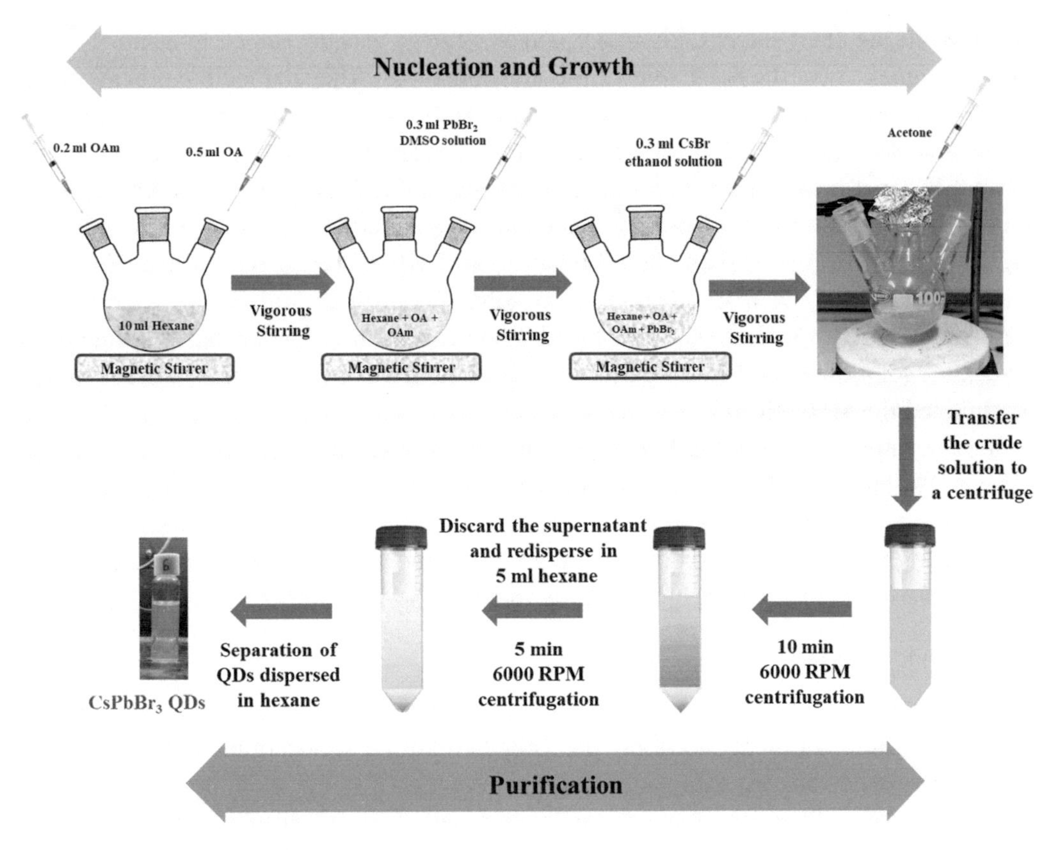

FIGURE 9.5 Schematic diagram for Emulsion LARP synthesis of $CsPbBr_3$. (Reproduced with permission from Ref. [11], © 2022, Elsevier B.V.)

to form NCs, following which the demulsifier is added. It is very difficult to differentiate between the LARP, emulsion LARP, or reverse microemulsion as the difference between these processes is very subtle [39]. The reverse microemulsion approach was first adopted by Schmidt et al. to prepare $MAPbBr_3$ NCs. In this method, MABr and $PbBr_2$ were dissolved in DMF, and octyl ammonium bromide was dissolved in a mixture of oleic acid and 1-octadecene, separately. Both the solutions were added to each other to form a reverse microemulsion. In the framework of reverse emulsion, nanocrystals were formed by the process of nucleation and growth. After adding acetone as a demulsifier, nanocrystals were precipitated [46]. By varying the molar concentration of the non-polar solvents and the capping ligands, $MAPbBr_3$ was prepared with improved optical properties and higher PLQY. Along with hybrid MHPs, perovskite-related structures such as Cs_4PbBr_6 were also reported to be prepared by reverse microemulsion by Chen et al. [47]. They used DMF as an aqueous phase and n-hexane as the oil phase.

9.2.5 Polar Solvent-Controlled Ionization

As discussed earlier, the LARP approach requires polar solvents to dissolve the metal ion precursors, and for scaling up the synthesis, polar solvents are required in large quantities. These polar solvents are not only toxic but affect the stability of MHP NCs as they get dissolved in the polar solvent. Therefore, a new strategy was developed by Fang et al., in which the quantity of the polar solvent was controlled [48]. They named the process 'polar solvent-controlled ionization'. In the LARP approach, metal ion precursors are dissolved in the polar solvent and the NCs are precipitated by dropping their aliquots in a non-polar solvent. But in a polar solvent-controlled ionization method, the metal ion precursors are added in a non-polar solvent first and then polar solvent is added to it, in a controlled manner, so that the precursors dissociate completely and MHPs start precipitating. Feng et al. synthesized $CsPbX_3$ (X = Halides) by polar solvent-controlled ionization method. They dissolved 0.5 ml of Cs-oleate, 0.5 ml of lead oleate and 0.5 ml of halogenated amine in 10 ml hexane solution. 0.5 ml isopropanol was added to this solution as polar solvent, which precipitated the MHP NCs.

Herein, some of the well-established methods for the synthesis of MHPs are discussed in brief. Although MHP NCs with controlled size and shape can be obtained by employing these protocols, there are issues with the stability and yield that need to be addressed while synthesizing these NCs. Therefore, lots of work is going on for designing a more feasible and reliable synthesis method, by which the stability and yield of MHPs can be improved.

9.3 APPLICATIONS

MHPs possess exceptional optical properties such as high absorption, color tunability, and color purity, among others and, therefore, they have found applications in light-emitting devices. Also, they possess exceptional light conversion efficiency and, hence, are used in photovoltaic cells. They have applications in photodetectors, transistors, lasers, and so on. Some of these applications are given below.

9.3.1 Perovskite Solar Cells

The MHPs are a class of semiconductor materials which are used for fabricating photovoltaic cells. The past ten years have witnessed vigorous research in developing perovskite solar cells with enhanced efficiency and they are regarded as the next generation photovoltaic materials. To date, the highest PCE achieved for a perovskite solar cell is 25.5% and it was obtained for $FAPbI_3$ [5]. Initially, perovskite NCs were used as a sensitizing layer in dye-sensitized solar cells (DSSCs) owing to their high absorption and light conversion efficiency. Kojima et al. devised a liquid electrolyte DSSC using $MAPbI_3$ and $MAPbBr_3$ as sensitizers, for which they achieved 3.81 and 3.13% PCE, respectively [49]. This DSSC cell consisted of a transparent conducting oxide (TCO) substrate, nano-porous TiO_2 scaffold, a perovskite layer as a sensitizer, an electrolyte (iodide (I^-)/triiodide ($I3^-$) redox couple) and a metal counter electrode as shown in Figure 9.6(a).

The working of this solar cell (Figure 9.6(a)) is as follows: (1) The sunlight passes through the TCO layer striking the perovskite layer. From there, it excites the perovskite material to produce photoelectrons. (2) Electrons flow to the conduction band of TiO_2 (n-type) semiconductor. Here, the semiconductor is used for the charge transport as well as for holding the sensitizer molecule. The perovskite layer is in the nano-dimensions, and the thick layer of the perovskite material is required to extract more photoelectrons. Therefore, the mesoporous TiO_2 is used as a scaffold to hold a greater number of perovskite molecules.

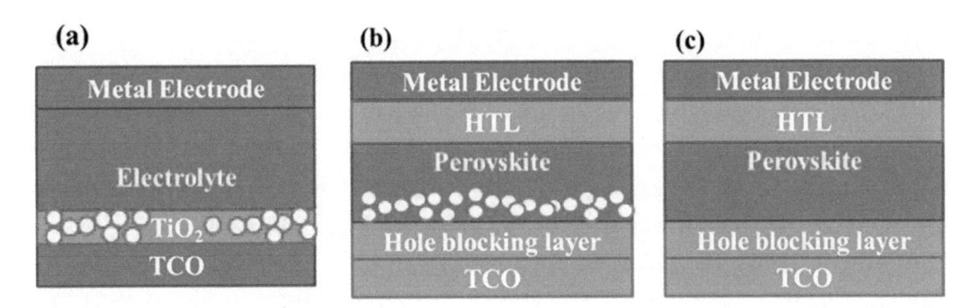

FIGURE 9.6 Structure of perovskite solar cell. (a) Liquid electrolyte dye-sensitized, (b) Solid-state mesoscopic heterojunction, (c) Planar n-i-p (HTL = hole transport layer, TCO = transparent conducting oxide) (Reproduced with permission from Ref. [50], Creative Commons license © 2016, WILEY-VCH Verlag GmbH & Co. KGaA, Weinheim.)

(3) Electrons from the TiO_2 layer will then flow towards the TCO (anode) and reach the metal electrode (cathode) producing current externally. Electrons in the perovskite are regenerated by the oxidation of I^-, which converts to I_3. Finally, (4) Electrons from the metal counter electrode (cathode) enter the cell again reducing I_3 to I^-, thus, completing the circuit. But the main drawback of this type of solar cells is the instability of the perovskite material in I^-/I^{-3} electrolyte, which reduces the efficiency of the solar cell. No other suitable liquid electrolyte has been found for perovskite-based DSSC.

To resolve this drawback, Kim et al. devised a perovskite solar cell with solid-state electrolyte material called a solid-state mesoscopic heterojunction solar cell. Heterojunction refers to the interface of two different types of semiconductors having unequal bandgaps. The combination of multiple heterojunctions is called the heterojunction structure of solar cell. In this solar cell, solid-state hole transport material (HTM) named spiro-OMeTAD (2,2′,7,7′-tetrakis (N,N-di-p-methoxyphenylamine)-9,9′-spirobifluorene) was used as an electrolyte, in which the perovskite remains stable. $CH_3NH_3PbI_3$ was used as an absorber on mesoporous TiO_2 scaffold. Figure 9.6(b) shows the design of a solid-state mesoscopic heterojunction. It consists of a TCO, a thin hole-blocking layer made of TiO_2, mesoscopic TiO_2, a perovskite absorber, spiro-OMeTAD as a hole transport layer (HTL) and a metal electrode. The hole-blocking layer is used for preventing contact between the TCO and the perovskite layer, to avoid recombination of charges by blocking hole transfer to the anode. This type of arrangement enhanced the PCE up to 9.7% [7]. Yang et al. chose formamidinium lead iodide ($FAPbI_3$) over $MAPbI_3$ and achieved more than 20% of PCE with the same type of structure [51]. A disadvantage of this type of solar cell is the cost and stability of spiro-OMeTAD, due to which its commercialization has not taken place yet. Lots of research has been done is finding a substitute for spiro-OMeTAD HTL. Researchers have developed both organic and inorganic HTM. PEDOT:PSS [poly(3,4-ethylenedioxythiophene): polystyrene sulfonate] and P3HT [poly(3-hexylthiophene-2,5-diyl)] are some of the organic HTMs that are cheaper than spiro-OMeTAD. However, they have stability issues that can degrade the overall performance of the solar cells. Inorganic materials such as Cu_2O, CuI, CuSCN, and MoS_2 have also been researched to be used as HTM [52]. A cell structure without mesoporous scaffold is called as planar heterojunction solar cell. This cell structure has two types, *viz.* n-i-p and p-i-n, based on the sequence of functional layer present in the device. Figure 9.6(c) shows the n-i-p based solar cell, consisting of a hole-blocking layer above TCO. The hole-blocking layer is made up of a compact TiO_2 or ZnO. Ball et al. were the first to introduce planar solar cell structure with $CH_3NH_3PbI_{3-x}Cl_x$ as absorber and recorded PCE up to 12.3% [53]. In a p-i-n planar solar cell, HTL is above the TCO layer. In this type of cell, PEDOT:PSS was used as the HTL, and a fullerene derivative was used as the ETL (electron transport layer). A p-i-n solar cell have some advantages over a n-i-p solar cell, such as low-temperature processing and compatibility with organic electronic manufacturing. One difficulty with the p-i-n planar solar cell is fabricating a pinhole-free perovskite layer on a TCO to avoid the leakage current. Both HTL- and ETL-free designs for perovskite solar cell have also been proposed. The highest PCE of 22.26% was achieved by Zang et al. for quasi-2D perovskite structure.

9.3.2 Perovskite Light-Emitting Diodes

The MHP NCs possess intense luminescence emission and color tunability, which makes them suitable for fabricating LEDs. LEDs prepared with perovskite as an emitting material called as perovskite LEDs (PeLEDs). The structure of PeLEDs is similar to that of OLEDs. PeLEDs consist of a perovskite light-emitting layer sandwiched between an n-type electron injection layer (EIL) and a p-type hole injection layer (HIL). Under a forward bias condition, charge carriers are injected in the perovskite layer, where they recombine radiatively to emit light. Inorganic-organic hybrid perovskites have proved to be superior to OLEDs and quantum dot LEDs (QLEDs) because: (1) the internal crystal structure of perovskites acts as a multiple quantum well due to which the light emitted by the perovskite has a narrow bandwidth (around 20 nm) and hence, this gives high color purity, (2) the ionization energy of a perovskite material is comparable to a common hole-injection material, and (3) perovskites are solution-processible and low-cost materials, and their emission color can be easily tuned by altering the chemical composition [54]. PeLEDs have demonstrated high external quantum efficiency up to 23.4% and current efficiency of 108 cd A^{-1} [55]. PeLEDs are solution processed and have a single-layered or multi-layered structure. Thin film formation of perovskites is the important factor in improving the properties of PeLEDs. Smooth and uniform thin films are required for good performance. Tan et al. were the first who reported a 3D PeLED [56]. They fabricated a device with (ITO)/TiO_2/$CH_3NH_3PbI_{3-x}Cl_x$/F8/MoO_3Ag structure as shown in Figure 9.7.

Here, the ITO functions as a cathode while MoO_3Ag works as an anode. A three-layered structure, that is, TiO_2/$CH_3NH_3PbI_{3-x}Cl_x$/F8, was sandwiched in between ITO and MoO_3Ag. They placed the perovskite layer between two large bandgap materials to maximize the confinement of charges in the perovskite layer. Here, TiO_2 injected the electrons in the perovskite and blocked the holes. TiO_2 was deposited on the ITO-coated glass substrate and a thin layer of perovskite was deposited on TiO_2. The perovskite layer needs to be very thin so that the spatial confinement of charges can be obtained. The thinner the layer, the more will be the confinement and radiative recombination. The perovskite layer was capped by F8 polymer, to form type-I heterojunction with perovskite. F8 polymer has low electron affinity and high ionizing potential; hence, electron and holes are blocked and

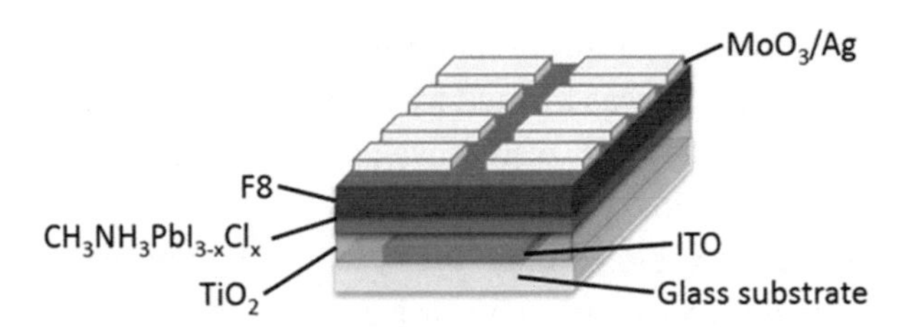

FIGURE 9.7 Schematic diagram for PeLED device. (Reproduced with permission from Ref. [56], © 2014, Nature publishing group.)

remain confined to the perovskite quantum well. A high work-function MoO_3/Ag anode layer is finally deposited on F8 polymer which injects holes in the device. The perovskite material used here was $CH_3NH_3PbI_{3-x}Cl_x$ for which the external quantum efficiency was 0.1%. Though, they did not achieve high efficiency, their research work has proved to be pioneering for further development of PeLEDs.

Organic-inorganic hybrid perovskite materials possess deep ionization potential and long exciton diffusion length, due to which there remains a hole injection barrier, which results in the quenching of light emission. Kim et al. introduced an HIL in PeLED structure to increase the hole injection in the perovskite layer [54]. They introduced buffer HIL (Buf-HIL) made up of PEDOT:PSS and PFI (perfluorinated polymeric acid, tetrafluoroethylene-perfluoro-3,6-dioxa-4-methyl-7-octene-sulfonic acid copolymer). With this additional layer, external quantum efficiency of the device was increased up to 0.125%. One of the reasons for getting low quantum efficiency in the above-mentioned PeLEDs was non-uniform coating of the perovskite layer. Spin coating resulted in the formation of non-uniform deposition of perovskite material with large grain-sized particles. Due to this non-uniform surface, leakage current increased and exciton diffusion length increased, which reduced the current efficiency. In the case of the perovskite solar cell, the requirements are different. Here, a thick layer of perovskite is needed to get facile exciton diffusion and dissociation. But in PeLEDs, a fine layer of perovskite is needed so that the exciton diffusion length is reduced and dissociation of exciton into the charge carrier is prevented. Cho et al. significantly improved the performance of PeLEDs, by employing a nanocrystal pinning process during spin coating to get pinhole-free perovskite layer. Also, they prevented the formation of metallic Pb which quenches the excitons by increasing the molar concentration of MABr. They reported an external quantum efficiency increase up to 8.35% [57]. Zang et al. prepared PeLEDs with mixed cation perovskite $Cs_{0.87}MA_{0.13}PbBr_3$, which emitted green color light. They introduced hydrophilic insulating polyvinyl pyrrolidine polymer on an EIL. They got high brightness light (91000 cd m^{-2}) and external quantum efficiency of 10.4% [58]. A quasi-2D structure of perovskite LEDs was introduced by Byun et al. [59]. A 2D perovskite structure resulted by substituting the methyl ammonium (MA) cations with larger organic ammonium (OA) cations. In 2D perovskites, excitons are more strongly bounded that increase the exciton binding energy, which is a requisite for efficient PeLEDs. But PeLEDs with 2D structure have a limitation of low current due to inhabitation of charge transport by insulating the OA group. To remove this disadvantage and retain the merits of 2D material, a quasi-2D structure of perovskite was introduced. The quasi-2D structure of perovskite was formed by mixing the 3D and 2D structures. In this series, perovskite quantum dot LEDs were developed by Song et al., and their quantum efficiency value reached up to 16.48% [60]. Recently, Kim et al. have reported the highest external quantum efficiency value to date, *viz.* 23.4%. They prepared PeLEDs based on $FAPbBr_3$ NCs. They doped zero-dipole guanidinium cation (CH_6N^{3+}; GA^+) in $FAPbBr_3$. The presence of GA-cation stabilized the surface and inhibited non-radiative recombination at the surface [61].

9.3.3 Lasers

Metal-halide perovskites are potential candidates as a gain medium for semiconductor laser applications, due to their high absorption coefficient, low defect density, high luminescence quantum yield, and low auger recombination rate. Color tunable lasing action was demonstrated from solution-processed $MAPbX_3$ (X = Cl, Br, I) thin films at room temperature. For lasers, all-inorganic $CsPbX_3$ NCs were found to be more stable compared to hybrid perovskites. Yakunin et al. reported a room temperature lasing from $CsPbX_3$ QDs in the visible region with low threshold and high value of model net gain [62]. Robust $CsPbX_3$ nanowires were also studied for their lasing action. Lasers based on perovskite materials exhibited ultra-stable amplified spontaneous emission at low threshold. Single crystal nanowires used for lasing attain 100% quantum yield due to less charge carrier trapping [63].

9.3.4 Photodetectors

In addition, MHPs are used to fabricate photodetectors. Dou et al. was the first who reported a solution-processed $CH_3NH_3PbI_{3-x}Cl_x$ perovskite-based photodetector [64]. The device structure for this photodetector was similar to a perovskite solar cell. The perovskite layer was placed in between the p-type HTL (PEDOT:PSS) and n-type ETL (PCBM-[6,6]-phenyl-C61-butyric acid methyl ester). This photodetector possessed fast photoresponse and large detectivity [64]. Lots of research has been done on perovskite-based photodetectors and efforts still continue to improve the existing photodetectors.

9.4 CHALLENGES

Certainly, MHPs have shown their superiority over all other existing photovoltaic and optical materials and are accepted by all as the next generation solar cell and lighting material for large-scale device manufacturing. However, the stability of perovskites over the long run remains a major issue of concern. Several parameters such as temperature, exposure to light, humidity and oxygen can affect the structural stability of perovskites and threaten the commercialization of perovskite-based devices. Perovskites, when exposed to oxygen, incline to the formation of superoxide O^{2-}, which fits in the iodine vacancy tending to deprotonate $CH_3NH_3^+$ ion. It leads to the dissociation of $MAPbI_3$ perovskite. Exposure to light and temperature can degrade the organic species leading to the formation of PbI_2. It impacts the operational stability of the device. Crystallographic defects induced during the fabrication and ion-migration-induced phase segregation is also responsible for the overall degradation [65]. For the commercialization of perovskite-based solar cells or LEDs, there must be stability during the device operation. Toxicity of lead is another disadvantage for perovskite-based devices, as the Pb-precursors can pollute the environment and cause serious damage to health. Developing lead-free perovskites that have optoelectronic properties like the lead-based perovskite is crucial.

9.5 CONCLUSION

Organic-inorganic metal-halide perovskites have gained widespread recognition owing to their solution processability, interesting structure and size-dependent properties, and application in optoelectronic and energy devices. This chapter gives a general overview on the perovskite materials, their structure, types of perovskite materials, various methods of synthesis of perovskite nanostructures, and their application. Challenges faced by perovskite materials are also discussed briefly. The past ten years have witnessed rapid development of perovskite materials from 0.1% efficiency to nearly 25% efficiency for both solar and LED devices. Although MHPs have demonstrated immense potential in numerous applications, certain challenges must be resolved for their ultimate commercialization. Stability, toxicity, ease of synthesis and efficiency are the main challenges which need to be overcome for the large-scale manufacturing of perovskite devices.

ACKNOWLEDGMENT

This work is supported by the South African Research Chairs Initiative of the Department of Science and Technology and National Research Foundation of South Africa (Grant 84415). The financial assistance from the University of the Free State is highly recognized.

REFERENCES

[1] X. Cheng, S. Yang, B. Cao, X. Tao, Z. Chen, Single crystal perovskite solar cells: Development and perspectives, *Adv. Funct. Mater.* 30 (2020) 1905021. https://doi.org/10.1002/adfm.201905021

[2] E.H. Jung, N.J. Jeon, E.Y. Park, C.S. Moon, T.J. Shin, T.Y. Yang, J.H. Noh, J. Seo, Efficient, stable and scalable perovskite solar cells using poly(3-hexylthiophene), *Nature.* 567 (2019) 511–515. https://doi.org/10.1038/s41586-019-1036-3

[3] K. Lin, J. Xing, L.N. Quan, F.P.G. de Arquer, X. Gong, J. Lu, L. Xie, W. Zhao, D. Zhang, C. Yan, W. Li, X. Liu, Y. Lu, J. Kirman, E.H. Sargent, Q. Xiong, Z. Wei, Perovskite light-emitting diodes with external quantum efficiency exceeding 20 per cent, *Nature.* 562 (2018) 245–248. https://doi.org/10.1038/s41586-018-0575-3

[4] Y. Cao, N. Wang, H. Tian, J. Guo, Y. Wei, H. Chen, Y. Miao, W. Zou, K. Pan, Y. He, H. Cao, Y. Ke, M. Xu, Y. Wang, M. Yang, K. Du, Z. Fu, D. Kong, D. Dai, Y. Jin, G. Li, H. Li, Q. Peng, J. Wang, W. Huang, Perovskite light-emitting diodes based on spontaneously formed submicrometre-scale structures, *Nature.* 562 (2018) 249–253. https://doi.org/10.1038/s41586-018-0576-2

[5] Y. Zhang, N.G. Park, Quasi-two-dimensional perovskite solar cells with efficiency exceeding 22%, *ACS Energy Lett.* 7 (2022) 757–765. https://doi.org/10.1021/acsenergylett.1c02645

[6] S. Adjokatse, H.H. Fang, M.A. Loi, Broadly tunable metal halide perovskites for solid-state light-emission applications, *Mater. Today.* 20 (2017) 413–424. https://doi.org/10.1016/j.mattod.2017.03.021

[7] H.S. Kim, C.R. Lee, J.H. Im, K.B. Lee, T. Moehl, A. Marchioro, S.J. Moon, R. Humphry-Baker, J.H. Yum, J.E. Moser, M. Grätzel, N.G. Park, Lead iodide perovskite sensitized all-solid-state submicron thin film mesoscopic solar cell with efficiency exceeding 9%, *Sci. Rep.* 2 (2012). https://doi.org/10.1038/srep00591

[8] N.-G. Park, Perovskite solar cells: An emerging photovoltaic technology, *Mater. Today.* 18 (2015) 65–72. https://doi.org/10.1016/j.mattod.2014.07.007
[9] A. Kojima, M. Ikegami, K. Teshima, T. Miyasaka, Highly luminescent lead bromide perovskite nanoparticles synthesized with porous alumina media, *Chem. Lett.* 41 (2012) 397–399. https://doi.org/10.1246/cl.2012.397
[10] W. Zhang, G.E. Eperon, H.J. Snaith, Metal halide perovskites for energy applications, *Nat. Energy.* 1 (2016). https://doi.org/10.1038/nenergy.2016.48
[11] G.B. Nair, S. Tamboli, R.E. Kroon, S.J. Dhoble, H.C. Swart, Facile room-temperature colloidal synthesis of $CsPbBr_3$ perovskite nanocrystals by the emulsion-based ligand-assisted reprecipitation approach: Tuning the color-emission by the demulsification process, *J. Alloys Compd.* 928 (2022) 167249. https://doi.org/10.1016/j.jallcom.2022.167249
[12] M.K. Kim, Z. Munkhsaikhan, S.G. Han, S.M. Park, H. Jin, J. Cha, S.J. Yang, J. Seo, H.S. Lee, C.-J. Choi, M. Kim, Structural engineering of single-crystal-like perovskite nanocrystals for ultrasensitive photodetector applications, *J. Mater. Chem. C.* 10 (2022) 11401–11411. https://doi.org/10.1039/D2TC01854C
[13] X. Lu, J. Li, Y. Zhang, Z. Han, Z. He, Y. Zou, X. Xu, Recent progress on perovskite photodetectors for narrowband detection, *Adv. Photonics Res.* 3 (2022) 2100335. https://doi.org/10.1002/adpr.202100335
[14] C. Wang, G. Dai, J. Wang, M. Cui, Y. Yang, S. Yang, C. Qin, S. Chang, K. Wu, Y. Liu, H. Zhong, Low-threshold blue quasi-2D perovskite laser through domain distribution control, *Nano Lett.* 22 (2022) 1338–1344. https://doi.org/10.1021/acs.nanolett.1c04666
[15] W. Yu, F. Li, L. Yu, M.R. Niazi, Y. Zou, D. Corzo, A. Basu, C. Ma, S. Dey, M.L. Tietze, U. Buttner, X. Wang, Z. Wang, M.N. Hedhili, C. Guo, T. Wu, A. Amassian, Single crystal hybrid perovskite field-effect transistors, *Nat. Commun.* 9 (2018). https://doi.org/10.1038/s41467-018-07706-9
[16] T. Matsushima, S. Hwang, A.S.D. Sandanayaka, C. Qin, S. Terakawa, T. Fujihara, M. Yahiro, C. Adachi, Solution-processed organic–inorganic perovskite field-effect transistors with high hole mobilities, *Adv. Mater.* 28 (2016) 10275–10281. https://doi.org/10.1002/adma.201603126
[17] S.P. Senanayak, M. Abdi-Jalebi, V.S. Kamboj, R. Carey, R. Shivanna, T. Tian, G. Schweicher, J. Wang, N. Giesbrecht, D. Di Nuzzo, H.E. Beere, P. Docampo, D.A. Ritchie, D. Fairen-Jimenez, R.H. Friend, H. Sirringhaus, A general approach for hysteresis-free, operationally stable metal halide perovskite field-effect transistors, *Sci. Adv.* 6 (2020). https://doi.org/10.1126/sciadv.aaz4948
[18] S.Y. Kim, J.M. Yang, E.S. Choi, N.G. Park, Layered $(C_6H_5CH_2NH_3)_2CuBr_4$ perovskite for multilevel storage resistive switching memory, *Adv. Funct. Mater.* 30 (2020). https://doi.org/10.1002/adfm.202002653
[19] E.J. Yoo, M. Lyu, J.-H. Yun, C.J. Kang, Y.J. Choi, L. Wang, Resistive switching behavior in organic-inorganic hybrid $CH_3NH_3PbI_{3-x}Cl_x$ perovskite for resistive random access memory devices, *Adv. Mater.* 27 (2015) 6170–6175. https://doi.org/10.1002/adma.201502889
[20] D. Panda, T.-Y. Tseng, Perovskite Oxides as Resistive Switching Memories: A Review, *Ferroelectrics.* 471 (2014) 23–64. https://doi.org/10.1080/00150193.2014.922389
[21] H. Nan, X. Hu, H. Tian, Recent advances in perovskite oxides for anion-intercalation supercapacitor: A review, *Mater. Sci. Semicond. Process.* 94 (2019) 35–50. https://doi.org/10.1016/j.mssp.2019.01.033
[22] F. Zhou, Z. Ren, Y. Zhao, X. Shen, A. Wang, Y.Y. Li, C. Surya, Y. Chai, perovskite photovoltachromic supercapacitor with all-transparent electrodes, *ACS Nano.* 10 (2016) 5900–5908. https://doi.org/10.1021/acsnano.6b01202
[23] B. Zhu, Y. Huang, L. Fan, Y. Ma, B. Wang, C. Xia, M. Afzal, B. Zhang, W. Dong, H. Wang, P.D. Lund, Novel fuel cell with nanocomposite functional layer designed by perovskite solar cell principle, *Nano Energy.* 19 (2016) 156–164. https://doi.org/10.1016/j.nanoen.2015.11.015
[24] H.-R. Xia, W.-T. Sun, L.-M. Peng, Hydrothermal synthesis of organometal halide perovskites for Li-ion batteries, *Chem. Commun.* 51 (2015) 13787–13790. https://doi.org/10.1039/C5CC05053G

[25] M. George, S.S. Patel, Synthesis gas production in oxy-carbondioxide reforming of methane over perovskite catalysts, *Inst. Technol. Nirma Univ. Ahmedabad.* 382 (2011) 8–10.
[26] C.N.R. Rao, Perovskites, in: *Encyclopedia of Physical Science and Technology*, Elsevier, 2003: pp. 707–714. https://doi.org/10.1016/B0-12-227410-5/00554-8
[27] R. Konta, T. Ishii, H. Kato, A. Kudo, Photocatalytic activities of noble metal ion doped $SrTiO_3$ under visible light irradiation, *J. Phys. Chem. B.* 108 (2004) 8992–8995. https://doi.org/10.1021/jp049556p
[28] H. Zhang, G. Chen, X. He, J. Xu, Electronic structure and photocatalytic properties of Ag–La codoped $CaTiO_3$, *J. Alloys Compd.* 516 (2012) 91–95. https://doi.org/10.1016/j.jallcom.2011.11.142
[29] K. Huang, R.S. Tichy, J.B. Goodenough, Superior perovskite oxide-ion conductor; strontium- and magnesium-doped $LaGaO_3$: I, phase relationships and electrical properties, *J. Am. Ceram. Soc.* 81 (2005) 2565–2575. https://doi.org/10.1111/j.1151-2916.1998.tb02662.x
[30] W.F. Zhang, J. Tang, J. Ye, Photoluminescence and photocatalytic properties of $SrSnO_3$ perovskite, *Chem. Phys. Lett.* 418 (2006) 174–178. https://doi.org/10.1016/j.cplett.2005.10.122
[31] H. Mizoguchi, H.W. Eng, P.M. Woodward, Probing the electronic structures of ternary perovskite and pyrochlore oxides containing Sn^{4+} or Sb^{5+}, *Inorg. Chem.* 43 (2004) 1667–1680. https://doi.org/10.1021/ic034551c
[32] M. Stefanski, M. Ptak, A. Sieradzki, W. Strek, Optical characterization of Yb^{3+}:$CsPbCl_3$ perovskite powder, *Chem. Eng. J.* 408 (2021) 127347. https://doi.org/10.1016/j.cej.2020.127347
[33] G.B. Nair, R. Krishnan, A. Janse van Vuuren, H.C. Swart, Synthesis of cesium lead bromide nanoparticles by the ultrasonic bath method: A polar-solvent-free approach at room temperature, *Dalt. Trans.* 52 (2023) 70–80. https://doi.org/10.1039/D2DT03689D
[34] C.C. Lin, A. Meijerink, R.S. Liu, Critical red components for next-generation white LEDs, *J. Phys. Chem. Lett.* 7 (2016) 495–503. https://doi.org/10.1021/acs.jpclett.5b02433
[35] L. Protesescu, S. Yakunin, M.I. Bodnarchuk, F. Krieg, R. Caputo, C.H. Hendon, R.X. Yang, A. Walsh, M.V. Kovalenko, Nanocrystals of cesium lead halide perovskites ($CsPbX_3$, X = Cl, Br, and I): Novel optoelectronic materials showing bright emission with wide color gamut, *Nano Lett.* 15 (2015) 3692–3696. https://doi.org/10.1021/nl5048779
[36] L.N. Quan, F.P. García de Arquer, R.P. Sabatini, E.H. Sargent, Perovskites for light emission, *Adv. Mater.* 30 (2018) 1–19. https://doi.org/10.1002/adma.201801996
[37] I. Borriello, G. Cantele, D. Ninno, Ab initio investigation of hybrid organic-inorganic perovskites based on tin halides, *Phys. Rev. B.* 77 (2008) 235214. https://doi.org/10.1103/PhysRevB.77.235214
[38] M.I. Saidaminov, J. Almutlaq, S. Sarmah, I. Dursun, A.A. Zhumekenov, R. Begum, J. Pan, N. Cho, O.F. Mohammed, O.M. Bakr, Pure Cs_4PbBr_6: Highly Luminescent zero-dimensional perovskite solids, *ACS Energy Lett.* 1 (2016) 840–845. https://doi.org/10.1021/acsenergylett.6b00396
[39] J. Shamsi, A.S. Urban, M. Imran, L. De Trizio, L. Manna, Metal halide perovskite nanocrystals: Synthesis, post-synthesis modifications, and their optical properties, *Chem. Rev.* 119 (2019) 3296–3348. https://doi.org/10.1021/acs.chemrev.8b00644
[40] L. Schade, A.D. Wright, R.D. Johnson, M. Dollmann, B. Wenger, P.K. Nayak, D. Prabhakaran, L.M. Herz, R. Nicholas, H.J. Snaith, P.G. Radaelli, Structural and optical properties of $Cs_2AgBiBr_6$ double perovskite, *ACS Energy Lett.* 4 (2019) 299–305. https://doi.org/10.1021/acsenergylett.8b02090
[41] L. Protesescu, S. Yakunin, M.I. Bodnarchuk, F. Bertolotti, N. Masciocchi, A. Guagliardi, M.V. Kovalenko, Monodisperse formamidinium lead bromide nanocrystals with bright and stable green photoluminescence, *J. Am. Chem. Soc.* 138 (2016) 14202–14205. https://doi.org/10.1021/jacs.6b08900
[42] F. Zhang, H. Zhong, C. Chen, X. Wu, X. Hu, H. Huang, J. Han, B. Zou, Y. Dong, Brightly luminescent and color-tunable colloidal $CH_3NH_3PbX_3$ (X = Br, I, Cl) quantum dots: Potential alternatives for display technology, *ACS Nano.* 9 (2015) 4533–4542. https://doi.org/10.1021/acsnano.5b01154

[43] S. Sun, D. Yuan, Y. Xu, A. Wang, Z. Deng, Ligand-mediated synthesis of shape-controlled cesium lead halide perovskite nanocrystals via reprecipitation process at room temperature, *ACS Nano.* 10 (2016) 3648–3657. https://doi.org/10.1021/acsnano.5b08193

[44] K.H. Wang, L. Wu, L. Li, H. Bin Yao, H.S. Qian, S.H. Yu, Large-scale synthesis of highly luminescent perovskite-related $CsPb_2Br_5$ nanoplatelets and their fast anion exchange, *Angew. Chemie - Int. Ed.* 55 (2016) 8328–8332. https://doi.org/10.1002/anie.201602787

[45] M. Leng, Y. Yang, K. Zeng, Z. Chen, Z. Tan, S. Li, J. Li, B. Xu, D. Li, M.P. Hautzinger, Y. Fu, T. Zhai, L. Xu, G. Niu, S. Jin, J. Tang, All-inorganic bismuth-based perovskite quantum dots with bright blue photoluminescence and excellent stability, *Adv. Funct. Mater.* 28 (2018). https://doi.org/10.1002/adfm.201704446

[46] L.C. Schmidt, A. Pertegás, S. González-Carrero, O. Malinkiewicz, S. Agouram, G. Mínguez Espallargas, H.J. Bolink, R.E. Galian, J. Pérez-Prieto, Nontemplate synthesis of $CH_3NH_3PbBr_3$ perovskite nanoparticles, *J. Am. Chem. Soc.* 136 (2014) 850–853. https://doi.org/10.1021/ja4109209

[47] D. Chen, Z. Wan, X. Chen, Y. Yuan, J. Zhong, Large-scale room-temperature synthesis and optical properties of perovskite-related Cs_4PbBr_6 fluorophores, *J. Mater. Chem. C.* 4 (2016) 10646–10653. https://doi.org/10.1039/C6TC04036E

[48] F. Fang, W. Chen, Y. Li, H. Liu, M. Mei, R. Zhang, J. Hao, M. Mikita, W. Cao, R. Pan, K. Wang, X.W. Sun, Employing polar solvent controlled ionization in precursors for synthesis of high-quality inorganic perovskite nanocrystals at room temperature, *Adv. Funct. Mater.* 28 (2018) 1706000. https://doi.org/10.1002/adfm.201706000

[49] A. Kojima, K. Teshima, Y. Shirai, T. Miyasaka, Organometal halide perovskites as visible-light sensitizers for photovoltaic cells, *J. Am. Chem. Soc.* 131 (2009) 6050–6051. https://doi.org/10.1021/ja809598r

[50] C. Zuo, H.J. Bolink, H. Han, J. Huang, D. Cahen, L. Ding, Advances in perovskite solar cells, *Adv. Sci.* 3 (2016) 1500324. https://doi.org/10.1002/advs.201500324

[51] W.S. Yang, J.H. Noh, N.J. Jeon, Y.C. Kim, S. Ryu, J. Seo, S.I. Seok, High-performance photovoltaic perovskite layers fabricated through intramolecular exchange, *Science.* 348 (2015) 1234–1237. https://doi.org/10.1126/science.aaa9272

[52] S. Li, Y.L. Cao, W.H. Li, Z.S. Bo, A brief review of hole transporting materials commonly used in perovskite solar cells, *Rare Met.* 40 (2021) 2712–2729. https://doi.org/10.1007/s12598-020-01691-z

[53] J.M. Ball, M.M. Lee, A. Hey, H.J. Snaith, Low-temperature processed meso-superstructured to thin-film perovskite solar cells, *Energy Environ. Sci.* 6 (2013) 1739–1743. https://doi.org/10.1039/c3ee40810h

[54] Y.-H. Kim, H. Cho, J.H. Heo, T.-S. Kim, N. Myoung, C.-L. Lee, S.H. Im, T.-W. Lee, Multicolored organic/inorganic hybrid perovskite light-emitting diodes, *Adv. Mater.* 27 (2015) 1248–1254. https://doi.org/10.1002/adma.201403751

[55] G. Pacchioni, Highly efficient perovskite LEDs, *Nat. Rev. Mater.* 6 (2021) 108–108. https://doi.org/10.1038/s41578-021-00280-5

[56] Z.-K. Tan, R.S. Moghaddam, M.L. Lai, P. Docampo, R. Higler, F. Deschler, M. Price, A. Sadhanala, L.M. Pazos, D. Credgington, F. Hanusch, T. Bein, H.J. Snaith, R.H. Friend, Bright light-emitting diodes based on organometal halide perovskite, *Nat. Nanotechnol.* 9 (2014) 687–692. https://doi.org/10.1038/nnano.2014.149

[57] H. Cho, S.-H. Jeong, M.-H. Park, Y.-H. Kim, C. Wolf, C.-L. Lee, J.H. Heo, A. Sadhanala, N. Myoung, S. Yoo, S.H. Im, R.H. Friend, T.-W. Lee, Overcoming the electroluminescence efficiency limitations of perovskite light-emitting diodes, *Science* 350 (2015) 1222–1225. https://doi.org/10.1126/science.aad1818

[58] L. Zhang, X. Yang, Q. Jiang, P. Wang, Z. Yin, X. Zhang, H. Tan, Y. Yang, M. Wei, B.R. Sutherland, E.H. Sargent, J. You, Ultra-bright and highly efficient inorganic based perovskite light-emitting diodes, *Nat. Commun.* 8 (2017) 15640. https://doi.org/10.1038/ncomms15640

[59] J. Byun, H. Cho, C. Wolf, M. Jang, A. Sadhanala, R.H. Friend, H. Yang, T.-W. Lee, Efficient visible quasi-2D perovskite light-emitting diodes, *Adv. Mater.* 28 (2016) 7515–7520. https://doi.org/10.1002/adma.201601369

[60] J. Song, T. Fang, J. Li, L. Xu, F. Zhang, B. Han, Q. Shan, H. Zeng, Organic–inorganic hybrid passivation enables perovskite QLEDs with an EQE of 16.48%, *Adv. Mater.* 30 (2018) 1805409. https://doi.org/10.1002/adma.201805409

[61] Y. H. Kim, S. Kim, A. Kakekhani, J. Park, J. Park, Y.H. Lee, H. Xu, S. Nagane, R.B. Wexler, D.H. Kim, S.H. Jo, L. Martínez-Sarti, P. Tan, A. Sadhanala, G.S. Park, Y.W. Kim, B. Hu, H.J. Bolink, S. Yoo, R.H. Friend, A.M. Rappe, T.W. Lee, Comprehensive defect suppression in perovskite nanocrystals for high-efficiency light-emitting diodes, *Nat. Photonics.* 15 (2021) 148–155. https://doi.org/10.1038/s41566-020-00732-4

[62] S. Yakunin, L. Protesescu, F. Krieg, M.I. Bodnarchuk, G. Nedelcu, M. Humer, G. De Luca, M. Fiebig, W. Heiss, M.V. Kovalenko, Low-threshold amplified spontaneous emission and lasing from colloidal nanocrystals of caesium lead halide perovskites, *Nat. Commun.* 6 (2015). https://doi.org/10.1038/ncomms9056

[63] Y. Zhao, K. Zhu, Organic-inorganic hybrid lead halide perovskites for optoelectronic and electronic applications, *Chem. Soc. Rev.* 45 (2016) 655–689. https://doi.org/10.1039/c4cs00458b

[64] L. Dou, Y.M. Yang, J. You, Z. Hong, W.H. Chang, G. Li, Y. Yang, Solution-processed hybrid perovskite photodetectors with high detectivity, *Nat. Commun.* 5 (2014). https://doi.org/10.1038/ncomms6404

[65] P. Pandey, S. Cho, S. Hayase, J. Sang Cho, D.-W. Kang, New strategies to develop high-efficiency lead-free wide bandgap perovskite solar cells, *Chem. Eng. J.* 448 (2022) 137622. https://doi.org/10.1016/j.cej.2022.137622

CHAPTER 10

Solar Cells with Recent Improvements and Energy-Saving Strategies for the Future World

Yatish R. Parauha

Shri Ramdeobaba College of Engineering and Management, Nagpur, India

B. Vengadaesvaran

Um Power Energy Dedicated Advanced Centre (UMPEDAC), University of Malaya – City Campus, Kuala Lumpur, Malaysia.

Nirupama S. Dhoble

Sevadal Mahila Mahavidhalaya, Sakkardara, Nagpur, India

Sanjay J. Dhoble

Rashtrasant Tukadoji Maharaj Nagpur University, Nagpur, India

10.1 INTRODUCTION

Renewable energy is critical for meeting international and national long-term climate and energy goals, such as shifting the energy supply from fossil-based to zero-carbon production. Renewable energy is also important for maintaining the environment and human health. With the growing challenges of global warming and the developing problems with environmental implications associated with fossil fuels, renewable energy and energy efficiency objectives and strategies are indispensable in addressing these challenges. A wide range of resources can be used to produce renewable energy such as wind, solar, water, tidal, geothermal, and biomass [1]. One of humanity's most pressing tasks is to create a sustainable energy source as the world's limited fossil fuel reserves are in direct conflict with rising energy demand. Solar energy is considered one of the most promising sustainable energy sources due to its limitless supply, universality, high efficiency, and environmental

DOI: 10.1201/9781003315261-13

friendliness [2]. However, natural solar radiation is scattered, sporadic, and constantly changing. As a result, it is still very difficult to effectively use solar energy in a way that is economical, easy, and clean. Therefore, this chapter focuses on solar energy as it is one of the most widely used forms of energy to maintain and achieve environmental sustainability. Solar energy is sustainable energy, which plays a crucial role in lowering greenhouse gas emissions and mitigating climate change, and is important for the protection of humans, wildlife and ecosystems. Solar power can help to enhance air quality and reduce water consumption associated with energy production.

There are numerous ways to use solar energy to generate primarily electrical, thermal, or mechanical energy. For instance, solar panels with photovoltaics simply capture solar energy and convert it to electrical energy, which may then be utilized to power devices or stored in a battery for later use [3]. These solar panel types also enable off-grid life by supplying power for a variety of uses. On the other hand, heat from the sun can be absorbed by thermal solar panels and converted into thermal energy or transferred to a medium such as water. Consumers can then receive hot water, for example, which eliminates the need to use fossil fuels for this reason. As an alternative, it has also been discovered that concentrated solar panels are a potential solar technology that can be used to create energy effectively by generating large amounts of heat [4].

Solar energy has attracted a great deal of attention in recent years as a major resource and because of increasing environmental concerns. Solar tower power plants offer the most potential of all solar energy infrastructure types because they can produce massive amounts of electricity at the lowest possible cost. Renewable and green energy sources are expected to account for more than half of the global electricity generation by 2035 [5, 6]. Photovoltaics are one of the renewable energies with the highest potential because they provide practically endless access to clean solar power. Due to the abundance of manufacturing materials and the maturity of the underlying technology, crystalline silicon (c-Si) solar cells are currently the mainstay of the photovoltaic industry. Currently, these devices account for 80% of all the photovoltaic devices used globally [7]. Various solar cell technologies as well as the emergence of a new class of photovoltaic absorber materials have led to substantial improvements in solar cell efficiency in recent years. In particular, two main current developments have shaped the research field of photovoltaics. First, crystalline although Si photovoltaics have reduced cost, they have very low efficiency, and, as a result, research must focus on high-efficiency concepts as opposed to low-cost and low-efficiency approaches. The second development is the search for a novel material for photovoltaic absorber material. Therefore, various absorbent materials have been investigated in recent years. This research work and investigation shows that solar cell efficiency can be increased using a variety of approaches. This chapter focuses on the development and challenges of solar cells.

10.2 SOLAR ENERGY: A MAJOR OPPORTUNITY FOR SOCIETY

Solar cells are also referred to as photovoltaic cells: device that converts the energy of light directly into electrical energy through the photovoltaic effect. This technology allows photons of solar radiation to be captured and converted into electrical energy, which is then immediately available to power a load or, if there is a surplus, can be stored for later use. Theoretically, the world consumption of a year can be covered with just one hour of sunshine. The vast majority of solar cells are made from silicon, ranging from amorphous (non-crystalline) to polycrystalline and crystalline (single-crystal) forms of silicon, with increasing efficiency and lower cost [8, 9]. The sun shines everywhere in the world and makes solar power usable everywhere. Because solar power can be combined with batteries for energy storage, solar power systems can operate independently of the public grid, making them cost-effective for remote locations. Solar panels have no moving parts, making maintenance costs low, and they are extremely reliable with a long lifespan of more than 25 years of guaranteed electricity. Solar power relies on the sun for a fuel source, so there is no need to drill for, refine, or ship petroleum-based fuels to site. As one can see, solar energy offers many advantages. Solar power systems and power plants do not emit greenhouse gases or pollute the air when they are in use. When solar energy replaces or reduces the use of other energy sources with lower environmental impacts, it can have a beneficial, indirect effect on the environment. But there are environmental concerns associated with the development and application of solar power systems. Photovoltaic cells and panels are made of hazardous chemicals, which need to be handled carefully to prevent release in the environment. Heavy metals are used in many photovoltaic cell technologies and when their useful life is over, these cells and photovoltaic panels need to be treated with care [10, 11]. In addition, some solar thermal systems use potentially hazardous liquids to transmit heat, and leaks of these substances can cause environmental damage. Thus, solar cells can also be dangerous for the environment, although the research community is trying to reduce all these effects, and, considering the benefits and future of solar cells, these side effects are negligible.

Advantages for Society:

- Solar energy is a renewable resource which is a great advantage. In years to come, we will have an infinite amount of sunlight available to us.
- Solar technology equipment requires no fuel, so it runs cleanly. There is not much emission of greenhouse gases or hazardous substances. We can significantly reduce environmental impact and make energy and financial savings by using solar power.
- Solar cookers provide a great alternative to cooking on a wood burning stove, and are a healthier and safer way to cook food and purify water.
- Solar energy is a great addition to the list of other renewable energy sources such as wind or hydroelectric energy.
- Solar panels installed on homes or businesses can actually generate surplus electricity. These home or business owners can sell energy back to the utility company, reducing or even eliminating their electricity expenses.

10.3 FUNDAMENTAL ASPECTS OF SOLAR CELLS

10.3.1 Construction and Working Principle

Solar cells are electrical components that convert sunlight directly into electrical energy. They are based on the photovoltaic effect at a boundary between the positively and negatively doped regions of a semiconducting material. Solar cells for generating electrical energy are therefore referred to as photovoltaic cells.

Solar cell is basically a p-n junction diode, although its construction is slightly different from that of a conventional p-n junction diode. Figure 10.1 shows a schematic representation of a photovoltaic solar cell. The principal layer of this cell includes an anti-reflective cover glass that protects the semi-conductor material against sunlight. The cell has small grid patterns with minor metal strips under the glass, so the top layer of the cell can be made using glass, metallic strips, and anti-reflective coats. The most important part of the cell is the middle layer where solar energy can be formed by the influence of photovoltaics. It consists of two semiconductor layers made of p-type and n-type materials. The base layer of this cell consists of two parts: a rear metallic electrode is below a p-type semiconductor, and it works with the metal grid to generate electric current in the principal layer. The last layer is called the reflective layer [12, 13].

The basic strategy behind the mechanism of electricity generation by a photovoltaic cell lies in the fact that these intense radiations, when they hit a solar panel, lead to the mobilization of electrons, thereby creating a potential difference that becomes visible in the form of direct electric current. There are two important processes involved in photovoltaic energy conversion in solar cells. First, a material is needed, which generates electron-hole pairs when light is absorbed. In the second process, the device's structure splits electrons and holes, sending the electrons to the negative electrode and the holes to the positive electrode thus producing electrical power [15].

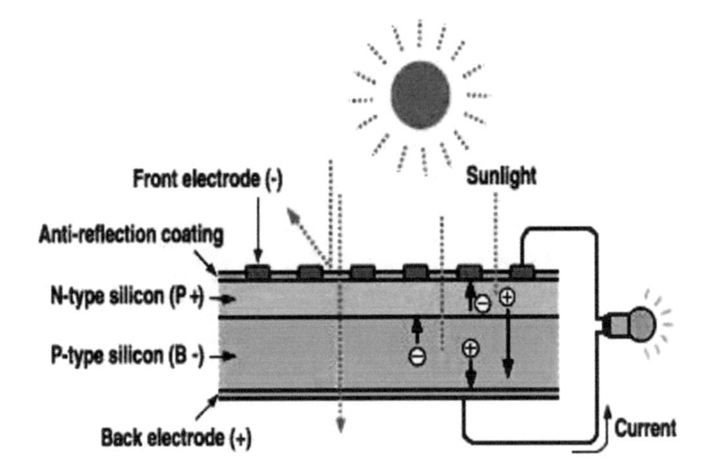

FIGURE 10.1 Basic diagram of a photovoltaic solar cell. (Reproduced from Ref. [14], under the Creative Commons License.)

When light enters the p-n junction, photons of light can readily enter the junction due to the very thin p-type layer. Light energy, in the form of photons, provides enough energy to the junction to build a number of electron-hole pairs. The incoming light disrupts the junction's thermal balance. The free electrons in the depletion zone might swiftly reach the junction's n-type side. Similarly, the depletion holes can soon reach the p-type side of the junction. Because of the barrier potential of the junction, freshly produced free electrons cannot cross the junction once they reach the n-type side [16]. Similarly, once the newly created holes come to the p-type side they cannot further cross the junction became of the same barrier potential of the junction. As the concentration of electrons becomes higher on one side, that is, the n-type side of the junction, and the concentration of holes increases on the other side, that is, the p-type side of the junction, the p-n junction will behave like a small battery cell. A voltage is set up which is known as photo voltage [16]. If we connect a small load across the junction, there will be a tiny current flowing through it [12, 13].

10.3.2 Basic Terms Related to Solar Cells

10.3.2.1 Short Circuit Current (J_{SC})

Short-circuit current is the current flowing through a solar cell when the voltage across it is 0 (i.e., when the solar cell is short-circuited). Usually, a short-circuit current is written as J_{sc}, as seen in Figure 10.2. The production and collection of light-generated carriers causes the short-circuit current. The short-circuit current and light-generated current are the same for a complete solar cell with the lightest resistive loss mechanism. Consequently, the short-circuit current is the maximum current that can be drawn from a solar cell.

The incident light, bandgap, and absorption coefficient of the semiconductors, the effectiveness of charge collection, and the active area of the solar cell all directly affect the photocurrent produced by a solar cell under illumination at the short circuit. It is typical to

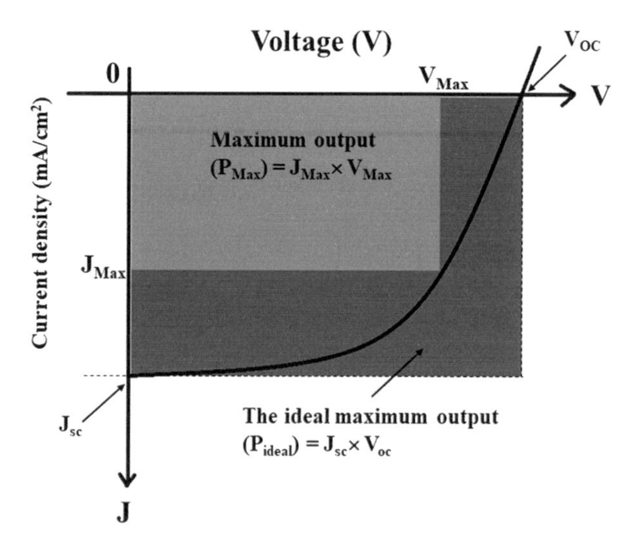

FIGURE 10.2 Schematic of the current-vs-voltage relationship for a solar cell.

(Source: Authors.)

mention the short-circuit current density (J_{sc} in mA/cm^2) rather than the short circuit current in order to eliminate the dependency on the solar cell area. The short-circuit current density increases with decreasing bandgap and increasing absorption coefficient. For an ideal solar cell, the equation for the short-circuit current density can be approximated as in Equation (10.1) [17]:

$$J_{sc} = q\int b_s(\mathrm{E})\mathrm{QE}(\mathrm{E})\mathrm{dE} \tag{10.1}$$

Where, q = unit charge, b_s(E) = incident spectral photon flux density. QE is determined by the solar cell materials' absorption coefficient, the efficiency of charge separation, and the efficiency of charge collecting in the device [17].

10.3.2.2 Open Circuit Voltage (V_{oc})

The highest voltage a solar cell can produce with no current flowing through it is known as the open-circuit voltage (V_{oc}). In the equivalent circuit, there will be a potential difference across the terminals of the cell. This potential difference generates a current that acts in the opposite direction to the photocurrent, and the net current is reduced by it short-circuit current. The open-circuit voltage refers to the amount of forward bias applied to the solar cell due to the biasing of the junction with the current produced by the light. The net current in the solar cell equation is set to zero to obtain the following expression for V_{oc} in Equation (10.2):

$$V_{OC} = \frac{nkT}{q}\ln\left(\frac{I_L}{I_0}+1\right) \tag{10.2}$$

V_{oc} increases with the increasing bandgap of the semiconductor. An upper limit to the open circuit voltage of a solar cell is the bandgap at unit charge.

10.3.2.3 Solar Cell Fill Factor (FF)

The maximum current and voltage from a solar cell are the short-circuit current and open-circuit voltage, respectively. However, the power from the solar cell is 0 at both of these operational points. P reaches a maximum at the cell's operating point (i.e., the maximum power point) at a voltage V_m with a corresponding current density J_m. A solar cell's power density is given by Equation (10.3) [17]:

$$P_{max} = J_{max} \times V_{max} \tag{10.3}$$

The 'fill factor' (FF), is a quantity that, along with V_{oc} and I_{sc}, defines the maximum power output of a solar cell. The FF is defined as the ratio of the maximum power produced by a solar cell to the product of V_{oc} and I_{sc}, as given in Equation (10.4).

$$\mathrm{FF} = \frac{P_{max}}{V_{OC} \times I_{SC}} \tag{10.4}$$

10.3.2.4 Solar Cell Efficiency

The most commonly used criterion for comparing the performance of one solar cell to another is efficiency. Efficiency (η) can be defined as the ratio of the energy output from a solar cell to the input energy from the sun. The efficiency depends on the spectrum, intensity, and solar cell temperature of the incident light, as given in Equation (10.5) [17].

$$\eta = \frac{J_{SC} V_{OC} FF}{P_s} \tag{10.5}$$

These four parameters, J_{sc}, V_{oc}, FF, and η, are the most important aspects of a solar cell.

10.3.2.5 External and Internal Quantum Efficiency

The quantum efficiency is the ratio of the number of carriers that are collected by the solar cell to the number of photons incident on the solar cell and can be expressed as a function of the wavelength or the photon energy [18]. There are two types of quantum efficiency: internal and external.

The internal quantum efficiency (IQE) is an important parameter quantifying the performance of a photodetector, as in Equation (10.6):

$$\text{IQE} = \frac{\text{Electron collected as photocurrent per second}}{\text{Photon absorbed per second}} \tag{10.6}$$

External quantum efficiency (EQE) is defined as the number of free electrons (produced by incident photons) collected in the device's external circuitry per photon incident on the device, as in Equation (10.7):

$$\text{EQE} = \frac{\text{Electron collected as photocurrent per second}}{\text{Photon incident per second}} \tag{10.7}$$

10.4 TYPES OF SOLAR CELLS

The ultimate aim of researchers in energy, physics, chemistry and other related fields has been to produce electricity using cleaner and more affordable energy resources. The scientific community is making steady efforts in this direction, especially in the current period of sustainability-focused enquiry, where energy has emerged as an important and central topic of study. Figure 10.3 shows that efforts in the field of solar cells can be divided into three generations. A detailed description of these solar cell generations is provided in Figure 10.3 [19].

10.4.1 First-Generation Solar Cells

In 1954, Bell Laboratories invented the first silicon solar cell, which had an efficiency of 6% [20]. Traditional silicon solar cells, first introduced for commercial applications, are now the most efficient solar cells for home use and are used in over 80% of all solar panels sold globally. In the case of single-cell photovoltaic systems, silicon solar cells are the most

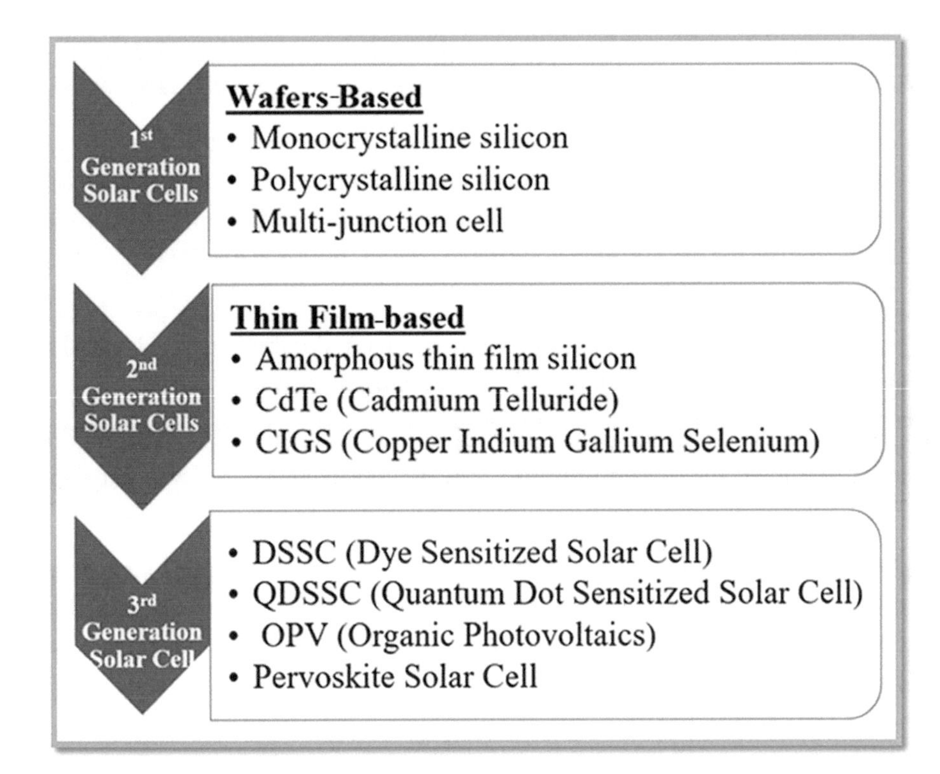

FIGURE 10.3 Different generations of solar cells.

(Source: Authors.)

efficient, and silicon is the second most abundant element on Earth after oxygen. With an energy bandgap of 1.1 eV, it is a suitable semiconductor material for solar energy applications [20].

Crystalline silicon cells are classified into three main types depending on how the Si wafers are made. The types are based on the type of silicon used, specifically: monocrystalline, polycrystalline, and multi-junction solar cells.

I. Monocrystalline silicon solar cells:

These are also known as mono-Si solar cells. They are one of the most widely used and efficient types of solar cells available today. Here are some key characteristics and features of monocrystalline silicon solar cells:

- Composition: Monocrystalline silicon solar cells are made from a single crystal structure of silicon. The silicon used is highly purified and has a very high degree of crystalline order, which enhances the cell's efficiency.
- Efficiency: Monocrystalline silicon solar cells have one of the highest conversion efficiencies among commercially available solar cell technologies. Typically, they have efficiency levels ranging from 15% to 22%. Higher efficiency means that more sunlight can be converted into electricity.

- Appearance: Monocrystalline silicon solar cells have a uniform black color. They are easily recognizable by their rounded edges and a consistent dark appearance.
- Manufacturing Process: The production of monocrystalline silicon solar cells involves the growth of a single crystal ingot. The ingot is then sliced into thin wafers, which are further processed and treated to form the solar cells. This manufacturing process contributes to the high cost of mono-Si solar cells.
- Space Efficiency: Monocrystalline silicon solar cells have a high power output per unit area. This makes them an excellent choice for situations where limited space is available for solar installations, such as residential rooftops.
- Durability and Longevity: Monocrystalline silicon solar cells have a long lifespan and are known for their durability. They can withstand harsh weather conditions and typically come with warranties ranging from 20 to 25 years.
- Performance in Sunlight: Monocrystalline silicon solar cells perform well in direct sunlight. They have a high light absorption capacity and are efficient in converting sunlight into electricity, making them suitable for a wide range of applications.
- Cost: Monocrystalline silicon solar cells tend to be more expensive compared to other types of solar cells, such as polycrystalline or thin-film cells. However, the higher efficiency and long-term performance often justify the initial investment.
- Market Dominance: Monocrystalline silicon solar cells have been dominant in the solar industry for many years. They hold a significant market share due to their efficiency, reliability, and established manufacturing processes.

It's important to note that while monocrystalline silicon solar cells are highly efficient, there are other solar cell technologies available, each with its own advantages and disadvantages. The choice of solar cell type depends on factors such as cost, available space, efficiency requirements, and specific project needs. In recent years, research work has continue to improve the performance of monocrystalline silicon solar cells.

Researchers are continuously working on increasing the energy conversion efficiency of monocrystalline silicon solar cells. This involves optimizing the material quality, reducing defects, and improving light-trapping techniques to enhance light absorption and reduce energy losses. Surface recombination can limit the performance of monocrystalline silicon solar cells. Researchers are investigating various passivation techniques to reduce surface recombination velocity and enhance the electrical properties of the cell surfaces. This includes the use of advanced passivation layers, such as silicon nitride (SiNx), aluminum oxide (Al_2O_3), and silicon oxides. Researchers are exploring different methods to create texturized surfaces on monocrystalline silicon wafers, such as chemical etching, reactive ion etching, and laser processing. Additionally, advanced antireflection coatings are being developed to further minimize reflection losses [21–23].

II. Polycrystalline silicon solar cells:

These are also known as multicrystalline silicon solar cells. They are made from polycrystalline silicon, which is a form of silicon consisting of multiple small crystal structures. They are one of the most widely used solar cell technologies due to their relatively high efficiency and lower production costs compared to other types of solar cells. The production process involves melting silicon and then cooling it quickly, resulting in the formation of multiple crystal structures. This rapid cooling creates grain boundaries between the crystals, which can affect the efficiency of the solar cell.

Polycrystalline silicon solar cells are less expensive to manufacture than monocrystalline silicon solar cells because the production process is simpler and uses less energy. However, they are generally less efficient in converting sunlight into electricity. The grain boundaries in polycrystalline silicon can impede the flow of electrons, reducing the overall efficiency of the cell.

Despite their lower efficiency, polycrystalline silicon solar cells are widely used in the solar industry due to their lower cost and adequate performance in various applications. They are commonly used in residential, commercial, and utility-scale solar power systems to generate renewable energy. Ongoing research and technological advancements continue to improve the efficiency of polycrystalline silicon solar cells, making them a viable option for solar energy generation. Researchers have been working on improving the quality of the polycrystalline silicon material and the manufacturing processes to enhance the overall efficiency of the solar cells. This includes reducing impurities, defects, and grain boundaries within the material, which can affect the performance.

III. Multi-junction solar cells:

In addition to these solar cells, there are also solar cells with more than three junctions, called multi-junction solar cells. These multi-junction solar cells contain additional semiconductor layers, each designed to capture a specific range of wavelengths. By stacking multiple junctions, these solar cells can achieve even higher efficiency.

They are mainly used in special applications such as satellite power systems and advanced concentrator photovoltaics. It is important to note that while multi-junction solar cells offer higher efficiency, they are generally more expensive to manufacture and are primarily used in specialized applications where space or performance constraints are critical. In most commercial and residential applications, the more cost-effective single-junction solar cells are still widely used [24–26].

Researchers have been working on improving the efficiency of multi-junction solar cells. By optimizing the bandgap energies of the different semiconductor layers and developing novel materials, researchers have achieved impressive efficiencies exceeding 46% [25]. These high-efficiency cells are primarily used in space applications where high-power generation is required. Researchers have also been exploring novel materials for multi-junction solar cells to improve their performance. For example, they have been investigating emerging materials like perovskites and quantum dots, which have shown promising optoelectronic properties and can be integrated into multi-junction structures.

10.4.2 Second-Generation Solar Cells

Second-generation solar cells refer to a category of solar cells that utilize silicon (Si) as the semiconductor material. These cells are distinct from the first generation of Si solar cells, which are typically made of single-crystal silicon. Second-generation Si solar cells are designed to address some of the limitations of first-generation cells, such as high manufacturing costs and the requirement of high-purity silicon. They are also known as thin-film silicon solar cells because they employ a thin layer of silicon as the active material. This generation can also be divided into three types of solar cells: amorphous thin-film, cadmium telluride (CdTe), and copper indium gallium selenium (CIGS) solar cells.

I. Amorphous thin-film:

Amorphous thin-film silicon solar cells are a type of solar cell technology that utilizes a non-crystalline form of silicon as the semiconductor material. Unlike traditional crystalline silicon solar cells, which require a thick and highly pure silicon wafer, amorphous thin film cells can be made using much thinner layers of silicon. These solar cells are typically fabricated using techniques such as chemical vapor deposition or sputtering, which allow for the deposition of silicon films onto various substrates, such as glass or flexible materials [27, 28].

The amorphous nature of the silicon film allows for greater flexibility in terms of the substrate and cell design. Amorphous thin-film silicon solar cells have several advantages over crystalline silicon cells. They are lightweight, flexible and can be produced at a lower cost due to their thinner silicon layers. Additionally, they have a higher tolerance to shade, allowing them to maintain a reasonable level of power output even when partially shaded.

However, amorphous thin-film silicon solar cells also have some disadvantages. They generally have lower efficiency compared to crystalline silicon cells, meaning they convert a smaller percentage of sunlight into electricity. This lower efficiency limits their use in applications where space or weight constraints are not significant factors.

Despite their lower efficiency, amorphous thin-film silicon solar cells have found applications in various settings. They are commonly used in portable electronic devices, such as calculators, watches, and small solar panels. They are also employed in building-integrated photovoltaics, where solar cells are integrated into windows or façades of buildings [29]. Researchers continue to explore ways to improve the efficiency of amorphous thin-film silicon solar cells through advancements in materials, device structures, and fabrication processes. These efforts aim to make this technology more competitive with other solar cell technologies and expand its range of applications in the future. Additionally, researchers have been investigating alternative materials for amorphous thin-film solar cells to overcome some of the limitations of amorphous silicon. For example, emerging materials like perovskites have gained significant attention in recent years due to their high efficiency and low-cost potential. Perovskite-based thin-film solar cells are being actively researched, and they have shown promising results in terms of efficiency and stability. Furthermore, researchers are exploring different deposition techniques and device architectures to optimize the performance of amorphous thin-film solar cells. Techniques such as plasma-enhanced chemical vapor deposition (PECVD), sputtering, and atomic layer deposition (ALD) are being used to deposit high-quality thin films with controlled properties. Novel device architectures, such as tandem solar cells and multi-junction structures, are also being investigated to achieve higher efficiencies.

II. Cadmium telluride (CdTe) solar cells:

CdTe solar cells are a type of thin-film solar cell that are known for their high efficiency and low manufacturing costs. They are made using a thin layer of cadmium telluride, a semiconductor material, which is sandwiched between two other layers of different materials to create a photovoltaic (PV) structure. CdTe solar cells have several advantages.

They have a high light-absorption coefficient, meaning they can efficiently convert a large amount of sunlight into electricity. CdTe solar cells have achieved high conversion efficiencies, with some commercial modules reaching efficiencies of around 22% [30]. This makes them one of the most efficient thin-film solar technologies available.

They also have a relatively low production cost compared to other solar cell technologies, making them cost-effective for large-scale installations.

One key feature of CdTe solar cells is their flexibility, which allows for their integration into various applications and surfaces [30]. They can be manufactured in large sizes, enabling easier installation and reducing the number of connections required and also perform well in low-light conditions, making them suitable for regions with less sunlight.

However, there are some concerns regarding the toxicity of cadmium, a component of CdTe. Proper handling and disposal of CdTe solar cells are necessary to prevent environmental contamination. Manufacturers have implemented measures to mitigate these concerns and ensure the safe use and recycling of CdTe solar cells.

Overall, CdTe solar cells offer a promising and cost-effective option for generating clean energy, contributing to the growth of renewable energy sources and reducing greenhouse gas emissions. CdTe solar cells have gained significant commercial success in recent years, with several companies manufacturing and deploying large-scale installations based on this technology. They are commonly used in utility-scale solar power plants, as well as in smaller residential and commercial solar systems.

III. Copper indium gallium selenium (CIGS) solar cells:

CIGS solar cells are a type of thin-film solar cell technology that converts sunlight into electricity. They are composed of a thin layer of copper, indium, gallium, and selenium compounds deposited on a flexible or rigid substrate.

CIGS solar cells have gained attention due to their high efficiency and potential for low-cost manufacturing. They offer several advantages, such as better performance in low light conditions, good temperature coefficient, and the ability to be integrated into various applications and surfaces. The CIGS layer absorbs sunlight and generates electrons, which are then collected and converted into usable electrical energy. The thin-film structure enables flexibility and lightweight characteristics, making them suitable for a range of applications, including building-integrated photovoltaics (BIPV), portable devices, and solar panels.

Overall, CIGS solar cells offer a promising alternative to traditional silicon-based solar cells, with the potential for higher efficiency and lower production costs. Ongoing research and development aim to further improve their performance and expand their application in the renewable energy sector.

10.4.3 Third-Generation Solar Cells

Third-generation solar cells refers to a category of advanced solar cell technologies that aim to overcome the limitations of first- and second-generation solar cells. While first-generation solar cells, such as crystalline silicon cells and second-generation solar cells, such as thin-film cells, have been successful in harnessing sunlight to generate electricity, third-generation solar cells explore new materials and concepts to further enhance efficiency and functionality. Here are some notable third-generation solar cell technologies.

I. Organic photovoltaic (OPV) cells:

OPV cells use organic materials, typically polymers or small molecules, to convert sunlight into electricity. The organic materials used in OPV cells offer advantages such as flexibility, light weight, and low-cost manufacturing. They can be processed using solution-based techniques, allowing for large-scale production and integration into various surfaces, including flexible substrates [31, 32]. However, OPV cells generally have lower conversion efficiencies compared to silicon-based solar cells. Efforts are ongoing to improve their efficiency and stability through material optimization, device engineering, and interface modifications. Overall, OPV cells represent a promising technology for solar energy conversion, particularly in applications that require lightweight, flexible, or customizable solar panels [31, 32].

II. Dye-sensitized solar cells (DSSCs):

DSSCs use a layer of light-absorbing dye to capture sunlight and transfer the energy to an electrolyte that generates an electric current. They have been successful in achieving high efficiencies for indoor or low-light conditions and can be manufactured using low-cost materials and processes. They consist of several key components: a semiconductor layer, a dye molecule, an electrolyte, and a counter electrode [33]. The semiconductor layer is typically composed of titanium dioxide (TiO_2) nanoparticles, which have a large surface area and high light absorption capability. The dye molecule also known as a sensitizing dye or photosensitizer, is adsorbed onto the surface of the semiconductor layer. The dye absorbs sunlight and transfers the energy to the TiO_2, creating electron-hole pairs. The electrolyte is a solution that fills the space between the dye-sensitized semiconductor layer and the counter electrode [34, 35]. It typically contains a redox couple, which helps to transport electrons between the semiconductor and the counter electrode. When sunlight hits the DSSC, the dye absorbs the photons and generates excited electrons. These electrons are injected into the TiO_2 semiconductor layer and move toward the counter electrode through the electrolyte. The counter electrode, usually made of a conductive material like platinum, acts as a catalyst to facilitate the transfer of electrons from the electrolyte back into the dye, completing the circuit. As the electrons flow through the external circuit from the semiconductor to the counter electrode, they generate an electric current that can be harnessed for various applications. This current can power electronic devices or be stored in batteries for later use.

DSSCs have several advantages over traditional silicon-based solar cells, including lower production costs, flexibility, and the ability to generate electricity under low-light conditions [34, 35]. However, their efficiency is generally lower than silicon solar cells, and they are more sensitive to environmental factors such as moisture and temperature.

III. Perovskite solar cells (PSCs):

PSCs use a class of materials called perovskites, which have shown great potential for high efficiency and low-cost solar cells. These cells can be fabricated using solution-based processes and have achieved rapid efficiency improvements in recent years.

However, stability issues, such as sensitivity to moisture and heat, remain a challenge for commercialization. The term 'perovskite' refers to a specific crystal structure that can be composed of various elements.

The key advantage of perovskite solar cells is their high efficiency in converting sunlight into electricity. Perovskite materials have demonstrated the potential to achieve efficiencies comparable to traditional silicon-based solar cells, but at a lower cost and with easier manufacturing processes [36, 37]. Typically, PSCs are made by depositing a thin layer of perovskite material onto a conductive substrate. When sunlight hits the perovskite layer, it generates electron-hole pairs, where the electrons and holes are separated at the interface between the perovskite layer and the substrate [38]. The separated charges are then collected by different electrodes, creating an electric current.

One of the challenges in the development of perovskite solar cells is their relatively short lifespan and sensitivity to moisture and heat. Researchers are actively working to improve the stability and durability of perovskite materials to make them more suitable for long-term use. Despite these challenges, PSCs hold great promise for the future of solar energy due to their high efficiency potential, low manufacturing costs, and the possibility of flexible and lightweight applications [36, 37]. Ongoing research and development efforts aim to overcome the limitations and make perovskite solar cells a commercially viable and widespread renewable energy technology.

IV. Quantum dot solar cells:

A quantum dot solar cell is a type of solar cell that utilizes quantum dots, which are tiny semiconductor particles, to convert sunlight into electricity. These cells are an emerging technology in the field of photovoltaics and have the potential to improve the efficiency and performance of solar cells. In a traditional solar cell, photons from sunlight create electron-hole pairs in a semiconductor material, generating an electric current. Quantum dot solar cells work on a similar principle but with some unique properties of quantum dots.

Quantum dots have a property called the 'quantum confinement effect', which means that their electronic properties are strongly influenced by their size and composition [39]. By adjusting the size and composition of the quantum dots, it is possible to tune their energy levels and absorption properties. This feature allows quantum dot solar cells to capture a broader range of the solar spectrum, including both visible and infrared light, which traditional solar cells may not efficiently capture. The construction of a quantum dot solar cell typically involves embedding quantum dots into a thin film or layer of a semiconductor material. When photons from sunlight hit the quantum dots, they excite electrons within the dots, creating electron-hole pairs [40]. These charge carriers can then be harvested as an electric current.

Quantum dot solar cells have several advantages over traditional solar cells. They have the potential for higher efficiency due to their ability to capture a wider range of solar wavelengths. Additionally, quantum dots can be synthesized using low-cost materials and solution-based processes, which could lead to cost-effective manufacturing [41, 42].

However, quantum dot solar cells are still in the early stages of development, and there are challenges to overcome. One of the main challenges is improving the stability and lifespan of the quantum dots, as they can degrade over time due to exposure to heat, light, and other environmental factors. Researchers are actively working on improving the durability and performance of quantum dot solar cells to make them commercially viable. Overall, quantum dot solar cells hold promise for enhancing solar energy conversion efficiency and expanding the capabilities of solar technology. Ongoing research and development efforts are focused on optimizing their performance, stability, and scalability for widespread adoption [41, 42].

10.5 EMERGING MATERIALS FOR SOLAR CELLS

In recent years, significant progress has been made in the field of solar cells, especially in the development of emerging materials that can enhance the efficiency, cost-effectiveness, and sustainability of solar energy conversion. It is worth noting that while these emerging materials hold promise for improvements in solar cell technologies, many are still in the research and development phase. Efforts are ongoing to address the challenges related to stability, scalability and commercial viability. Nonetheless, these materials may represent exciting avenues for enhancing the efficiency and sustainability of solar energy conversion in the future. Next, we briefly discuss some of the notable materials that have been reported by researchers over the years.

Lee et al. [43] prepared amorphous InZnO thin film and reported thermally evaporated InZnO thin film can be used as a transparent conducting oxide (TCO) layer to decrease of shunt resistance in the solar cell. In particular, an a-IZO film with Zn content x of 0.27 at 200°C yields excellent electrical and optical properties; m of 29 $cm^2 V^{-1} s^{-1}$ at N of 3.9 × 1020 cm^{-3} and average T above 80% obtained from 400 nm to 1500 nm. The shows better performance with a-IZO than ITO layer.

Li et al. [44] have reported $CsPbBr_3$ film prepared by a vapor-assisted solution technique for Perovskite solar cells (PSCs). The authors described a unique intermediate energy level of MnS used as a hole transport layer (HTL) in a $CsPbBr_3$ PSC, as shown in Figure 10.4. The power conversion efficiency (PCE) of the optimized $CsPbBr_3$ PSC based on all inorganic transport layers is 10.45%, while it is 8.16% for the device without the intermediate layer, which is one of the highest PCEs achieved among $CsPbBr_3$-based PSCs to date. Furthermore, the optimized device maintained 80% PCE of its original efficiency at 85°C at 80% RH over 90 days, demonstrating a strong environmental tolerance promoting the commercial use of low-cost, efficient, and stable all-inorganic PSCs.

Ablekim and co-workers reported 19% reproducible efficiency of CdTe solar cell, using a thermal evaporation technique [45]. In this investigation, Ablekim et al. deposited CdSe and CdTe bilayers on commercially coated SnO_2 glasses, after $CdCl_2$ treatment with and

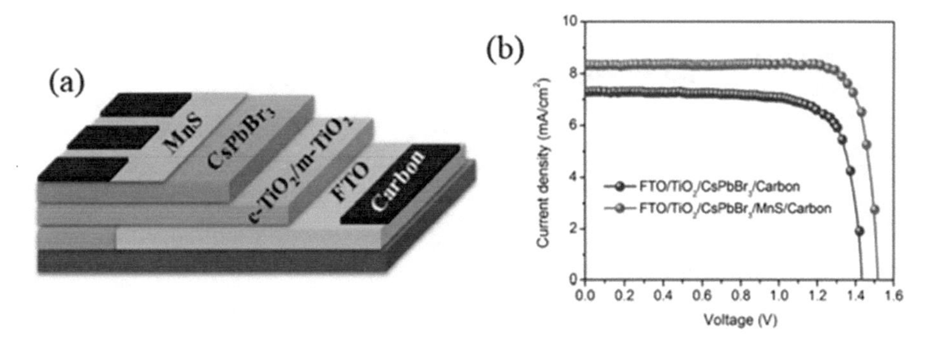

FIGURE 10.4 (a) Schematic diagram of prepared PSC structure, and (b) J-V curve of prepared $CsPbBr_3$ film with and without MnS intermediate layer. (Reproduced with permission from Ref. [44], © 2019, American Chemical Society).

without MgZnO. The characterization results show good absorber structural and electro-optical quality of the prepared solar cell.

Dhapodkar et al. reported a downconversion phosphate-based glass sample doped with three different rare-earth ions (Dy^{3+}, Tb^{3+}, Eu^{3+}) and studied energy transfer behavior, color tunability as well as I-V characteristics [46]. The proposed $Mg_{21}Ca_4Na_4(PO_4)_{18}$:$Dy^{3+}$, Tb^{3+}, Eu^{3+} glass material shows strong absorption in the UV and near-UV region, which can overcome solar cell energy losses in this region and enhance solar cell efficiency. The prepared material was coated on the surface of Si-solar cell and I-V characteristics were analyzed as shown in Figure 10.5. The I-V characteristics results represent a 43.33% solar cell efficiency enhancement as compared to the blank solar cell. The overall study shows proposed material has huge potential for solar cell application. Similar to this work, Parauha et al. [47] reported color tunable $La(PO_4)$:Dy^{3+},Eu^{3+} phosphors and suggested that prepared material may be useful for WLEDs and solar cell application. The authors report energy transfer from Dy^{3+} to Eu^{3+} and I-V characteristics of prepared samples. The highly intense samples were coated on the surface of Si-solar cell and found solar cell efficiency enhanced around 17.34% and 12.76% under direct sunlight and solar simulator, respectively. Similar to these, I-V characteristics of Ce^{3+}, Eu^{2+} doped Na_2CaSiO_4 phosphor show around 16.48 and 20.89 % solar cell efficiency enhancement under solar simulator and direct sunlight, respectively [48]. All these studies suggested that rare-earth doped phosphor is also useful for solar cell application.

Bagavathi et al. [49] reported iron oxide–carbon black (Fe_3O_4–CB) nanocomposite material as an alternative counter electrode in DSSC. The nanocomposite materials were

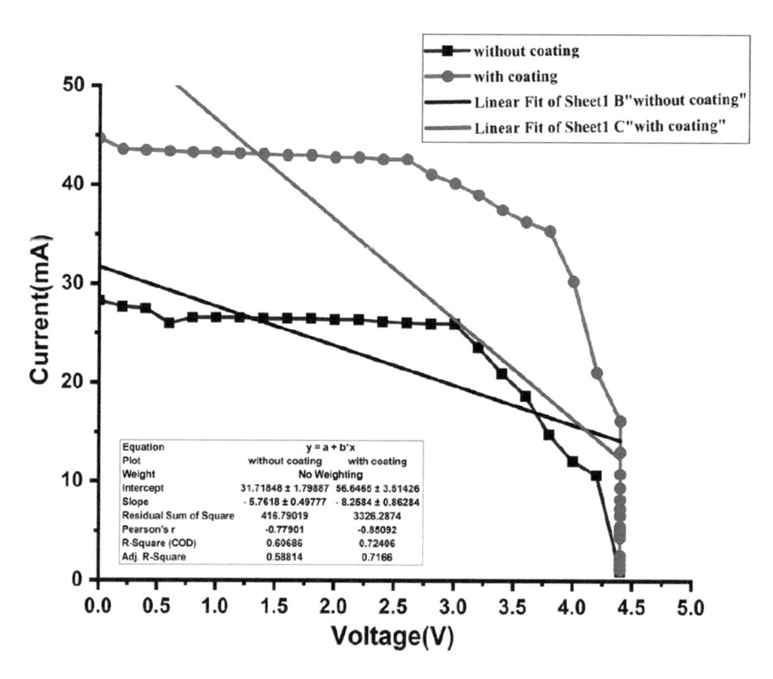

FIGURE 10.5 I-V characteristics of Dy^{3+}-Tb^{3+}-Eu^{3+} triple-doped $Mg_{21}Ca_4Na_4(PO_4)_{18}$ glass sample. (Reproduced with permission from Ref. [46]. © 2022 Elsevier B.V.)

synthesized by a simple solution-mixing process with different ratios of Fe_3O_4–CB. Under illumination of 80 mW cm^{-2}, photovoltaic characterization of the DSSCs revealed that the Fe_3O_4-CB (1:2) nanocomposite had a PCE (6.1%) that was higher than the sputtered Pt CE (4.1%). The overall results suggest that the simple preparation, excellent electrocatalytic activity, and low-cost nature of Fe_3O_4-CB (1:2) nanocomposite CEs indicate them as highly effective, reasonable cost replacements in DSSCs compared to conventional Pt CEs.

Karthick et al. [50] reported aggregated Nb_2O_5 nano-assemblies as anode material for DSSC. In this investigation, chain-like assembly is prepared with Nb_2O_5 particles size around 7 ± 2 nm and diameter 1 ± 0.3 μm. The prepared DNA-doped Nb_2O_5 nano-assemblies have been used in two different energy applications. They are used as anode sample in DSSC and electrochemical supercapacitor studies. The prepared sample shows 0.72% PCE for DSSC.

Singh et al. [51] reported $KFeO_2$ nanoparticles (NPs) prepared by sol-gel method for DSSC. The size of prepared NPs is around 12 nm with a spherical shape. The PCE performance of fabricated DSSC were evaluated and found that the short circuit current–density (J_{sc}) is 0.84 V, open-circuit voltage (V_{oc}) is 4.37 mA/cm^2, fill factor (FF) is 62 and PCE is 2.27 %, respectively. The bandgap of prepared material was 1.88 eV. The obtained results suggested that prepared NPs have potential for DSSC application.

Yan et al. [52] developed a choline chloride monolayer on the surface of SnO_2 as electron transport layer (ETL) and tried to improve the performance of $CH_3NH_3PbI_3$ ($MAPbI_3$) PSCs along with enhancing the high-output voltage. Figure 10.6(a) shows constructed $CH_3NH_3PbI_3$ ($MAPbI_3$) PSCs with a choline chloride monolayer as ETL. The as-prepared

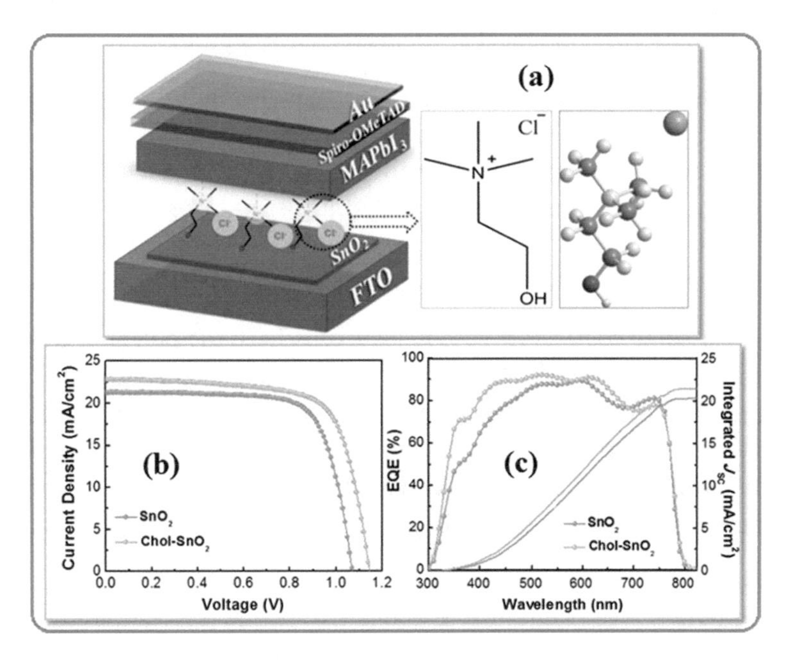

FIGURE 10.6 (a) Fabricated $CH_3NH_3PbI_3$ ($MAPbI_3$) PSCs with choline chloride monolayer as ETL, (b) *J-V* curves, and (c) EQE spectrum of PSC with SnO_2 and Chol-SnO_2 as ETL. (Reproduced with permission from Ref. [52] © 2020, American Chemical Society).

ETL enhances the PCE to 18.90% under a solar illumination, which is much higher than unmodified devices as shown in Figure 10.6(b). More significantly, the MAPbI3 PSC based on Chol-SnO_2 ETL demonstrates a higher open-circuit voltage (V_{oc}) of 1.145 V compared to the control device (1.071 V).

Li et al. [53] studied $CuInSe_2$ and $CuInSe_2$-ZnS material for quantum dots sensitized solar cells (QDSC). As already reported, $CuInS_2$ quantum dots have huge potential for QDSC and are widely serving as light-harvesting sensitizer materials. Inspired by this, Li et al. fabricated $CuInSe_2$-ZnS QD as liquid-junction QDSCs using a linker-molecule-assisted postsynthesis assembly approach. The obtained results from the sandwich $CuInSe_2$-ZnS QD-based champion solar cells exhibit an efficiency of 6.79% (J_{sc} = 22.61 mA/cm^2, V_{oc} = 0.583 V, FF = 0.515) under AM 1.5 G full one sun irradiation.

10.6 RESEARCH ADVANCES AND FUTURE PLANS

Solar cell technology has seen significant advancements in recent years, and there are several ongoing research efforts and future plans to further improve its efficiency, durability, and cost-effectiveness. Let's explore some of these developments in detail:

- **Advanced silicon solar cells**: Silicon-based solar cells dominate the market, and ongoing research focuses on improving their efficiency and reducing production costs. Some key advancements include the development of passivation techniques to reduce surface recombination, the implementation of advanced light-trapping structures, and the integration of silicon nanowires or quantum dots to enhance light absorption. Additionally, bifacial solar cells, which can generate electricity from both sides, are gaining attention for their increased energy yield.
- **Tandem solar cells**: Tandem solar cells, also known as multi-junction solar cells, aim to increase efficiency by stacking multiple layers of different semiconductors with complementary absorption properties. Researchers are investigating various materials and designs for tandem solar cells, including perovskite/silicon, perovskite/copper indium gallium selenide (CIGS), and perovskite/perovskite combinations. The goal is to achieve high efficiency while maintaining long-term stability and cost-effectiveness.
- **Transparent solar cells**: Transparent solar cells have the potential to be integrated into windows and other transparent surfaces, allowing them to generate electricity without obstructing the view. Researchers are developing materials that can absorb ultraviolet and infrared light while maintaining high transparency in the visible spectrum. Additionally, novel device architectures are being investigated to maximize efficiency while balancing transparency and power generation.
- **Perovskite solar cells**: Perovskite solar cells have gained immense attention due to their high-power conversion efficiency potential and low fabrication costs. Researchers are actively working on enhancing their stability and durability, as perovskite materials are sensitive to moisture and light degradation. Strategies such as encapsulation techniques, interface engineering, and material composition optimization are being explored to address these challenges.

- **Organic solar cells**: Organic solar cells utilize organic semiconductors, offering advantages such as flexibility, light weight, and low-cost manufacturing. Researchers are working on enhancing the efficiency and stability of organic solar cells by improving material properties, device architecture, and interface engineering. Tandem structures combining organic and inorganic materials are also being explored to achieve higher efficiency levels.
- **Emerging materials**: Apart from perovskites and organic semiconductors, researchers are exploring various emerging materials with unique properties for solar cells. These include quantum dots, carbon-based materials (such as graphene), and 2D materials (such as molybdenum disulfide). These materials offer advantages like tunable bandgaps, high carrier mobility, and excellent mechanical properties, paving the way for next-generation solar cell technologies.

In terms of future plans, the solar cell industry aims to achieve several milestones:

- **Higher efficiency**: The primary focus is on achieving higher conversion efficiencies for solar cells. The goal is to break the theoretical efficiency limits and develop technologies that can surpass the Shockley-Queisser limit for single-junction solar cells, which is approximately 33.7%.
- **Improved durability**: Enhancing the durability and stability of solar cells is crucial for their long-term performance. Researchers are working on developing robust encapsulation techniques, advanced materials with improved resistance to environmental factors, and degradation-resistant device architectures.
- **Cost reduction**: Making solar energy more cost-effective is essential for its widespread adoption. The industry aims to reduce the overall cost of solar cell production through advanced manufacturing techniques, material optimization, and economies of scale.
- **Integration and smart applications**: Solar cells are increasingly being integrated into various products, such as building materials, wearable devices, and even vehicles. Future plans involve further integration into infrastructure and smart grids, enabling efficient energy generation.

10.7 SUMMARY

In conclusion, solar cells are an important and promising technology with the potential to play a major role in addressing global energy challenges. Here are some key points to summarize their significance:

- **Renewable energy**: Solar cells generate electricity using sunlight, a renewable resource that is abundant and available almost everywhere on the planet. This makes solar energy a sustainable alternative to fossil fuels, reducing greenhouse gas emissions and combating climate change.

- **Clean and environmentally friendly**: Solar cells produce electricity without emitting harmful pollutants or greenhouse gases. They have a minimal impact on the environment compared to traditional energy sources such as coal or natural gas. Solar energy helps to improve air quality and mitigate the negative effects of climate change.
- **Cost savings**: While the initial installation cost of solar panels can be high, solar energy offers significant long-term cost savings. Once installed, solar cells require little maintenance, and sunlight is free, allowing for reduced energy bills and potential savings over the system's lifetime.
- **Energy independence**: Solar cells provide individuals, businesses, and communities with a degree of energy independence. By generating electricity on-site, solar energy reduces dependence on the grid and fossil fuel-based power sources. This resilience is particularly valuable during power outages or in remote areas where grid access is limited.
- **Diverse applications**: Solar cells are versatile and can be used in various applications, from large-scale solar farms to small residential installations and portable devices. Solar power can be integrated into buildings, powering homes, offices, and even electric vehicles, further expanding its potential impact.
- **Technological advancements**: Research and development efforts continue to enhance solar cell efficiency and reduce costs. Advancements in materials, manufacturing processes, and system designs have led to more efficient and affordable solar panels, making solar energy increasingly accessible to a wider range of users.

While solar cells have made significant strides, there are still challenges to address, such as energy storage and intermittency. However, continued investment in research and development, along with supportive policies and incentives, can drive further advancements in solar technology and maximize its potential. Overall, solar cells offer a clean, sustainable, and economically viable solution for meeting our energy needs while reducing environmental impact. With ongoing innovation and adoption, solar energy has the potential to transform the global energy landscape and contribute to a more sustainable future.

REFERENCES

[1] L. Ahmad, N. Khordehgah, J. Malinauskaite, H. Jouhara, Recent advances and applications of solar photovoltaics and thermal technologies, *Energy.* 207 (2020) 118254. https://doi.org/10.1016/j.energy.2020.118254

[2] J. Gong, C. Li, M.R. Wasielewski, Advances in solar energy conversion, *Chem. Soc. Rev.* 48 (2019) 1862–1864. https://doi.org/10.1039/c9cs90020a

[3] S.A.S. Eldin, M.S. Abd-Elhady, H.A. Kandil, Feasibility of solar tracking systems for PV panels in hot and cold regions, *Renew. Energy.* 85 (2016) 228–233. https://doi.org/10.1016/j.renene.2015.06.051

[4] M.K. Hairat, S. Ghosh, 100 GW solar power in India by 2022 – A critical review, *Renew. Sustain. Energy Rev.* 73 (2022) 1041–1050. https://doi.org/10.1016/j.rser.2017.02.012

[5] W.J. Ho, J.J. Liu, B.X. Ke, Characterization of luminescent down - Shifting spectral conversion effects on silicon solar cells with various combinations of Eu - Doped phosphors, *Materials (Basel).* 15 (2022) 452.
[6] T. Kober, H. Schiffer, M. Densing, E. Panos, Global energy perspectives to 2060 – WEC's World Energy Scenarios 2019, *Energy Strateg. Rev.* 31 (2020) 100523. https://doi.org/10.1016/j.esr.2020.100523
[7] C. Battaglia, A. Cuevasb, S. De Wolfc, High-efficiency crystalline silicon solar cells: Status and perspectives, Energy *Strateg. Rev.* 9 (2016) 1552–1576. https://doi.org/10.1039/c5ee03380b
[8] S. Roy, M.S. Baruah, S. Sahu, B.B. Nayak, Materials today: Proceedings computational analysis on the thermal and mechanical properties of thin film solar cells, *Mater. Today Proc.* 44 (2021) 1207–1213. https://doi.org/10.1016/j.matpr.2020.11.241
[9] C.W. Teplin, D.S. Ginley, H.M. Branz, A new approach to thin film crystal silicon on glass: Biaxially-textured silicon on foreign template layers, *J. Non. Cryst. Solids.* 352 (2006) 984–988. https://doi.org/10.1016/j.jnoncrysol.2006.01.024
[10] M.M. Aman, K.H. Solangi, M.S. Hossain, A. Badarudin, G.B. Jasmon, H. Mokhlis, A.H.A. Bakar, S.N. Kazi, A review of Safety, Health and Environmental (SHE) issues of solar energy system, *Renew. Sustain. Energy Rev.* 41 (2015) 1190–1204. https://doi.org/10.1016/j.rser.2014.08.086
[11] T. Tsoutsos, N. Frantzeskaki, V. Gekas, Environmental impacts from the solar energy technologies, *Energy Policy.* 33 (2005) 289–296. https://doi.org/10.1016/S0301-4215(03)00241-6
[12] M.H. Alaaeddin, S.M. Sapuan, M.Y.M. Zuhri, E.S. Zainudin, F.M.A.- Oqla, Photovoltaic applications: Status and manufacturing prospects, Renew. *Sustain. Energy Rev.* 102 (2019) 318–332. https://doi.org/10.1016/j.rser.2018.12.026
[13] M.H.A. Khatibi, Ali, Fatemeh Razi Astaraei, Generation and combination of the solar cells: A current model review, *Energy Sci. Eng.* 7 (2019) 305–322. https://doi.org/10.1002/ese3.292
[14] Q. Su, G. Zhang, J. Lai, S. Feng, W. Shi, Green solar electric vehicle changing the future lifestyle of human, *World Electr. Veh.* 4 (2010) 128–132. https://doi.org/10.3390/wevj4010128
[15] J. Day, S. Senthilarasu, T.K. Mallick, Improving spectral modi fi cation for applications in solar cells: A review, *Renew. Energy.* 132 (2019) 186–205. https://doi.org/10.1016/j.renene.2018.07.101
[16] Z.L.W. Shiquan Lin, The tribovoltaic effect, *Materialstoday.* 62 (2023) 111–128. https://doi.org/10.1016/j.mattod.2022.11.005
[17] F. Wang, X. Liu, F. Gao, Chapter 1 – Fundamentals of solar cells and light-emitting diodes in: *Advanced Nanomaterials for Solar Cells and Light Emitting Diodes*, Elsevier Inc., 2019. 1–35. https://doi.org/10.1016/B978-0-12-813647-8.00001-1
[18] M.Z.F. Khaleda, B. Vengadaesvaran, N.A. Rahim, Chapter 18 - Spectral response and quantum efficiency evaluation of solar cells: A review, in: *Energy materials fundamentals to applications.* Elsevier, (2021) 525–566 https://doi.org/10.1016/B978-0-12-823710-6.00014-5
[19] K.S. Ahmad, S.N. Naqvi, S.B. Jaffri, Systematic review elucidating the generations and classifications of solar cells contributing towards environmental sustainability integration, *Rev. Inorg. Chem.* 41 (2021) 21–39.
[20] A. Goetzberger, C. Hebling, Photovoltaic materials, past, present, future, *Sol. Energy Mater. Sol. Cells.* 62 (2000) 1–19.
[21] Y. Wang, J. Ye, Review and development of crystalline silicon solar cell with intelligent materials, *Adv. Mater. Res.* 321 (2011) 196–199. https://doi.org/10.4028/www.scientific.net/AMR.321.196
[22] H. Liu, Y. Du, X. Yin, M. Bai, W. Liu, Micro / nanostructures for light trapping in monocrystalline silicon solar cells, *J. Nanomater.* 2022 (2022) 8139174. https://doi.org/10.1155/2022/8139174
[23] S.A. Kalogirou, Chapter – 9 Photovoltaic Systems. in: *Solar Energy Engineering*, Elsevier, (2009) 469–519. https://doi.org/10.1016/B978-0-12-374501-9.00009-1
[24] N. Asim, K. Sopian, S. Ahmadi, K. Saeedfar, M.A. Alghoul, O. Saadatian, S.H. Zaidi, A review on the role of materials science in solar cells, *Renew. Sustain. Energy Rev.* 16 (2012) 5834–5847. https://doi.org/10.1016/j.rser.2012.06.004

[25] S.P. Philipps, F. Dimroth, A.W. Bett, Chapter 1-4-B High efficiency III V multijunction solar cells in: *McEvoy's Handbook of Photovoltaics*, Elsevier Ltd, (2018) 439–472. https://doi.org/10.1016/B978-0-12-809921-6.00012-4
[26] V. Avrutin, N. Izyumskaya, H. Morkoç, Superlattices and microstructures semiconductor solar cells : Recent progress in terrestrial applications, *Superlattices Microstruct.* 49 (2011) 337–364. https://doi.org/10.1016/j.spmi.2010.12.011
[27] S. Issue, Thin-film solar cells: An overview, *Prog. Photovolt Res. Appl.* 92 (2004) 69–92. https://doi.org/10.1002/pip.541
[28] A. Shah, J. Meier, A. Buechel, U. Kroll, J. Steinhauser, F. Meillaud, Towards very low-cost mass production of thin-film silicon photovoltaic (PV) solar modules on glass, *Thin Solid Films.* 502 (2006) 292 – 299. https://doi.org/10.1016/j.tsf.2005.07.299
[29] M. Li, F. Igbari, Z. Wang, L. Liao, Indoor thin-film photovoltaics: Progress and challenges, *Adv. Energy Mater.* 2000641 (2020) 1–25. https://doi.org/10.1002/aenm.202000641
[30] M.A. Scarpulla, B. Mccandless, A.B. Phillips, Y. Yan, M.J. Heben, C. Wolden, G. Xiong, W.K. Metzger, D. Mao, D. Krasikov, I. Sankin, S. Grover, A. Munshi, W. Sampath, J.R. Sites, A. Bothwell, D. Albin, M.O. Reese, A. Romeo, M. Nardone, R. Klie, J.M. Walls, T. Fiducia, A. Abbas, S.M. Hayes, Solar energy materials and solar cells CdTe-based thin film photovoltaics: Recent advances, current challenges and future prospects, *Sol. Energy Mater. Sol. Cells.* 255 (2023) 112289. https://doi.org/10.1016/j.solmat.2023.112289
[31] H. Yao, J. Wang, Y. Xu, S. Zhang, J. Hou, Recent progress in chlorinated organic photovoltaic materials, *Acc. Chem. Res.* 53 (2020) 822–832. https://doi.org/10.1021/acs.accounts.0c00009
[32] Yong Cui, Ling Hong, and Jianhui Hou, Organic photovoltaic cells for indoor applications: opportunities and challenges, *ACS Appl. Mater. Interfaces.* 12 (2020) 38815–38828. https://doi.org/10.1021/acsami.0c10444
[33] A. Agrawal, S.A. Siddiqui, A. Soni, G.D. Sharma, Advancements, frontiers and analysis of metal oxide semiconductor, dye, electrolyte and counter electrode of dye sensitized solar cell, *Sol. Energy.* 233 (2022) 378–407. https://doi.org/10.1016/j.solener.2022.01.027
[34] T. Mudiyanselage, W. Jayalath, B. Jayamaha, M. Chandi, H. Federico, A review of textile dye - Sensitized solar cells for wearable electronics, *Ionics (Kiel).* (2022) 2563–2583. https://doi.org/10.1007/s11581-022-04582-8
[35] N. Bennett, B. Chen, H. Upadhayaya, K. Raghava, D.K. Kumar, J. Kr, V. Sadhu, Functionalized metal oxide nanoparticles for efficient dye-sensitized solar cells (DSSCs): A review ˇíz, *Mater. Sci. Energy Technol.* 3 (2020) 472–481. https://doi.org/10.1016/j.mset.2020.03.003
[36] M. Shahbazi, H. Wang, Progress in research on the stability of organometal perovskite solar cells, *Sol. Energy.* 123 (2016) 74–87. https://doi.org/10.1016/j.solener.2015.11.008
[37] N.H. Tiep, Z. Ku, H.J. Fan, Recent Advances in Improving the stability of perovskite solar cells, *Adv. Energy Mater.* 6 (2016) 1501420. https://doi.org/10.1002/aenm.201501420
[38] P. Basumatary, P. Agarwal, A short review on progress in perovskite solar cells, *Mater. Res. Bull.* 149 (2022) 111700. https://doi.org/10.1016/j.materresbull.2021.111700
[39] R.S. Sudar, D. Pukazhselvan, C.K. Mahadevan, Studies on the synthesis of cubic ZnS quantum dots, capping and optical – electrical characteristics, *J. Alloys Compd.* 517 (2012) 139–148. https://doi.org/10.1016/j.jallcom.2011.12.060
[40] J. Yuan, A. Hazarika, Q. Zhao, X. Ling, T. Moot, W. Ma, J.M. Luther, Review metal halide perovskites in quantum dot solar cells: Progress and prospects, *Joule.* 4 (2020) 1160–1185. https://doi.org/10.1016/j.joule.2020.04.006
[41] I.J. Kramer, E.H. Sargent, The architecture of colloidal quantum dot solar cells: Materials to devices, *Chem. Rev.* 114 (2013) 863–882. https://doi.org/10.1021/cr400299t
[42] H.K. Jun, M.A. Careem, A.K. Arof, Quantum dot-sensitized solar cells — Perspective and recent developments: A review of Cd chalcogenide quantum dots as sensitizers, *Renew. Sustain. Energy Rev.* 22 (2013) 148–167. https://doi.org/10.1016/j.rser.2013.01.030

[43] W. Lee, D. Cho, Y. Do, M. Choi, J. Chul, Y. Chung, Thermally evaporated amorphous InZnO thin fi lm applicable to transparent conducting oxide for solar cells, *J. Alloys Compd.* 806 (2019) 976–982. https://doi.org/10.1016/j.jallcom.2019.07.321

[44] X. Li, Y. Tan, H. Lai, S. Li, Y. Chen, S. Li, P. Xu, J. Yang, Evaporation-assisted deposition and setting intermediate energy levels efficiency by evaporation-assisted deposition and setting, *ACS Appl. Mater. Interfaces.* 11 (2019) 29746–29752. https://doi.org/10.1021/acsami.9b06356

[45] T. Ablekim, J.N. Duenow, X. Zheng, H. Moutinho, J. Moseley, C.L. Perkins, W. Johnston, P.O. Keefe, E. Colegrove, D.S. Albin, M.O. Reese, W.K. Metzger, Thin film solar cells with 19 % efficiency by thermal evaporation of CdSe and CdTe, *ACS Energy Lett.* 5 (2020) 892–896. https://doi.org/10.1021/acsenergylett.9b02836

[46] T.S. Dhapodkar, A.R. Kadam, N. Brahme, S.J. Dhoble, Efficient white light-emitting $Mg_{21}Ca_4Na_4(PO_4)_{18}$: Dy^{3+}, Tb^{3+}, Eu^{3+} triple-doped glasses: A multipurpose glasses for WLEDs, solar cell efficiency enhancement, and smart windows applications, *Mater. Today Chem.* 24 (2022) 100938. https://doi.org/10.1016/j.mtchem.2022.100938

[47] Y.R. Parauha, S.J. Dhoble, Color–tunable luminescence, energy transfer behavior and I-V characteristics of Dy^{3+}/Eu^{3+} co–doped $La(PO_4)$ phosphors for WLEDs and solar applications, *New J. Chem.* 46 (2022) 6230–6243. https://doi.org/10.1039/d2nj00232a

[48] Y.R. Parauha, S.J. Dhoble, Enhancement of photoluminescence and tunable properties for Ce^{3+}, Eu^{2+} activated Na_2CaSiO_4 downconversion phosphor: A novel approach towards spectral conversion, *J. Lumin.* 251 (2022) 119173. https://doi.org/10.1016/j.jlumin.2022.119173

[49] M. Bagavathi, A. Ramar, R. Saraswathi, Fe_3O_4–carbon black nanocomposite as a highly efficient counter electrode material for dye-sensitized solar cell, *Ceram. Int.* 42 (2016) 13190–13198. https://doi.org/10.1016/j.ceramint.2016.05.111

[50] K. Karthick, U. Nithiyanantham, S.R. Ede, S. Kundu, DNA aided formation of aggregated Nb_2O_5 nano-assemblies as anode material for dye sensitized solar cell (DSSC) and supercapacitor applications DNA aided formation of aggregated Nb_2O_5 nano-assemblies as anode material for dye sensitized solar cell (D), *ACS Sustain. Chem. Eng.* 4 (2016) 3174–3188. https://doi.org/10.1021/acssuschemeng.6b00200

[51] G. Singh, T.P. Kaur, A.K. Tangra, Novel $KFeO_2$ nanoparticles for dye-sensitized solar cell, *Mater. Res. Express.* 6 (2019) 1150f5.

[52] Jingjing Yan, Zhichao Lin, Qingbin Cai, Xiaoning Wen, and Cheng Mu, Choline chloride-modified SnO 2 achieving high output voltage in $MAPbI_3$ perovskite solar cells, *ACS Appl. Energy Mater.* 3 (2020) 3504–3511. https://doi.org/10.1021/acsaem.0c00038

[53] Wenjie Li, Zhenxiao Pan, and Xinhua Zhong, $CuInSe_2$ and $CuInSe_2$-ZnS based high efficiency "green" quantum dot sensitized solar cells, *J. Mater. Chem. A.* 3 (2015) 1649–1655. https://doi.org/10.1039/C4TA05134C

IV

Sensors and Detectors

CHAPTER 11

Energy-Saving Materials for Self-Powered Photodetection

Nupur Saxena

Indian Institute of Technology, Jammu, India

Tania Kalsi

Central University of Jammu, Jammu, India

Ashok Bera

Indian Institute of Technology, Jammu, India

Pragati Kumar

Central University of Jammu, Jammu, India

11.1 INTRODUCTION

Photodetectors (PDs) are devices that detect and convert optical signals into electrical ones. Hence, they find tremendous applications in space, defense, household electrical appliances, energy sectors, electronic devices, and environmental monitoring [1]. PDs with spectral as well as broadband response, faster speed, improved sensitivity, and stability are the requirement of the hour [2]. Simultaneously, the growing energy requirement in the world is to be limited as much as possible. Numerous self-powered devices and sensors have been developed so far in different areas like temperature sensors, gas sensors [3], pressure sensors, solar cells [4], light-emitting diodes (LEDs) as well as health-monitoring sensors [5], among others.

Self-powered devices based on different naturally occurring phenomena in various materials are a way to save energy that can be used for other purposes. Particularly, the requirement of external bias for the operation of fast PDs with high photosensitivity limits their uses in different environments and areas [6]. Hence, self-powered PDs (SPPDs) are an essential step towards energy saving and harvesting for future generations and the extension of applications of PDs in multiple fields [7]. In this chapter, an introduction to types of PDs, their performance parameters, and photo-sensing mechanism are given, followed by a discussion on various effects that serve as the basis for energy saving. A detailed

DOI: 10.1201/9781003315261-15

description of different materials used for SPPDs will be presented in the next section, before concluding the chapter.

11.1.1 Types of Photodetectors

Conventionally, PDs are classified in various ways depending upon their architecture, geometry, mechanism, detection of spectrum and so forth. A detailed classification of PDs based on their basic structure and working principles is presented in [8, 9]. However, in this chapter we will discuss only the types of SPPDs.

Designing an SPPD that can work effortlessly and effectively is a dedicated research problem for next-generation nanodevices and has attracted enormous interest worldwide. An ideal SPPD works without any external bias and meets the expectations of miniaturization, weight, and speed. In this context, it is imperative to mention here that there are various ways to design an SPPD based on the mechanism as well as the architecture. These mechanisms will be discussed in a later subsection. Broadly, SPPDs are categorized into two groups on the basis of their methods of energy conversion; one exploits the photovoltaic effect (built-in potential (BP)), and the other utilizes the energy-harvesting unit integrated with photoconductive devices [10]. Commonly, the figure of merits (FOMs) of photoconductive devices are better than that of the photovoltaic-type devices due to integrated energy system [10]. Further, each category of SPPDs is subdivided according to the nature of the junction and integrated energy source, respectively, as will be discussed in the next section along with their mechanisms.

11.1.2 Performance Parameters

The performance of a PD is evaluated by a number of parameters that include photosensitivity (S) [11–13], photoresponsivity (R) [13–15], external quantum efficiency (EQE) [11, 13–15], specific detectivity (D^*) [11, 14, 16], gain (G) [14, 16], response time/speed (R_λ), and noise equivalent power (NEP) [14, 16] which are mathematically expressed by Equations (11.1)–(11.6):

$$S = \frac{I_{\text{Ph}} - I_d}{I_d} \tag{11.1}$$

$$R_\lambda = \frac{I_{\text{Ph}} - I_d}{P_{\text{in}}}\left(AW^{-1}\right) \tag{11.2}$$

$$\text{EQE}\left(\eta_{\text{ex}}\right) = R_\lambda \frac{hc}{e\lambda} = R_\lambda \frac{1.24}{\lambda(\text{nm})} \times 10^3 \tag{11.3}$$

$$D^* = R_\lambda \sqrt{\frac{A}{2qI_d}}\left(\text{Jones or cmHz}^{1/2}\text{W}^{-1}\right) \tag{11.4}$$

$$\text{NEP} = \frac{\sqrt{A\Delta f}}{D^*} = \frac{I_N}{R_\lambda} \tag{11.5}$$

$$G = \frac{\tau_l}{\tau_t} = \frac{\tau_l}{L^2}\mu V_{DS} \tag{11.6}$$

where I_d and I_{ph} are the dark and photocurrents, P_{in} and A indicate the light intensity of the source and effective area of the device, Δf and I_N are electrical bandwidth and noise current respectively, τ_l is lifetime of trapped carrier say hole, τ_t is transit time of the other carrier say electron, L is the channel length, μ is the carrier mobility, and V_{DS} is the applied bias voltage.

Alternative critical performance parameter 'f_c' is defined as in [11–14, 16] and can be estimated as [15];

$$f_c = \frac{1}{2\pi\tau} \tag{11.7}$$

In addition to the above-mentioned characteristic parameters, the open-circuit voltage (V_{oc}) and short-circuit current (I_{sc}) are turned out as new figures of merit for SPPDs, especially in the detection of low light signals (<100 μWcm^{-2}) [17]. Typically, V_{oc} is defined as the voltage developed by a potential difference across a very large load resistor when the diode is illuminated, that is, it is the maximum voltage available from a potential difference when PD is drawing zero current. Conversely, I_{sc} is the I_{ph} generated by a PD into a short circuit, that is, it is the maximum I_{ph} flowing through the PD at zero bias voltage. I_{sc} depends strongly on the intensity and spectral distribution of the light source.

11.1.3 Photo-Sensing and Self-Powering Mechanism(s)

In 1887, the experimental demonstration of the photoelectric effect by Heinrich Hertz, and the theoretical explanation of the said effect in 1905 by Albert Einstein led to the development of a light-detection mechanism. More than two decades later, in 1930, the commercial product of photon detection, that is, the photomultiplier tube (PMT) was invented.

In the photogeneration process, the generation of electron–hole pairs (EHPs) in the active region of semiconducting material takes place via absorption of incident photons. Further, these electrons and holes are collected as photocurrent at the electrodes. The BP across the depletion region or space charge layer (SCL), or active region of a homo/heterojunction PD, plays a critical role in the operation of PDs. Photodetectors can be operated either with or without external power supply. When a PD is reverse biased (connected to an external power supply), reverse biasing encourages and accelerates the photogenerated charge carriers, which enhances the photocurrent and augments the performance of the device. However, photocurrent in the PD circuit may be produced via BP-assisted splitting, and drifting of the photogenerated EHPs can take place without adding an external power supply. The former is the mechanism of conventional PDs, whereas the latter is the response mechanism of one kind of SPPD that utilizes the photovoltaic effect (PE) [18, 19]. The detected signals themselves feedback the power to the PDs for self-operation in these types of SPPDs. Thus, the fabricated PDs must have the ability to detect the signals and produce measurable electric energy from these detected signals simultaneously [20]. Further, SPPDs

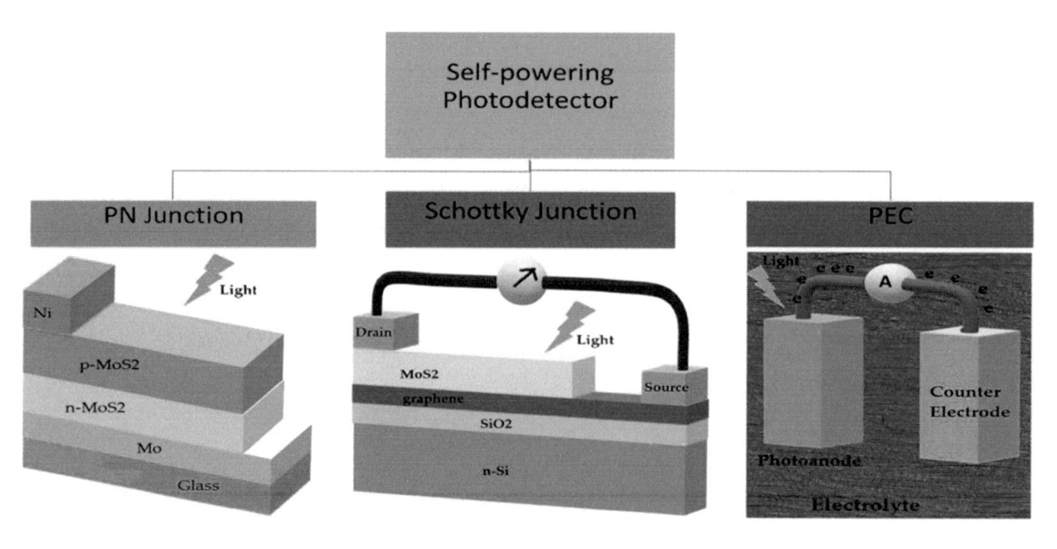

FIGURE 11.1 Types of SPPDs utilizing the photovoltaic effect (PE). (Reproduced from Ref. [18], © U. Sundararaju et al. under the Creative Commons license.)

based on the PE can be categorized into three types based on the charge separation features of the interface. These are: (1) Schottky junction, (2) p–n junction, and (3) photoelectrochemical (PEC) cell. The first two types of PDs exploit the BP to separate the photogenerated carriers, whereas charge separation for the PEC type PDs depends on the energy barrier between the electrode materials and electrolyte. The device schematics for the aforementioned SPPDs are shown in Figure 11.1.

Another kind of SPPD is an integrated self-powered nanosystem. Commonly, a nanosystem for SPPD contains three components: (1) a photo sensor, (2) a unit for power supply, and (3) a system for electrical measurement. The power unit could either be an energy-harvesting unit (a nanogenerator or solar cell) or a storage unit (battery or supercapacitor) [19–22]. The energy-harvesting unit of such systems collects energy from nature and drives the light sensor to achieve detection. Principally, an integrated power unit in this type of SPPD is essential for the generation of the driving force that restrains the recombination of the photogenerated EHPs [23, 24]. Indeed, integration of a power source offers a potential difference (PD) that controls the directional movement of the photogenerated electron and hole, and thus separates the photogenerated EHPs in the photoresponsing nanosystem.

Usually, nanogenerator (NG) technology employs three distinctive approaches named piezoelectric, triboelectric, and thermoelectric or pyroelectric [20, 25–28]. Piezoelectric nanogenerators (PENGs) employ piezoelectric materials to convert random mechanical energy into electric energy. An external strain generates piezoelectric potential that drives electrons in the external load [29, 30]. Triboelectric NGs (TENGs) are different from the PENG in design, and a very common phenomenon (i.e. electrostatic charge) of daily life is associated. The generation of electricity from electrostatic charges/potential via tribological process is a brilliant idea. Primarily, the triboelectric effect is the process of electrification in contact via mechanical deformation. Under the application of an external mechanical deformation, the two stacked insulating materials are touched and rubbed with each other to generate triboelectric charges of opposite signs over the internal surfaces of the two materials [25]. Further, the release of deformation results in separation of the opposite triboelectric charges with an air gap and formation of a dipole moment that establishes an electric potential difference between the two planar electrodes [25]. To compensate for this potential difference and achieve equilibrium, electrons flow from lower to higher potential, which leads to the accumulation of electrostatically induced charges on the electrodes. The reapplication of deformation, the dipole moment, either disappears or is decreased in magnitude so that the two materials remain in contact. As a result, the flow of electrons in the electrode is reversed due to a reduction in electric potential difference which results in the vanishing of the accumulated induced charges. The electrons can flow through the external load in an alternating manner by the repetition of the above processes. The TENGs have advantages over the others because of their capability of energy harvesting from human activities, such as rotating tires, ocean waves, or mechanical vibrations. Figure 11.2(a) and (b) demonstrate the integration of ZnO PENG and test circuit diagram SPPD system using paper-based TENG energy source respectively. On the other hand, the Seebeck effect, that is, a temperature difference across a device, leads to the diffusion of charge carriers in conventional thermoelectric generators (THENGs) as shown in Figure 11.2(d) [27, 31–33]. In spite of the necessity of a temperature difference for the

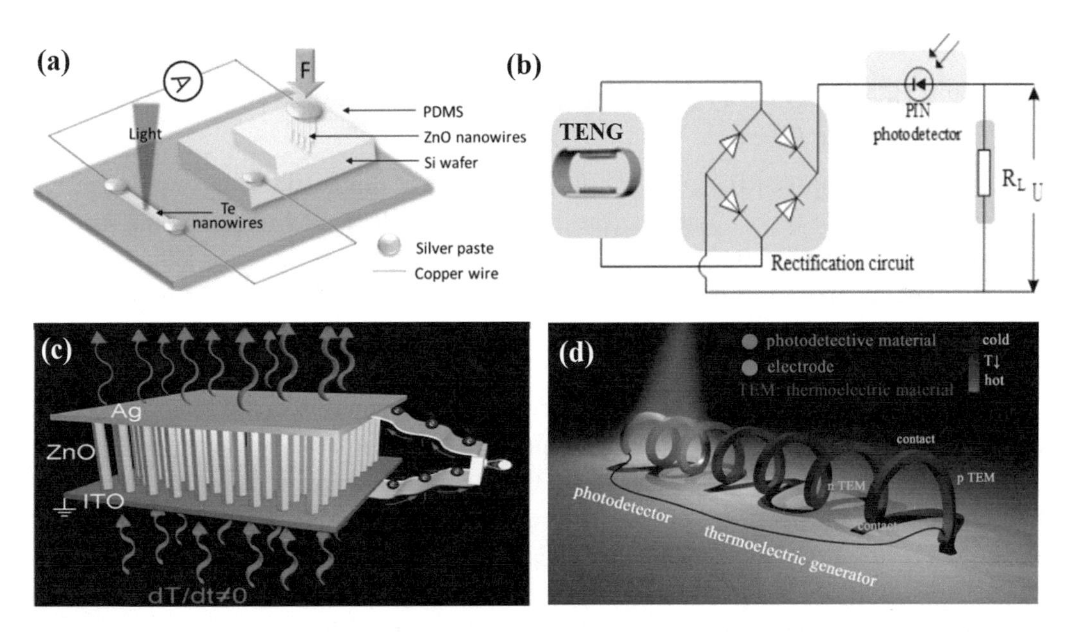

FIGURE 11.2 (a) The design scheme of a self-powered nanosystem with ZnO nanowires PENG as the power source. (Reproduced with permission from Ref. [34], © 2021 Elsevier.) (b) Test circuit diagram SPPD system using TENG energy source. (Reproduced with permission from Ref. [35], © 2020 Elsevier.) (c) Schematic diagram showing the structure of the PNG. (Reproduced with permission from Ref. [27], © 2012 American Chemical Society.) (d) Design principle of the self-adjustable and self-powered integral device: a photodetector and THENG (where color difference represents temperature difference) are connected in series on a three-dimensional helical polymer substrate. (Reproduced with permission from Ref. [36], © 2019 Royal Chemical Society.)

operation of such devices, they cannot work in the spatially uniform environmental temperature and under a time-dependent temperature fluctuation. In such a situation, the pyroelectric effect can be the choice, which is due to spontaneous polarization in certain anisotropic solids owing to a transient temperature gradient. The core of the pyroelectric nanogenerator (PNG) is to utilize the polarization electric field and charge separation created along the nanowire as a result of the time-dependent change in temperature, as illustrated in Figure 11.2(c).

The operation and mechanism of a solar cell (an electrical device that transforms solar energy directly into electricity) are now well known. In addition to being a renewable and clean energy source, a solar cell can be integrated into nanosystems to self-drive them. Akin to conventional externally powered PDs, integrated systems also exhibit light detection ability. Indeed, solar energy is intermittent which results in variances in time and space, and, therefore, the interruption in the harvesting of energy leads to unstable output power in solar cells and NGs. Alternatively, energy-storage devices like lithium-ion batteries and supercapacitors can deliver a steady and durable output [37, 38] and thereby can be integrated with light sensors to form SPPDs. In contrast, microbial fuel cells (MFCs) that convert chemical energy into electricity have a number of advantages like more rapid recharging and much higher energy storage density over conventional batteries [20]. Thus, it is the MFCs of sizes in the nanometer range that can be integrated with a conventional PD to build a high-performance SPPD.

11.2 MULTIFARIOUS EFFECTS AS ENERGY-SAVING BOOSTERS

11.2.1 Piezoelectric Effect

The crystals with non-centrosymmetry show induction of dipole moments or polarization of charges under the application of stress that develop a piezopotential along the stressed direction in the micro/nanostructures at the heterojunction or interface [39]. Typically, non-mobile and non-annihilated ionic charges generate a strain-induced inner-crystal field, which is termed as piezopotential. Thus, piezo-polarization of charges and piezopotential persist until the removal/disappearance of strain. Piezopotential can support electrons' flow in the external load in case of mechanical deformation of the materials; this is the basic principle of the NGs [21, 40]. Piezopotential can also modify the band alignment of the heterojunction effectively and control the separation, transport, and/or recombination of charge in optoelectronic phenomena, and serve as the basis of the piezo-phototronic effect (PPE) [20, 41]. As shown in Figure 11.3, investigations on the distribution of piezopotential based on two typical configurations (transverse and axial) of nanowire (NW) devices suggest that transverse deflection of NWs have utility in energy-harvesting applications (NGs) [42], while the axially strained NWs have efficacy in piezotronic applications on flexible substrates [43]. Besides, theoretical studies have predicted the generation of a strong piezoelectric field induced inside the NW heterostructures due to lattice mismatch [44, 45]. The electric potential in laterally bent NW is usually independent of vertical height except for the regions very close to the fixed end of the NW, whereas the electrostatic potential is a function of the aspect ratio of the NW in place of its absolute dimension [41]. The induced piezopotential is proportional to the maximum deflection at the NW tip

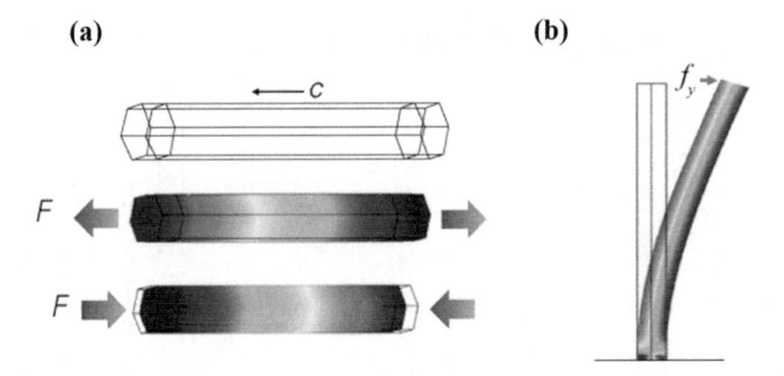

FIGURE 11.3 (a) The distribution of piezopotential along a ZnO NW under axial strain calculated by numerical methods and (b) the distribution of piezopotential along a transversely deflected ZnO NW calculated by numerical methods. The color gradient represents the distribution of piezopotential in which red indicates positive piezopotential and blue indicates negative piezopotential. The growth direction of the NW is along the c-axis. (Reproduced with permission from Ref. [41], © 2014 Oxford Academic.)

for an NW with a fixed aspect ratio. Conversely, upon axial straining, the piezopotential drops, and electron energy increases continuously from one end to the other end of the NW. Meanwhile, the Fermi level remains flat all over the NW at equilibrium. Accordingly, the electron energy barrier between NW and metal electrodes is raised and lowered at two ends respectively, which leads to asymmetry in the *I-V* characteristics of the NW device. It serves as the fundamental of piezotronics and piezo-phototronics [46].

11.2.2 Pyroelectric Effect

The change in dipole moment per unit volume (P_s) of polar or anisotropic materials due to a thermal fluctuation is referred as pyroelectric effect. A temporal temperature gradient is essential for pyroelectic effect instead of the spatial temperature gradient that is needed for thermoelectric effect. Typically, pyroelectric materials are a subclass of piezoelectric materials, and, hence, this class of materials must retain a unique polar axis, which may or may not be switched by an external electric field, in addition to a non-centrosymmetric crystal structure. Every unit cell of the pyroelectric material has its own dipole moment and thereby shows spontaneous polarization [39]. Generally, heating or cooling induces pyroelectric polarization charges (pyrocharges) at two polar ends and generates an instantaneous potential in pyroelectric materials.

Under the exposure of light illuminations, a rapid rise in temperature is naturally induced within pyroelectric nanomaterials (preferably 1D), resulting in a distribution of pyroelectric potentials throughout the crystal with polarization pyrocharges presented at both ends of the NW. Thus, pyroelectric polarizations manipulate the charge–carrier transport process across the interface during device operation [7, 47, 48]. Indeed, this pyroelectric effect is coupled with the PE of conventional PDs and results in two kinds of transient short-circuit current curves upon light illumination [7]. Just after illumination, a sharp rise in current occurs due to the coupling of pyroelectric effect with the PE. This part of current curve can be labelled as photovoltaic–pyroelectric current I_1 on *I-t* graph (Figure 11.4(a)).

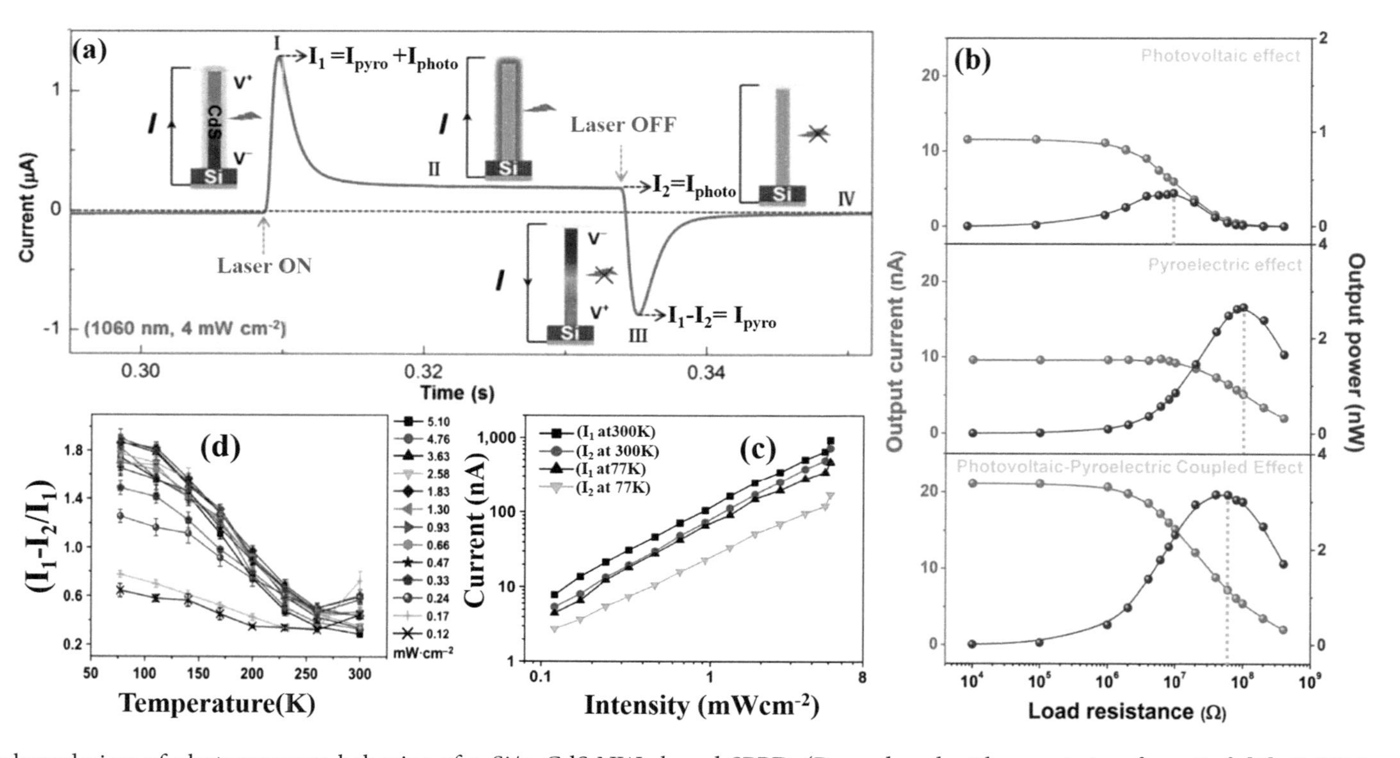

FIGURE 11.4 (a) Enlarged view of photoresponse behavior of p-Si/n-CdS NWs based SPPD. (Reproduced with permission from Ref. [7], © 2018 WILEY-VCH.) (b) Output current and output power of ITO/BTO/Ag device regarding PE, pyroelectric effect, and photovoltaic–pyroelectric coupled effect as a function of loading resistance. (Reproduced with permission from Ref. [47], © 2017 WILEY-VCH.) (c) Photocurrent responses to the pyro-phototronic and PE as a function of UV illumination intensity at 77 and 300 K. (d) variation of output-signal enhancement caused by the pyro-phototronic effect with temperature under different UV light intensities. (Reproduced with permission from Ref. [48], © 2016 Tsinghua University Press and Springer-Verlag Berlin Heidelberg.)

The pyroelectric current is gradually reduced and disappears very shortly due to stabilization of temperature induced by the light. However, the short-circuit current (photovoltaic current I_2) remains constant until turning off the light. As soon as the light is turned off, the current attains a peak in negative direction immediately due to the pyroelectric effect that results from the temporal change in temperature [47]. The pyroelectric current gradually reduces again and stabilizes around zero, which suggests that the peak signal is due solely to the pyroelectric effect. The time of surface temperature change rate dT/dt from peak value to near zero can be defined as the decay time of pyroelectric current. The extended view of photoresponse behavior of the flexible SPPD based on p-Si/n-CdS nanowires under 1060 nm laser illumination with the power density of 4 mW cm^{-2} under zero bias is shown in Figure 11.4(a). The insets of Figure 11.4(a) depict schematics of the basic mechanism of the pyro-phototronic effect corresponding to various steps, representing the respective schematic circuit with pyroelectric potential and light-induced heating effect in CdS NW. Figure 11.4(b) demonstrates the maximum output current and power recorded at matched loading resistance under 405 nm illumination (25.4 mW cm^{-2}) in ITO/BTO/Ag device for the pyroelectric effect (I_{pyro}), photovoltaic effect (I_{photo}), and photovoltaic–pyroelectric ($I_{pyro + photo}$) coupled effect [47]. Figure 11.4(c) displays the enhancement in the output currents induced by the coupling of the pyroelectric effect with the photovoltaic effect as a function of UV illumination intensity at 77 and 300 K for the ZnO/perovskite heterostructured (ZPH) SPPDs, while the ratio of pyrocurrent to the photovoltaic current as a function of temperature at different intensities of UV illumination are given in Figure 11.4(d). It is obvious that both $I_{pyro + photo}$ and I_{photo} are proportional with the light intensity at 77 and 300 K, and the light self-induced temperature changes within ZnO increase at most temperatures (77–260 K) with the increase in the UV light intensity leading to enhanced I_{pyro} and $I_{pyro+photo}$ [48].

11.2.3 Triboelectric Effect

Akin to piezo and pyroelectric effects, the triboelectric effect may also be exploited for self-powered devices. Here, some special design is used for the fabrication of self-powering (SP) devices. In particular, an SPPD comprises two separate parts, as demonstrated in the schematic of the device construction (Figure 11.5(a)). The bottom part may be constructed on a transparent substrate like ITO or fluorine-doped tin oxide (FTO), by deposition of a compact hole-blocking layer (HBL), an electron transport layer (ETL) of organic/inorganic materials, and a light-absorption layer (LAL) of suitable material. Subsequently, two pieces of conducting polymer films with desired dimensions are synthesized and connected to the substrates at opposite edges. They provide a gap between the two contact surfaces by supporting them. In contrast, the upper part consists of a metal layer (ML) deposited on an elastic buffer layer to increase the effective contact area between the ML and the LAL. The ML and the conducting oxide layer on the substrate act as the top and the bottom electrode, respectively, that may be coupled with the positive and the negative terminals of an electrometer for data acquisition [49, 50]. Typically, this category of SPPD is essentially a TENG that can convert mechanical energy into electrical energy. The periodic and repeated contact of the top copper electrode to the perovskite layer generates opposite

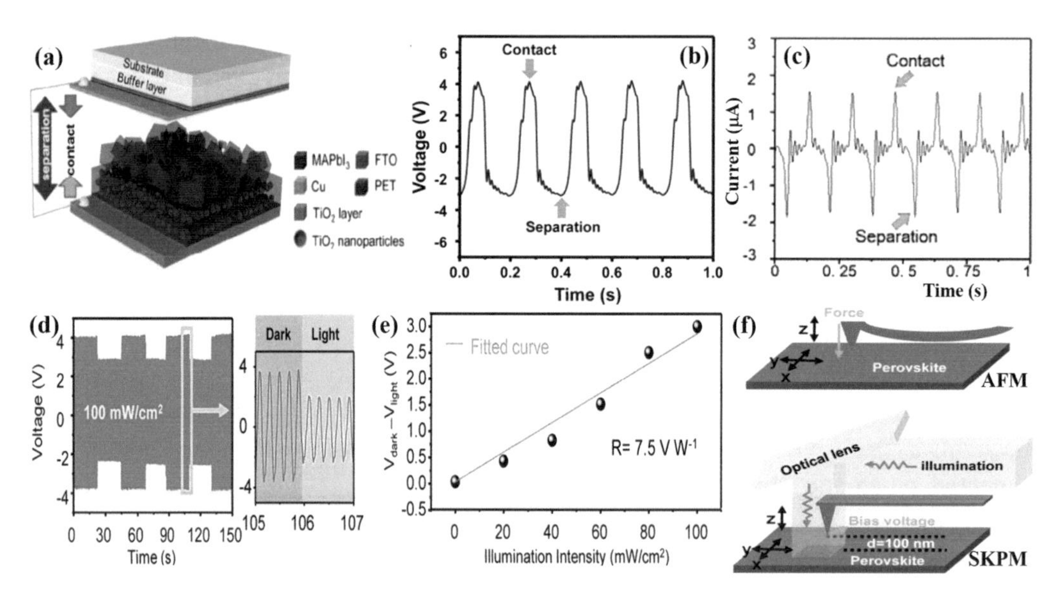

FIGURE 11.5 (a) Device structure of SPPD with $MAPbI_3$ as the photosensitive and triboelectric layer, (b) open-circuit voltage (V_{oc}) and (c) short-circuit current (I_{sc}) output of the $MAPbI_3$ and $Pb(Zr,Ti)O_3$ (PZT) based SPPD respectively without illumination, (d) change in magnitude of V_{oc} under light intensity of 100 mW/cm², (e) light intensity dependency of V_{oc} change and its linearly fitted curve, and (f) schematic illustrations of (top) contact-mode AFM and (bottom) SKPM mode with applied illumination. (Figure 11.5 (a)–(b) and (d)–(f) reproduced with permission from Ref. [50] © 2015 American Chemical Society, and (c) reproduced with permission from Ref. [49] © 2017 Elsevier.)

charges (triboelectric charges) on the contact surfaces that induces an oscillating open-circuit voltage (V_{oc}) between the two electrodes or flow a short-circuit current (I_{sc}) as shown in Figure 11.5(b) and (c) respectively. Indeed, such electricity is generated by the coupling effect of triboelectrification and electrostatic induction. Remarkably, the obtained V_{oc}/I_{sc} is essentially the voltage/current that is applied/flows onto/through the inner resistance of a TENG, as the LAL is a semiconductor with finite conductivity.

Under the dark, the V_{oc}/I_{sc} has certain peak-to-peak. On light illumination, the amplitude of V_{oc}/I_{sc} immediately drops and recovers speedily to the original value instantly when illumination is off, as illustrated in Figure 11.5(d). Usually, the variation of V_{oc}/I_{sc} amplitude upon illumination increases on increasing light intensity and provides the average value of responsivity as in Figure 11.5(e). Commonly, both surface and bulk properties of the LAL are believed to be the cause of sudden reduction of the V_{oc}/I_{sc} upon illumination, and thus SPPD possesses a dual sensing mechanism. In general, atomic force microscopy in contact mode (CAFM) and scanning Kevin probe microscopy (SKPM) are two of the best techniques to estimate the surface properties and were used for the same purpose [49]. During the characterization, the LAL surface was rubbed by conducting tip of AFM in the contact mode (top Figure 11.5(e)). Then SKPM is employed to monitor the surface potential (a direct indicator of the surface charge density) of the rubbed area at a certain bias voltage, while the rubbed surface is illuminated (bottom Figure 11.5(e)). The ±ve surface potential

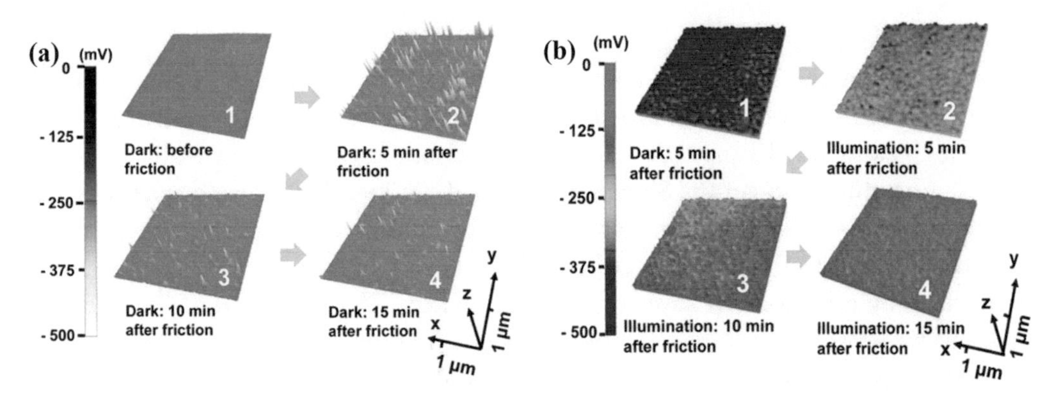

FIGURE 11.6 3D mappings of the surface potential as a function of friction time: (a) without the illumination, and (b) with the illumination applied 5 min after the friction. (Reproduced with permission from Ref. [50] © 2015 American Chemical Society.)

induced by the rubbing process correspond to generate ±ve triboelectric charges respectively on the LAL surface (Figure 11.6(a)). This surface potential gradually reduces with time in the absence of illumination after the rubbing due to gentle dissipation of triboelectric charges into the air, and eventually vanishes for an extended period. On the contrary, feeding of illumination after rubbing leads to the spontaneous upswing in the surface potential, which indicates the momentary drop in surface charge density. However, the subsequent reduction of the surface potential is again associated with the slow dissipation of the charges (Figure 11.6(b)).

11.3 ENERGY-SAVING MATERIALS FOR SELF-POWERED PHOTODETECTORS

A wide variety of materials, including elemental semiconductors (Si, Ge), compound semiconductors (metal oxides, metal chalcogenides), perovskite, 2D materials (graphene, metal dichalcogenides), polymers, and metal organic frameworks (MOFs), among others, and their composites/heterostructures have been used for the development of BP-based SPPDs. Usually, the choice of material depends on the spectral region to be detected. Typically, extra wide bandgap materials like Ga_2O_3 (4.4–5.3 eV), Gd_2O_3 (~5.2 eV), AlN (~6.2 eV), diamond (~5.5 eV), BN (4.5–5.5 eV), and their alloys such as $Al_xGa_{1-x}N$ and Zn or In doped MgO (7.7 eV) are suitable for deep UV (DUV) or solar blind (SB) detection (UV-C, 280–200 nm), whereas wide bandgap materials such as ZnO (~3.3 eV), TiO_2 (~3.2 eV), SnO_2 (~3.6 eV), NiO (3.6–4.0 eV) and so forth, are well suited for UV-A (400–315 nm) and UV-B (315–280 nm). In contrast, the best candidates for infrared (IR) or near-infrared (NIR) detection are narrow bandgap materials like PbS (~0.41 eV), PbSe (~0.26 eV), HgS (~2.0 eV), and intermediate bandgap materials such as CdS (2.42 eV), CdO (~2.3 eV), oxides of copper (1.2–2.5 eV), and cobalt (1.5–2.6 eV), among others, displayed extremely high detection capability for visible radiations. However, the spectral range of detection can be tailored by various means, including bandgap and interface engineering. Here, we present a short overview of the various energy-saving materials under two broad categories: non-2D materials and 2D materials, for the design of SPPDs working on the BP mechanism only.

11.3.1 Non-2D Materials

A single germanium NW was employed to fabricate the metal–semiconductor–metal (MSM) type SPPD. The device showed a significant photoresponse (responsivity~10^3-10^5 A W^{-1}) in the wavelength range 300–1100 nm [51]. Also, a ZnO NWs array with Pt Schottky contacts was utilized to get the SP action for UV detection. The device displayed a high sensitivity of 475 without external bias [52]. In both studies, the self-powering feature was ascribed as a result of asymmetrical height of Schottky barrier, while a UV-SPPD was designed using a photoactive layer of single-crystalline ZnS nanotubes (NTs) and electrodes of Ag NWs network. The device exhibited a high on/off ratio, sensitivity, responsivity, and detectivity of 19173, 19172, 2.56 A/W, and 1.67×10^{10} cm $Hz^{1/2}W^{-1}$ respectively with fast response ($\tau_r = 0.09$ s, $\tau_f = 0.07$ s) at 0 V. They believed that SP action depends on the contact area of symmetrical MSM structure as demonstrated for three devices in Figure 11.7(a)–(c) and their corresponding *I-V* characteristics in Figure 11.7(d)–(f) [53]. A GaN-based MSM-type UV-SPPD was designed over the sapphire substrate by employing a lateral polarity structure (LPS). The SP action and efficient carrier separation were associated with the in-plane internal electric field and different Schottky barrier heights at a metal/semiconductor interface. Without biasing the device, dark current of 6.8 nA/cm^2 and detectivity of 1.0×10^{12} jones were recorded, whereas under −10 V, a high photo-to-dark current ratio of 1.2×10^4 with peak responsivity of 933.7 mA/W were achieved [54]. The MSM architecture based on epitaxial β-Ga_2O_3 layer showed an ultra-low dark current of 800 fA at 0 V and detection of DUV/SB radiation (265 nm) with peak responsivity of 11.6 μA/W under illumination intensity of 75 μW/cm^2 [55]. Jia et al. developed planar-type MSM SPPDs using $CH_3NH_3PbI_3$ to take the advantage of its property of spontaneous polarization, which can generate an extra polarization in the electric field. The Au/$CH_3NH_3PbI_3$/Cu device exhibited maximum responsivity of 0.288 A W^{-1} and detectivity of 4.69×10^{12} jones in 400–840 nm under the influence of BP caused by spontaneous polarization and asymmetric Schottky junction. Furthermore, the operation stability (photocurrent) of the unencapsulated photodetector remained 92% after 15 days in normal ambience [56]. Cao et al. designed a broadband (250–1500 nm) SPPD based on an electronically conductive MOF. The device demonstrated high external quantum efficiency (84%) along with high I_{on}/I_{off} (~10^3) and short rise (7 ms), and fall time (30 ms) [57].

A UV SPPD developed using p-n junction of NiO/Ga_2O_3 displayed the responsivity, detectivity, and I_{on}/I_{off} ratio of 57 μA/W, 5.45×10^9 Jones, and 122, respectively, under the irradiation of 254 nm light at 0 V [58]. In comparison, the responsivity of SPPD was improved to 420 mA/W by utilization of α-Ga_2O_3 nanorod array/Cu_2O microsphere heterojunction under the same illumination [59]. Whereas the detectivity and other parameters were improved by employing small-molecule hole transport materials (SMHTMs) with β-Ga_2O_3. In that work, four types of SMHTMs viz; 4,4′-Bis(N-carbazolyl)-1,1′-biphenyl (CBP), methylcyclopropene (MCP), butylphthalide (3-n-butylphthalide or NBP), and 4-(1 -cyclohexyl)-N,N-bis(4-methylphenyl)aniline (TAPC) were used to construct an SPPD with high I_{on}/I_{off} ratio of about >10^5. The best PD was designed with β-Ga_2O_3/TAPC heterojunction with a dark current of about 20 fA, an I_{on}/I_{off} ratio of 5.9×10^5, and a detectivity of 1.02×10^{13} Jones at 0 V [60]. Further, the sensing parameters of Ga_2O_3 based p-n type SPPD

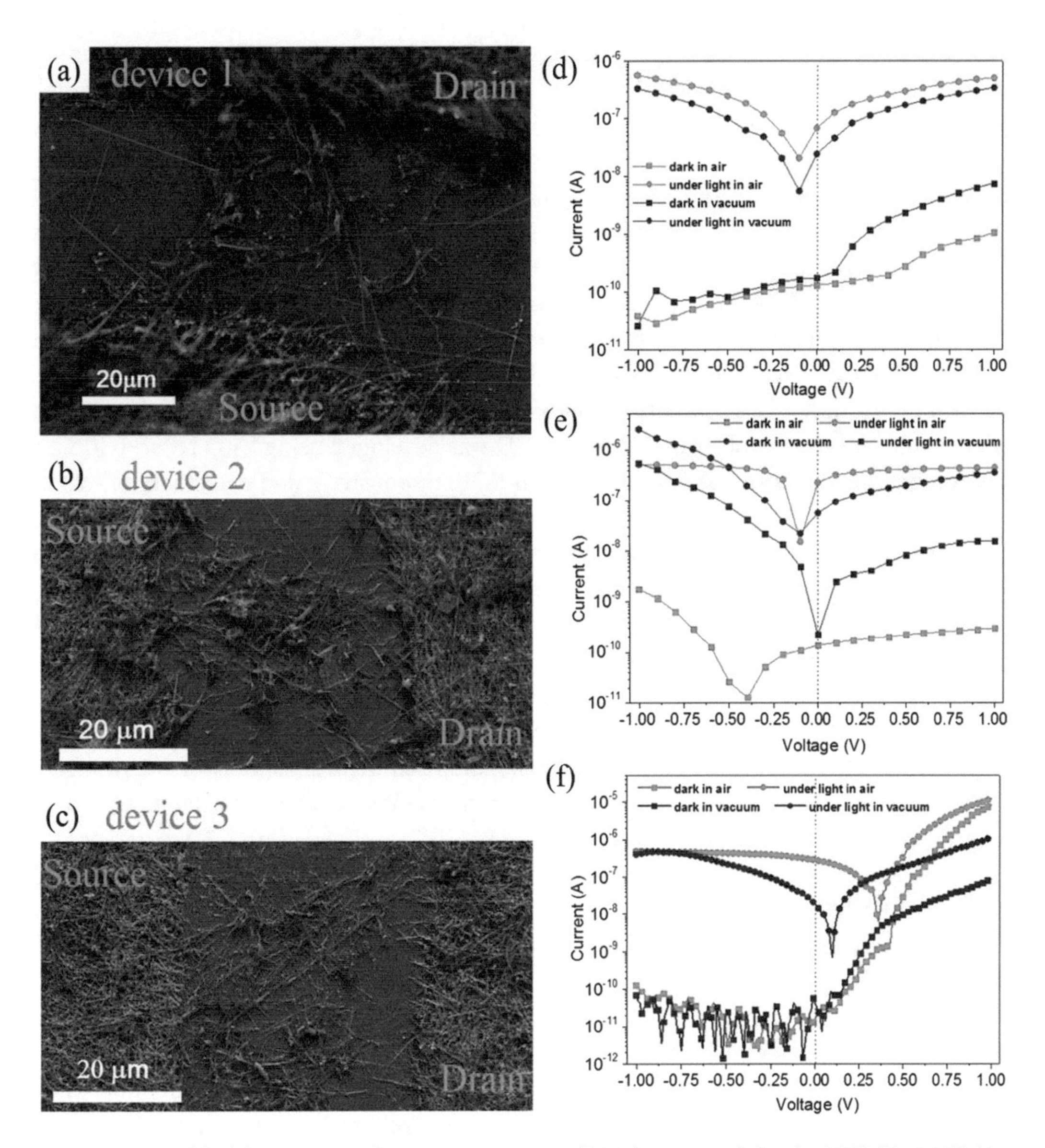

FIGURE 11.7 (a)–(c) Scanning electron microscopy (SEM) image of the Ag NWs/ZnS NTs/Ag NWs (MSM) SPPD devices 1–3, and (d)–(f) corresponding *I–V* characteristics under dark and UV light in air and vacuum. (Reproduced from Ref. [53], © 2017 Q. An et al. under the Creative Commons license.)

were improved by exploiting GaN/Sn:Ga_2O_3 heterojunction. This device showed high responsivity, UV/visible rejection ratio of $R_{254\ nm}/R_{400\ nm}$, detectivity, and I_{on}/I_{off} ratio of 3.05 A/W, 5.9 × 10^3, 1.69 × 10^{13} Jones, and ~10^4 respectively, under the same radiation with a very low dark current (18 pA) and fast response time of 18 ms [61]. Chen et al. designed an SPPD based on depolarization field (E_{dp}) of ferroelectric materials and BP at p-n junction ($Ep\text{-}_n$). Their device consisted of NiO/$Pb_{0.95}La_{0.05}Zr_{0.54}Ti_{0.46}O_3$ (PLZT) heterojunction, taking advantage of energy level alignments that favored the electron extraction and exhibited a tunable performance upon varying the polarization direction of PLZT. In a poling down state of PLZT layer, the device illustrated a higher responsivity, detectivity, and faster response speed of $1.8 \pm 0.12 \times 10^{-4}$ A/W, $3.69 \pm 0.2 \times 10^{9}$ Jones and ~ 0.36 ± 0.02 s, with a lower dark current (1.3 ± 0.19 pA) under 254 nm at zero bias than the sole PLZT-based device [62]. The inorganic/organic hybrid p-n junction of Sb_2S_3/2,2',7,7'-tetrakis-(N,N-di-pmethoxyphenylamine)-9,9'-spirobifluorene (Spiro-OMeTAD) was used to develop visible (470 nm) SPPD. The device showed a rectification ratio of ~10^2 with responsivity and response time of 0.087 A/W and < 25 ms, respectively [63]. Sb_2Se_3/VO_2 heterojunction was used to fabricate visible (520 nm) SPPD that responded within 360 ms with a responsivity of 244 mA/W [64]. The PbS/porous silicon (PS) heterojunction was also employed to fabricate SPPD that showed a sensitivity of 5.66×10^4 % under 10 mW/cm^2 with a response time of 1.1 s [65]. A broadband (350–700 nm) SPPD was developed using ZnO NRs/PbS/RGO heterojunction. The device showed maximum EQE, responsivity, and detectivity of ~78%, ~ 0.25 A/W, and ~ 82944.55 $Hz^{1/2}$/W with fast response of 74 ms at 400 nm under zero bias. Besides, the device showed significant photovoltaic effect with V_{OC} ~ 0.27 V, I_{SC} ~1.03 mA and PEC ~ 0.73%. The observed superior photoresponse and photovoltaic behavior was attributed to the formation of p-n junction between the p-PbS/n-ZnO NRs interface and favorable carrier transport at RGO electrode [66]. A zeolitic imidazolate framework-8 (ZIF-8)@H:ZnO core-shell nanorods arrays/Si heterojunction SPPD enhanced photoresponsive characteristics due to combination of hydrogenation and ZIF-8 passivation. The device exhibited superior detectivity, high responsivity and prominent sensitivity of ~ 2.14×10^{16} Jones, ~ 7.07×10^4 mA/W, ~ 2.08×10^{12} cm^2/W with broadband detection ranging from UV to NIR [67]. Table 11.1 lists materials employed in SPPD grouped as non-2D materials.

11.3.2 2D Materials

Since the successful separation of monoatomic graphene from bulk graphite in 2004 [95], 2D materials have secured persistent attraction globally due to their excellent performance in various fields. A number of 2D materials such as transition metal chalcogenides [96, 97], hexagonal boron nitride (h-BN) [98], black phosphorus, MXene [99–102], and metal oxides [84, 103] among others, have been synthesized with the advancement of theory and technology. Similar to graphene, other 2D materials have facilitated many novel submissions in electronics as well as optoelectronics owing to properties like good electron mobility [85], suitable bandgap [104], good light absorption, and photoresponse [105]. 2D materials possess unique properties like 2D planar and stable atomic structures with large specific surface areas and tunable electronic and optical properties with changes in thickness/layers compared with traditional semiconductor materials. Recently, a number of SPPDs based on 2D materials with excellent photodetection performance have been reported, as illustrated below.

An MSM monolayer WS_2-based SPPD exhibited linear photoresponse with a responsivity of 1.4 mA/W, detectivity of 10^{10} Jones, dark current level of 0.1 pA and response speed of tens of milliseconds under 532 nm [106]. However, the performance of the device could be improved by using irregular WS_2. The device demonstrated a fast response speed of 7.8/37.2 ms with high responsivity, I_{on}/I_{off} ratio, and detectivity of 777 mA/W, 104 and

TABLE 11.1 SPPD Based on Various Non-2D Materials Along with Their Characteristic Parameters

Material	λ (nm)	V_{OC} (mV)/I_{SC} (μA/cm²)	S (%)/R (mA/W)	D (Jones)	EQE (%)/Gain	τ_r/ τ_f (ms)	References
AgNWs/NiO/TiO_2/FTO	**365**	130/2.441	—/136	1.11×10^9	—	7.6/15	**[51]**
Ge: β-Ga_2O_3	**230**	—	—/90	—	52/—	—	**[53]**
AgNWs/Cu_4O_3/TiO_2/FTO	**365**	300/0.25	—/187	—	—	0.439/0.423	**[54]**
SnO_2-TiO_2 nanomace arrays	**365**	—	—/145	—	49/—	37/25	**[57]**
ZnO NW/$CuCrO_2$ core–shell	**395**	16/38.5	—/5.87	—	—	0.032 /0.035	**[64]**
SnO_2 NS-Ti_2O-TiO_2 core-shell	**365**	560/709.5	563/600	—	—	20/4	**[65]**
ZIF-8@H:ZnO NRs/Si	**365**	—	—/70700	6.354×10^{13}	—	0.252/0.607	**[67]**
ITO/SnO_2/In_2O_3/SnS_2/FTO/Glass	**365**	—	—/2.9	5.9×10^7	—	0.059/0.079	**[68]**
Gradient O-doped CdS/perovskite	**700**	—	—/480	2.1×10^{13}	—	0.54/ 2.21	**[69]**
p-GaSe/n-InSe	**470**	—	—/21	3.7×10^{12}	9.3/—	0.0018/0.0020	**[70]**
rGO/n-Si	**600**	—	—/1.52	—	—	0.002/0.0037	**[71]**
ZnO/Cu_2O BHAs	**355**	—/107	525/19.3	—	—	140/360	**[72]**
NiO:Cu(1.0%)	**325**	—	—/202	5.9×10^9	47/—	2900/3200	**[73]**
p-NiO/n-ZnO@Al_2O_3 core-shell structure	**380**	—	2400/1.4	—	—	40	**[74]**
NiO/TiO_2NRs/TiO_x	**380**	—	—/5.66	2.50×10^{12}	—	<100	**[75]**
ZnO-Cu_2O core-shell nanowires	**VIS**	—	—/0.007	—	—	<90	**[76]**
ZnO NW–Co_3O_4	**Solar light**	—	—/21800	4.12×101^2	—	6	**[77]**
p-NiO/n-ZnO heterojunction	**370**	270/0.085	—	—	—	323/12	**[78]**
FTO/TiO_2/Co_3O_4/ NiO/Ag	**365**	—	10^4/3000	2.7×10^{11}	—	0.20/1	**[79]**
ZnO/Ga_2O_3 core/shell heterostructure	**251**	—	—/9.7	6.29 × 1012	—	0.1/0.9	**[80]**
Cu_2O/ZnO NRs heterojunctions	**450**	—	—/70	—	—	75/70	**[81]**
p-NiO/ZnO-NR array heterojunction	**355**	—	—/0.44	—	—	230/210	**[82]**
ITO/ZnO/Ag NWs	**365**	—	10^2/6	2.14×101^1	—	360 /1910	**[83]**
Co_3O_4/ZnO heterojunction	White light	165/76	4.57×10^6/12.9	5.58×10^{11}	—	0.081/0.178	**[84]**
MoO_3/ Co_3O_4/ZnO/ITO	**365**	—	—	—	—	35/82	**[85]**

(Continued)

TABLE 11.1 (CONTINUED) SPPD Based on Various Non-2D Materials Along with Their Characteristic Parameters

Material	**λ (nm)**	V_{OC} **(mV)/**I_{SC} **(μA/cm²)**	**S (%)/R (mA/W)**	**D (Jones)**	**EQE (%)/Gain**	τ_r/τ_f **(ms)**	**References**
NiO/ZnO/ITO	**365**	—	1944/0.02	7.2×10^{11}	—	0.041/0.071	**[86]**
Ag/NiO/TiO_2/FTO	**365**	400/—	—/430	8.6×10^{9}	—	3.9/4.1	**[87]**
n-ZnO/p-NiO core-shell NW arrays	**355**	—	—	—	—	0.01/0.03	**[88]**
ZnO homojunction nanofibers	**360**	—	2.5×10^{4}/1	—	—	3900/4700	**[89]**
ZnO/perovskite-heterostructured	**325**	—	—/26.7	4.0×10^{10}	--	0.053/0.063	**[90]**
CdSe QD	**522**	—	—/10.23	8.81×10^{9}	—	17.9/18	**[91]**
Zno-ZnS micro-structured composite	**365**	—	—/3.34	8.9×10^{12}	—	22500/45000	**[92]**
$Sn_{0.12}Zn_{0.88}S$ thin films	**280**	—	847/47.1	1.64×10^{9}	—/0.3	9.8/12.4	**[93]**
$Sn_{0.40}Zn_{0.60}S$ alloy thin films	**280**	1800 – 2400/—	2406/212	6.28×10^{9}	—/1.33	1.9/2.6	**[94]**

4.94×10^{11} Jones, respectively, under 405 nm illumination due to a Schottky barrier height difference of ≈ 50.2 mV through Fermi-level pinning effect and different contact area [107]. Further, the performance of WS_2-based SPPD was improved by replacing one of the Au electrodes with a graphene electrode. The device illustrated a high responsivity of 7.55 A/W, low dark current of ~10^{-2} pA, high detectivity of 3×10^{12} Jones, and very high I_{on}/I_{off} ratio of 10^8 under 650 nm illumination [108]. The SP activities of two different symmetric metals (Ag and Au) interdigitated MoS_2 PD were studied under 632 nm light, and noticed that the dark current of the device with Au electrode was ten times higher than the device with Ag electrode due to a lower barrier height for Au. The photosensing parameters of device with Au electrode were also found superior, with maximum responsivity and detectivity of 128.0 mA/W and 1.32×10^{10} Jones, respectively [109]. An MSM SPPD was fabricated based on ambipolar WSe_2 with two asymmetrical electrodes comprising 2D van der Waals metal Fe_3GeTe_2 and semimetal graphene. The device displayed broadband (450–800 nm) detection with maximum I_{sc}, V_{oc}, I_{on}/I_{off} ratio, EQE, detectivity, responsivity, and response time of 17 nA, −0.177 V, 106, 27.14, 3.4×10^{10} Jones, 116.38 mA/W and 370 μs, respectively, under 650 nm illumination with a dark current of 1 fA [110]. Similarly, a broadband (300–1000 nm) SPPD was designed by irregular 2D InSe, which demonstrated responsivity, detectivity, and I_{on}/I_{off} ratio of 0.103 A/W, 1.83×10^{10} Jones, and 10^4, respectively, with a response time of 1 ms [111]. A 2D Nb_2CT_x MXene@Bi quantum dots (QDs) Schottky heterojunction was used to develop a PEC-type SPPD that showed a response/recovery time (0.08/0.08 s), a photocurrent of 1.56 mA/cm^{-2}, a responsivity of 585.25 mA/W, and detectivity of 4.63×10^{12} Jones. Besides, the device exhibited SP capability after bending 180° for 1000 cycles [112].

Patil et al. constructed a 2D p-n van der Waals heterojunction (vdW HJ) SPPD by vertically stacking p-type and n-type InSe flakes. Such a device demonstrated a threefold enhancement in responsivity in the NIR spectral region (980 nm) with a dark current of few pA compared to PD based on only p- or n-type InSe [113]. Whereas the novel p-GeSe/n-$MoSe_2$ vdW HJ SPPD detected 850 nm radiation and demonstrated a high photoresponsivity, EQE detectivity, normalized I_{on}/I_{off} ratio and NEP of 465 mA/W, 670%, 7.3×10^9 Jones, 1.9×10^{10} W^{-1}, and 1.22×10^{-13} WHz$^{-1/2}$, respectively, with a short rise time of 180 ms, and a fall time of 360 ms [114]. While vdW HJ SPPD was developed by replacing GeSe via GaSe to tune the detection from NIR to UV-Vis. (375–633 nm) and showed a large V_{oc} of 0.61 V, fast response of 5 ms, and high responsivity of 900 mA/W [115]. Further, the detection range was extended up to the IR region (375—1550 nm) by employing p-WSe_2/n-Bi_2Te_3 vdW HJ, with maximum I_{sc} and V_{oc} of 18 nA and 0.25 V, respectively, for 633 nm illumination. The device displayed a low response time (~210 μs) and high responsivity (20.5 A/W at 633 nm and 27 mA/W at 1550 nm) [116]. A 2D MOF ($Cu_3(HHTP)_2$, HHTP = 2,3,6,7,10, 11-hexahydroxytriphenylenehydrate)/ZnO-based UV SPPD showed ultrafast response time of 70 μs with a responsivity of 1.05 mA/W and detectivity of 1.31×10^8 Jones [117]. Recently, Guo et al. fabricated an SPPD based on 2D ferroelectric PMA_2PbCl_4 monocrystalline microbelt (MMB). The device with architecture Ag/Bi/2D PMA_2PbCl_4 MMB/Bi/Ag PDs showed a high photoresponsivity up to 9 A/W under 320 nm laser illumination with a response speed of hundreds of μs. It was noticed that the responsivity induced by the

TABLE 11.2 2D Materials-Based SPPDs and Their Figures of Merit

Material	λ (nm)	V_{OC} (mV)/I_{SC} (μA/cm²)	S (%)	R (mA/W)	D (Jones)	τ_r/ τ_f (ms)	References
WSe_2/InSe	**520**	470/11700	10^5	61	2.5×10^{11}	0.063/0.076	**[121]**
MoS_2/Sb_2S_3	**650**	—	10^3	>150	—	—	**[122]**
Ti_3C_2/ϵ-Ga_2O_3	**254**	—	2.5×10^6	15.5	2.15×10^{11}	43/145	**[123]**
MXene/GaN NRs	**382**	—	—	48.6	5.9×10^{12}	—	**[124]**
p-GaTe/n-MoS_2	**633**	—	—	1360	—	10	**[125]**
CdSe NB/ graphene	**633**	—	3.5×10^5	10200	—	0.082/0.18	**[126]**
CdSe nanobelt/ graphene	**633**	—	1.2×10^5	8700	—	0.070/0.137	**[127]**
$PtSe_2$/Si	**808**	—	—	520	3.26×10^{13}	55.3/170. 5	**[128]**
graphene/GaAs NW	**532**	430/3.16	—	15.4	—	0.071/.194	**[129]**
graphene/ ZnO:Al NR	**UV**	—	10^2	39	—	0.037/.330	**[130]**

ferro-pyro-phototronic effect is 128 times larger than that induced by the photovoltaic effect [118]. MXene ($Ti_3C_2T_x$)/GaN vdW HJ was employed to design the UV SPPD that demonstrated large responsivity (284 mA/W), high detectivity (7.06×10^{13} Jones), fast response speed (7.55 μs/1.67 ms) and PCE of 7.33 % under 355 nm [119]. SPPDs were developed using two vdW HJ of Cs-I-black phosphorus and Sn-I-black phosphorus. It was noticed that both Cs-I-black phosphorus and Sn-I-black phosphorus heterostructures have type-II band alignment with interface potential drop (E_P) of 6.02 eV and 5.64 eV, respectively [120]. Table 11.2 lists SPPDs developed employing 2D materials.

11.4 CONCLUSION

In summary, PDs are devices that convert light directly into electricity and are the backbone of many electronic devices. The SPPD is a special class of PD that operates without external power and consumes much less power. A number of materials, including conventional elemental and compound semiconductors as well as emerging 2D and perovskite materials, have been used as energy-saving materials to design SPPDs. Undoubtedly, out of two broad categories of SPPDs, namely, BP-based and integrated-power-source-based, the performance of the latter is superior to the former. But the latter one needs a complex structure with the integration of a power unit like a nanogenerator/solar cell or a storage unit like a battery/supercapacitor. Contrarily, the architectures of BP-based SPPDs are simple and economical. Moreover, their performance can be augmented via coupling of photoelectronics with piezo-, pyro- or tribo-electric effects. In the MSM type SPPDs, asymmetric Schottky barrier height/area at two contacts are the cause of BP and thus SP action, while in p-n junction type SPPDs, it is the junction width/potential height that governs the SP action.

REFERENCES

[1] M. Long, P. Wang, H. Fang, and W. Hu, Progress, challenges, and opportunities for 2D material based photodetectors. *Advanced Functional Materials*, 2018. **29**(19): p. 1803807. DOI: 10.1002/adfm.201803807

[2] G. Konstantatos and E.H. Sargent, Nanostructured materials for photon detection. *Nat Nanotechnol*, 2010. **5**(6): p. 391–400. DOI: 10.1038/nnano.2010.78

[3] Z. Wen, Q. Shen, and X. Sun, Nanogenerators for self-powered gas sensing. *Nanomicro Lett*, 2017. **9**(4): p. 45. DOI: 10.1007/s40820-017-0146-4

[4] T. Mori and S. Priya, Materials for energy harvesting: At the forefront of a new wave. *MRS Bulletin*, 2018. **43**(3): p. 176–180. DOI: 10.1557/mrs.2018.32

[5] L. Zhao, H. Li, J. Meng, and Z. Li, The recent advances in self-powered medical information sensors. *InfoMat*, 2019. **2**(1): p. 212–234. DOI: 10.1002/inf2.12064

[6] P. Guo, J. Xu, K. Gong, X. Shen, Y. Lu, Y. Qiu, et al., On-nanowire axial heterojunction design for high-performance photodetectors. *ACS Nano*, 2016. **10**(9): p. 8474–8481. DOI: 10.1021/acsnano.6b03458

[7] Y. Dai, X. Wang, W. Peng, C. Xu, C. Wu, K. Dong, et al., Self-powered Si/CdS flexible photodetector with broadband response from 325 to 1550 nm based on pyro-phototronic effect: An approach for photosensing below bandgap energy. *Adv Mater*, 2018. **30**(9): p. 1705893. DOI: 10.1002/adma.201705893

[8] N. Saxena, T. Kalsi, and P. Kumar, CdS-based photodetectors for visible-UV spectral region, in: *Handbook of II-VI semiconductor-based sensors and radiation detectors: Volume 1, materials, technologies and light detectors*, G. Korotcenkov, Editor. 2023, Springer. pp. 251–279.

[9] N. Saxena, S. Sharma, and P. Kumar, All metal oxide-based photodetectors, in: *Metal oxides for next generation optoelectronic, photonic, and photovoltaic applications*, V. Kumar, V. Sharma, and H.C. Swart, Editors. 2023, Elsevier.

[10] L. Su, W. Yang, J. Cai, H. Chen, and X. Fang, Self-powered ultraviolet photodetectors driven by built-in electric field. *Small*, 2017. **13**(45): p. 1701687. DOI: 10.1002/smll.201701687

[11] T. Kalsi and P. Kumar, Cd1-xMgxS CQD thin films for high performance and highly selective NIR photodetection. *Dalton Trans*, 2021. **50**(36): p. 12708–12715. DOI: 10.1039/d1dt01547h

[12] P. Kumar, N. Saxena, S. Dewan, F. Singh, and V. Gupta, Giant UV-sensitivity of ion beam irradiated nanocrystalline CdS thin films. *RSC Advances*, 2016. **6**(5): p. 3642–3649. DOI: 10.1039/c5ra21026g

[13] P. Kumar, N. Saxena, F. Singh, and V. Gupta, Ion beam assisted fortification of photoconduction and photosensitivity. *Sensors and Actuators A: Physical*, 2018. **279**: p. 343–350. DOI: 10.1016/j.sna.2018.06.037

[14] S. Miao and Y. Cho, Toward green optoelectronics: Environmental-friendly colloidal quantum dots photodetectors. *Frontiers in Energy Research*, 2021. **9**. DOI: 10.3389/fenrg.2021.666534

[15] K. Xu, W. Zhou, and Z. Ning, Integrated structure and device engineering for high performance and scalable quantum dot infrared photodetectors. *Small*, 2020. **16**(47): p. e2003397. DOI: 10.1002/smll.202003397

[16] S. Guo, L. Wang, C. Ding, J. Li, K. Chai, W. Li, et al., Tunable optical loss and multi-band photodetection based on tin doped CdS nanowire. *Journal of Alloys and Compounds*, 2020. **835**. DOI: 10.1016/j.jallcom.2020.155330

[17] M. Kielar, T. Hamid, M. Wiemer, F. Windels, L. Hirsch, P. Sah, et al., Light detection in open-circuit voltage mode of organic photodetectors. *Advanced Functional Materials*, 2019. **30**(9). DOI: 10.1002/adfm.201907964

[18] U. Sundararaju, M.A.S. Mohammad Haniff, P.J. Ker, and P.S. Menon, MoS(2)/h-BN/graphene heterostructure and plasmonic effect for self-powering photodetector: A review. *Materials (Basel)*, 2021. **14**(7). DOI: 10.3390/ma14071672

[19] W. Tian, Y. Wang, L. Chen, and L. Li, Self-powered nanoscale photodetectors. *Small*, 2017. **13**(45): p. 1701848. DOI: 10.1002/smll.201701848

[20] L. Peng, L. Hu, and X. Fang, Energy harvesting for nanostructured self-powered photodetectors. *Advanced Functional Materials*, 2014. **24**(18): p. 2591–2610. DOI: 10.1002/adfm.201303367

[21] Z.L. Wang, Towards self-powered nanosystems: from nanogenerators to nanopiezotronics. *Advanced Functional Materials*, 2008. **18**(22): p. 3553–3567. DOI: 10.1002/adfm.200800541

[22] W. Wu, S. Bai, M. Yuan, Y. Qin, Z.L. Wang, and T. Jing, Lead zirconate titanate nanowire textile nanogenerator for wearable energy-harvesting and self-powered devices. *ACS Nano*, 2012. **6**(7): p. 6231–6235.

[23] C.-H. Chen, K.-R. Wang, S.-Y. Tsai, H.-J. Chien, and S.-L. Wu, Nitride-based metal–semiconductor–metal photodetectors with InN/GaN multiple nucleation layers. *Japanese Journal of Applied Physics*, 2010. **49**(4S): p. 04dg06. DOI: 10.1143/jjap.49.04dg06

[24] Q. Yang, Y. Liu, Z. Li, Z. Yang, X. Wang, and Z.L. Wang, Self-powered ultrasensitive nanowire photodetector driven by a hybridized microbial fuel cell. *Angew Chem Int Ed Engl*, 2012. **51**(26): p. 6443–6446. DOI: 10.1002/anie.201202008

[25] F.-R. Fan, Z.-Q. Tian, and Z. Lin Wang, Flexible triboelectric generator. *Nano Energy*, 2012. **1**(2): p. 328–334. DOI: 10.1016/j.nanoen.2012.01.004

[26] X. Wang, J. Song, J. Liu, and Z.L. Wang, Direct-current nanogenerator driven by ultrasonic waves. *Science*, 2007. **316**(5821): p. 102–105. DOI: 10.1126/science.1139366

[27] Y. Yang, W. Guo, K.C. Pradel, G. Zhu, Y. Zhou, Y. Zhang, et al., Pyroelectric nanogenerators for harvesting thermoelectric energy. *Nano Lett*, 2012. **12**(6): p. 2833–2838. DOI: 10.1021/nl3003039

[28] D. Zhang, Y. Wang, and Y. Yang, Design, performance, and application of thermoelectric nanogenerators. *Small*, 2019. **15**(32): p. e1805241. DOI: 10.1002/smll.201805241

[29] X. Wang, J. Zhou, J. Song, J. Liu, N. Xu, and Z.L. Wang, Piezoelectric field effect transistor and nanoforce sensor based on a single ZnO nanowire. *Nano Lett*, 2006. **6**(12): p. 2768–2772.

[30] Z.L. Wang, Nanopiezotronics. *Advanced Materials*, 2007. **19**(6): p. 889–892. DOI: 10.1002/adma.200602918

[31] L.E. Bell, Cooling, heating, generating power, and recovering waste heat with thermoelectric systems. *Science*, 2008. **321**: p. 1457–1461.

[32] F.J. DiSalvo, Thermoelectric cooling and power generation. *Science*, 1999. **285**: p. 703–706.

[33] M. Zebarjadi, K. Esfarjani, M.S. Dresselhaus, Z.F. Ren, and G. Chen, Perspectives on thermoelectrics: From fundamentals to device applications. *Energy Environ. Sci.*, 2012. **5**(1): p. 5147–5162. DOI: 10.1039/c1ee02497c

[34] R. Wang, S. Liu, C.R. Liu, and W. Wu, Data-driven learning of process-property-performance relation in laser-induced aqueous manufacturing and integration of ZnO piezoelectric nanogenerator for self-powered nanosensors *Nano Energy*, 2021. **83**: p. 105820.

[35] J. Wang, K. Xia, J. Liu, T. Li, X. Zhao, B. Shu, et al., Self-powered silicon PIN photoelectric detection system based on triboelectric nanogenerator. *Nano Energy*, 2020. **69**: p. 104461. DOI: 10.1016/j.nanoen.2020.104461

[36] S. Cai, X. Xu, X. Wu, J. Chen, C. Zuo, R. Jin, et al., A self-adjustable wearable photodetector powered by flexible thermoelectric generators. *J. Mater. Chem. C*, 2019. **7**: p. 13097–13103.

[37] C.K. Chan, H. Peng, G. Liu, K. McIlwrath, X.F. Zhang, R.A. Huggins, et al., High-performance lithium battery anodes using silicon nanowires. *Nat Nanotechnol*, 2008. **3**(1): p. 31–35. DOI: 10.1038/nnano.2007.411

[38] J. Chmiola, C. Largeot, P.L. Taberna, P. Simon, and Y. Gogotsi, Monolithic carbide-derived carbon films for micro-supercapacitors. *Science*, 2010. **328**(5977): p. 480–483. DOI: 10.1126/science.1184126

[39] P. Kumar and L. Unnikrishnan, Pyroelectric and piezoelectric polymers, in: (eds. Inamuddin, Mohd Imran Ahamed, Rajender Boddula, Tariq A. Altalhi. *Polymers in energy conversion and storage*. 2022. CRC Press. p. 109–139.

[40] Z.L. Wang, R. Yang, J. Zhou, Y. Qin, C. Xu, Y. Hu, et al., Lateral nanowire/nanobelt based nanogenerators, piezotronics and piezo-phototronics. *Materials Science and Engineering: R: Reports*, 2010. **70**(3–6): p. 320–329. DOI: 10.1016/j.mser.2010.06.015
[41] Z.L. Wang and W. Wu, Piezotronics and piezo-phototronics: Fundamentals and applications. *National Science Review*, 2014. **1**(1): p. 62–90. DOI: 10.1093/nsr/nwt002
[42] Z.L. Wang and J. Song, Piezoelectric nanogenerators based on zinc oxide nanowire arrays. *Science*, 2006. **312**: p. 242–246.
[43] J. Zhou, P. Fei, Y. Gu, W. Mai, Y. Gao, R. Yang, et al., Piezoelectric-potential-controlled polarity-reversible schottky diodes and switches of ZnO wires. *Nano Lett*, 2008. **8**(11): p. 3973–3977.
[44] F. Boxberg, N. Sondergaard, and H.Q. Xu, Photovoltaics with piezoelectric core-shell nanowires. *Nano Lett*, 2010. **10**(4): p. 1108–1112. DOI: 10.1021/nl9040934
[45] F. Boxberg, N. Sondergaard, and H.Q. Xu, Elastic and piezoelectric properties of zincblende and wurtzite crystalline nanowire heterostructures. *Adv Mater*, 2012. **24**(34): p. 4692–4706. DOI: 10.1002/adma.201200370
[46] Z.L. Wang, Progress in piezotronics and piezo-phototronics. *Adv Mater*, 2012. **24**(34): p. 4632–4646. DOI: 10.1002/adma.201104365
[47] N. Ma, K. Zhang, and Y. Yang, Photovoltaic–pyroelectric coupled effect induced electricity for self-powered photodetector system. *Advanced Materials*, 2017. **29**(46): p. 1703694. DOI: 10.1002/adma.201703694
[48] W. Peng, R. Yu, X. Wang, Z. Wang, H. Zou, Y. He, et al., Temperature dependence of pyro-phototronic effect on self-powered ZnO/perovskite heterostructured photodetectors. *Nano Research*, 2016. **9**(12): p. 3695–3704. DOI: 10.1007/s12274-016-1240-5
[49] L. Su, H.Y. Li, Y. Wang, S.Y. Kuang, Z.L. Wang, and G. Zhu, Coupling of photoelectric and triboelectric effects as an effective approach for PZT-based high-performance self-powered ultraviolet photodetector *Nano Energy*, 2017. **31**: p. 264–269.
[50] L. Su, Z.X. Zhao, H.Y. Li, J. Yuan, Z.L. Wang, G.Z. Cao, et al., High-performance organolead halide perovskite-based self-powered triboelectric photodetector. *ACS Nano*, 2015. **9**(11): p. 11310–11316.
[51] S. Sett, S. Sengupta, N. Ganesh, K.S. Narayan, and A.K. Raychaudhuri, Self-powered single semiconductor nanowire photodetector. *Nanotechnology*, 2018. **29**(44): p. 445202. DOI: 10.1088/1361-6528/aada2d
[52] Z. Bai, X. Yan, X. Chen, H. Liu, Y. Shen, and Y. Zhang, ZnO nanowire array ultraviolet photodetectors with self-powered properties. *Current Applied Physics*, 2013. **13**(1): p. 165–169. DOI: 10.1016/j.cap.2012.07.005
[53] Q. An, X. Meng, K. Xiong, and Y. Qiu, Self-powered ZnS nanotubes/Ag nanowires MSM UV photodetector with high on/off ratio and fast response speed. *Sci Rep*, 2017. **7**(1): p. 4885. DOI: 10.1038/s41598-017-05176-5
[54] C. Guo, W. Guo, Y. Dai, H. Xu, L. Chen, D. Wang, et al., Self-powered ultraviolet MSM photodetectors with high responsivity enabled by a lateral n+/n− homojunction from opposite polarity domains. *Optics Letters*, 2021. **46**(13): p. 3203–3206.
[55] B.R. Tak, M.M. Yang, Y.H. Lai, Y.H. Chu, M. Alexe, and R. Singh, Photovoltaic and flexible deep ultraviolet wavelength detector based on novel beta-Ga(2)O(3)/muscovite heteroepitaxy. *Sci Rep*, 2020. **10**(1): p. 16098. DOI: 10.1038/s41598-020-73112-1
[56] C. Jia, H. Liu, X. Zhang, S. Wang, and X. Li, A high performance Au/CH3NH3PbI3/Cu planar-type self-powered photodetector. *Journal of Materials Chemistry C*, 2022. **10**(35): p. 12602–12609. DOI: 10.1039/d2tc02477b
[57] L.-A. Cao, M.-S. Yao, H.-J. Jiang, S. Kitagawa, X.-L. Ye, W.-H. Li, et al., A highly oriented conductive MOF thin film-based Schottky diode for self-powered light and gas detection. *Journal of Materials Chemistry A*, 2020. **8**(18): p. 9085–9090. DOI: 10.1039/d0ta01379j
[58] Y. Wang, C. Wu, D. Guo, P. Li, S. Wang, A. Liu, et al., All-Oxide NiO/Ga2O3p–n Junction for Self-Powered UV Photodetector. *ACS Applied Electronic Materials*, 2020. **2**(7): p. 2032–2038. DOI: 10.1021/acsaelm.0c00301

[59] C. He, D. Guo, K. Chen, S. Wang, J. Shen, N. Zhao, et al., α-Ga2O3 nanorod array–Cu2O microsphere p–n junctions for self-powered spectrum-distinguishable photodetectors. *ACS Applied Nano Materials*, 2019. **2**(7): p. 4095–4103. DOI: 10.1021/acsanm.9b00527

[60] C. Wu, F. Wu, C. Ma, S. Li, A. Liu, X. Yang, et al., A general strategy to ultrasensitive Ga2O3 based self-powered solar-blind photodetectors. *Materials Today Physics*, 2022. **23**: p. 100643. DOI: 10.1016/j.mtphys.2022.100643

[61] D. Guo, Y. Su, H. Shi, P. Li, N. Zhao, J. Ye, et al., Self-powered ultraviolet photodetector with superhigh photoresponsivity (3.05 A/W) based on the GaN/Sn:Ga(2)O(3) pn junction. *ACS Nano*, 2018. **12**(12): p. 12827–12835. DOI: 10.1021/acsnano.8b07997

[62] J. Chen, D. You, Y. Zhang, T. Zhang, C. Yao, Q. Zhang, et al., Highly sensitive and tunable self-powered UV photodetectors driven jointly by p-n junction and ferroelectric polarization. *ACS Appl Mater Interfaces*, 2020. DOI: 10.1021/acsami.0c15816

[63] A. Bera, A. Das Mahapatra, S. Mondal, and D. Basak, Sb(2)S(3)/spiro-OMeTAD inorganic-organic hybrid p-n junction diode for high performance self-powered photodetector. *ACS Appl Mater Interfaces*, 2016. **8**(50): p. 34506–34512. DOI: 10.1021/acsami.6b09943

[64] Y. Xin, J. Jiang, Y. Lu, H. Liang, Y.J. Zeng, and Z. Ye, Self-powered broad spectral photodetector with ultrahigh responsivity and fast response based on Sb 2 Se 3 /VO 2 heterojunction. *Advanced Materials Interfaces*, 2021. **8**(10). DOI: 10.1002/admi.202100058

[65] Z.A. Bashkany, I.K. Abbas, M.A. Mahdi, H.F. Al-Taay, and P. Jennings, A self-powered heterojunction photodetector based on a PbS nanostructure grown on porous silicon substrate. *Silicon*, 2016. **10**(2): p. 403–411. DOI: 10.1007/s12633-016-9462-4

[66] N. Deka, P. Chakraborty, D. Chandra Patra, S. Dhar, and S.P. Mondal, Self-powered broadband photodetection using PbS decorated ZnO nanorods/reduced graphene oxide junction. *Materials Science in Semiconductor Processing*, 2020. **118**. DOI: 10.1016/j.mssp.2020.105165

[67] T. Guo, C. Ling, X. Li, X. Qiao, X. Li, Y. Yin, et al., A ZIF-8@H:ZnO core–shell nanorod arrays/Si heterojunction self-powered photodetector with ultrahigh performance. *Journal of Materials Chemistry C*, 2019. **7**(17): p. 5172–5183. DOI: 10.1039/c9tc00290a

[68] S. Abbas, D.-K. Ban, and J. Kim, Functional interlayer of In2O3 for transparent SnO2/SnS2 heterojunction photodetector. *Sensors and Actuators A: Physical*, 2019. **293**: p. 215–221. DOI: 10.1016/j.sna.2019.04.049

[69] F. Cao, L. Meng, M. Wang, W. Tian, and L. Li, Gradient energy band driven high-performance self-powered perovskite/CdS photodetector. *Adv Mater*, 2019. **31**(12): p. e1806725. DOI: 10.1002/adma.201806725

[70] F. Yan, L. Zhao, A. Patane, P. Hu, X. Wei, W. Luo, et al., Fast, multicolor photodetection with graphene-contacted p-GaSe/n-InSe van der Waals heterostructures. *Nanotechnology*, 2017. **28**(27): p. 27LT01. DOI: 10.1088/1361-6528/aa749e

[71] G. Li, L. Liu, G. Wu, W. Chen, S. Qin, Y. Wang, et al., Self-powered UV-near infrared photodetector based on reduced graphene oxide/n-Si vertical heterojunction. *Small*, 2016. **12**(36): p. 5019–5026. DOI: 10.1002/smll.201600835

[72] Z. Bai and Y. Zhang, Self-powered UV–visible photodetectors based on ZnO/Cu2O nanowire/electrolyte heterojunctions. *Journal of Alloys and Compounds*, 2016. **675**: p. 325–330. DOI: 10.1016/j.jallcom.2016.03.051

[73] R. Balakarthikeyan, A. Santhanam, R. Anandhi, S. Vinoth, A.M. Al-Baradi, Z.A. Alrowaili, et al., Fabrication of nanostructured NiO and NiO:Cu thin films for high-performance ultraviolet photodetector. *Optical Materials*, 2021. **120**. DOI: 10.1016/j.optmat.2021.111387

[74] Z. Chen, B. Li, X. Mo, S. Li, J. Wen, H. Lei, et al., Self-powered narrowband p-NiO/n-ZnO nanowire ultraviolet photodetector with interface modification of Al2O3. *Applied Physics Letters*, 2017. **110**(12). DOI: 10.1063/1.4978765

[75] Y. Gao, J. Xu, S. Shi, H. Dong, Y. Cheng, C. Wei, et al., TiO(2) nanorod arrays based self-powered UV photodetector: Heterojunction with NiO nanoflakes and enhanced UV photoresponse. *ACS Appl Mater Interfaces*, 2018. **10**(13): p. 11269–11279. DOI: 10.1021/acsami.7b18815

[76] P. Ghamgosar, F. Rigoni, S. You, I. Dobryden, M.G. Kohan, A.L. Pellegrino, et al., ZnO-Cu2O core-shell nanowires as stable and fast response photodetectors. *Nano Energy*, 2018. **51**: p. 308–316. DOI: 10.1016/j.nanoen.2018.06.058

[77] P. Ghamgosar, F. Rigoni, M.G. Kohan, S. You, E.A. Morales, R. Mazzaro, et al., Self-powered photodetectors based on core-shell ZnO-Co(3)O(4) nanowire heterojunctions. *ACS Appl Mater Interfaces*, 2019. **11**(26): p. 23454–23462. DOI: 10.1021/acsami.9b04838

[78] M.R. Hasan, T. Xie, S.C. Barron, G. Liu, N.V. Nguyen, A. Motayed, et al., Self-powered p-NiO/n-ZnO heterojunction ultraviolet photodetectors fabricated on plastic substrates. *APL Mater*, 2015. **3**(10): p. 106101. DOI: 10.1063/1.4932194

[79] P. Mahala, M. Patel, D.-K. Ban, T.T. Nguyen, J. Yi, and J. Kim, High-performing self-driven ultraviolet photodetector by TiO2/Co3O4 photovoltaics. *Journal of Alloys and Compounds*, 2020. **827**. DOI: 10.1016/j.jallcom.2020.154376

[80] B. Zhao, F. Wang, H. Chen, L. Zheng, L. Su, D. Zhao, et al., An ultrahigh responsivity (9.7 mA W −1) self-powered solar-blind photodetector based on individual ZnO–Ga 2 O 3 heterostructures. *Advanced Functional Materials*, 2017. **27**(17). DOI: 10.1002/adfm.201700264

[81] C. Wang, J. Xu, S. Shi, Y. Zhang, Y. Gao, Z. Liu, et al., Optimizing performance of Cu 2 O/ZnO nanorods heterojunction based self-powered photodetector with ZnO seed layer. *Journal of Physics and Chemistry of Solids*, 2017. **103**: p. 218–223. DOI: 10.1016/j.jpcs.2016.12.026

[82] Y. Shen, X. Yan, Z. Bai, X. Zheng, Y. Sun, Y. Liu, et al., A self-powered ultraviolet photodetector based on solution-processed p-NiO/n-ZnO nanorod array heterojunction. *RSC Advances*, 2015. **5**(8): p. 5976–5981. DOI: 10.1039/c4ra12535e

[83] M.I. Saleem, S. Yang, A. Batool, M. Sulaman, Y. Song, Y. Jiang, et al., All-solution-processed, high-performance self-powered UVA photodetectors with non-opaque silver nanowires electrode. *Sensors and Actuators A: Physical*, 2021. **322**. DOI: 10.1016/j.sna.2021.112606

[84] H. Xie, Z. Li, L. Cheng, A.A. Haidry, J. Tao, Y. Xu, et al., Recent advances in the fabrication of 2D metal oxides. *iScience*, 2022. **25**(1): p. 103598. DOI: 10.1016/j.isci.2021.103598

[85] X. Jiang, S. Liu, W. Liang, S. Luo, Z. He, Y. Ge, et al., Broadband nonlinear photonics in few-layer MXene Ti3C2Tx (T = F, O, or OH). *Laser Photonics Rev.*, 2018. **12**(2): p. 1870013.

[86] M. Patel and J. Kim, Transparent NiO/ZnO heterojunction for ultra-performing zero-bias ultraviolet photodetector on plastic substrate. *Journal of Alloys and Compounds*, 2017. **729**: p. 796–801. DOI: 10.1016/j.jallcom.2017.09.158

[87] T.T. Nguyen, M. Patel, and J. Kim, Self-powered transparent photodetectors for broadband applications. *Surfaces and Interfaces*, 2021. **23**. DOI: 10.1016/j.surfin.2021.100934

[88] P.-N. Ni, C.-X. Shan, S.-P. Wang, X.-Y. Liu, and D.-Z. Shen, Self-powered spectrum-selective photodetectors fabricated from n-ZnO/p-NiO core–shell nanowire arrays. *Journal of Materials Chemistry C*, 2013. **1**(29). DOI: 10.1039/c3tc30525b

[89] Y. Ning, Z. Zhang, F. Teng, and X. Fang, Novel transparent and self-powered UV photodetector based on crossed ZnO nanofiber array homojunction. *Small*, 2018. **14**(13): p. e1703754. DOI: 10.1002/smll.201703754

[90] Z. Wang, R. Yu, C. Pan, Z. Li, J. Yang, F. Yi, et al., Light-induced pyroelectric effect as an effective approach for ultrafast ultraviolet nanosensing. *Nat Commun*, 2015. **6**: p. 8401. DOI: 10.1038/ncomms9401

[91] H. Kumar, Y. Kumar, B. Mukherjee, G. Rawat, C. Kumar, B.N. Pal, et al., Electrical and optical characteristics of self-powered colloidal CdSe quantum dot-based photodiode. *IEEE Journal of Quantum Electronics*, 2017. **53**(3): p. 1–8. DOI: 10.1109/jqe.2017.2696487

[92] K. Benyahia, F. Djeffal, H. Ferhati, A. Bendjerad, A. Benhaya, and A. Saidi, Self-powered photodetector with improved and broadband multispectral photoresponsivity based on ZnO-ZnS composite. *Journal of Alloys and Compounds*, 2021. **859**. DOI: 10.1016/j.jallcom.2020.158242

[93] S. Ebrahimi and B. Yarmand, Solvothermal growth of aligned SnxZn1-xS thin films for tunable and highly response self-powered UV detectors. *Journal of Alloys and Compounds*, 2020. **827**. DOI: 10.1016/j.jallcom.2020.154246

[94] S. Ebrahimi and B. Yarmand, Tunable and high-performance self-powered ultraviolet detectors using leaf-like nanostructural arrays in ternary tin zinc sulfide system. *Microelectronics Journal*, 2021. **116**. DOI: 10.1016/j.mejo.2021.105237

[95] K.S. Novoselov, A.K. Geim, S.V. Morozov, D. Jiang, Y. Zhang, S.V. Dubonos, et al., Electric field effect in atomically thin carbon films. *Science*, 2004. **306**(5696): p. 666–669.

[96] D. Kong, H. Wang, J.J. Cha, M. Pasta, K.J. Koski, J. Yao, et al., Synthesis of MoS2 and MoSe2 films with vertically aligned layers. *Nano Lett*, 2013. **13**(3): p. 1341–1347. DOI: 10.1021/nl400258t

[97] C. Lee, H. Yan, L.E. Brus, T.F. Heinz, J. Hone, and S. Ryu, Anomalous lattice vibrations of singleand few-layer MoS2. *ACS Nano*, 2010. **4**(5): p. 2695–2700.

[98] S. Wang, X. Wang, and J.H. Warner, All chemical vapor deposition growth of MoS2:h-BN vertical van der Waals heterostructures. *ACS Nano*, 2015. **9**(5): p. 5246–5254.

[99] M. Zhang, Q. Wu, F. Zhang, L. Chen, X. Jin, Y. Hu, et al., Black-phosphorous-based pulsed lasers: 2D black phosphorus saturable absorbers for ultrafast photonics. *Adv Opt Mater*, 2019. **7**(1): p. 1970001.

[100] J. Pang, A. Bachmatiuk, Y. Yin, B. Trzebicka, L. Zhao, L. Fu, et al., Applications of phosphorene and black phosphorus in energy conversion and storage devices. *Advanced Energy Materials*, 2018. **8**(8). DOI: 10.1002/aenm.201702093

[101] X.-D. Zhu, Y. Xie, and Y.-T. Liu, Exploring the synergy of 2D MXene-supported black phosphorus quantum dots in hydrogen and oxygen evolution reactions. *Journal of Materials Chemistry A*, 2018. **6**(43): p. 21255–21260. DOI: 10.1039/c8ta08374f

[102] M.-Q. Zhao, M. Torelli, C.E. Ren, M. Ghidiu, Z. Ling, B. Anasori, et al., 2D titanium carbide and transition metal oxides hybrid electrodes for Li-ion storage. *Nano Energy*, 2016. **30**: p. 603–613. DOI: 10.1016/j.nanoen.2016.10.062

[103] R. Kumar, X. Liu, J. Zhang, and M. Kumar, Room-temperature gas sensors under photoactivation: From metal oxides to 2D materials. *Nanomicro Lett*, 2020. **12**(1): p. 164. DOI: 10.1007/s40820-020-00503-4

[104] F. Xia, H. Wang, D. Xiao, M. Dubey, and A. Ramasubramaniam, Two-dimensional material nanophotonics. *Nat. Photonics*, 2014. **8**: p. 899–907.

[105] B. Wang, S.P. Zhong, Z.B. Zhang, Z.Q. Zheng, Y.P. Zhang, and H. Zhang, Broadband photodetectors based on 2D group IVA metal chalcogenides semiconductors. *Applied Materials Today*, 2019. **15**: p. 115–138. DOI: 10.1016/j.apmt.2018.12.010

[106] F. Liu, Y. Yan, D. Miao, J. Xu, J. Shi, X. Gan, et al., Gate-tunable self-driven photodetector based on asymmetric monolayer WSe2 channel. *Applied Surface Science*, 2023. **616**. DOI: 10.1016/j.apsusc.2023.156444

[107] W. Gao, S. Zhang, F. Zhang, P. Wen, L. Zhang, Y. Sun, et al., 2D WS2 based asymmetric schottky photodetector with high performance. *Advanced Electronic Materials*, 2020. **7**(7). DOI: 10.1002/aelm.202000964

[108] C. Zhou, S. Zhang, Z. Lv, Z. Ma, C. Yu, Z. Feng, et al., Self-driven WSe2 photodetectors enabled with asymmetrical van der Waals contact interfaces. *NPJ 2D Materials and Applications*, 2020. **4**(1). DOI: 10.1038/s41699-020-00179-9

[109] A.H. Abdullah Ripain, N.A.A. Zulkifli, C.L. Tan, W.H. Abd Majid, and R. Zakaria, Contributions of symmetric metal contacts on liquid exfoliation 2D-MoS2 flakes based MSM photodetector by spray pyrolysis: A CVD-free technique. *Optical and Quantum Electronics*, 2022. **54**(12). DOI: 10.1007/s11082-022-04253-y

[110] X.G. Liang, L. DanMin, L.J. Zhen, J.J. Li and Y.E.S. Shuai, Self-powered and bipolar photodetector based on a van der Waals metal-semiconductor junction: Graphene/WSe_2/Fe_3GeTe_2 heterojunction. *Science China*, 2022. **16**(6): p. 1263–1272. DOI: 10.1007/s11431-022-2031-7

[111] J. Chen, Z. Zhang, J. Feng, X. Xie, A. Jian, Y. Li, et al., 2D InSe self-powered schottky photodetector with the same metal in asymmetric contacts. *Advanced Materials Interfaces*, 2022. **9**(35). DOI: 10.1002/admi.202200075

[112] Y. Zhang, Y. Xu, L. Gao, X. Liu, Y. Fu, C. Ma, et al., MXene-based mixed-dimensional Schottky heterojunction towards self-powered flexible high-performance photodetector. *Materials Today Physics*, 2021. **21**. DOI: 10.1016/j.mtphys.2021.100479

[113] C. Patil, C. Dong, H. Wang, B.M. Nouri, S. Krylyuk, H. Zhang, et al., Self-driven highly responsive p-n junction InSe heterostructure near-infrared light detector. *Photonics Research*, 2022. **10**(7): p. A97–A105. DOI: 10.1364/prj.441519

[114] M. Hussain, S.H.A. Jaffery, A. Ali, C.D. Nguyen, S. Aftab, M. Riaz, et al., NIR self-powered photodetection and gate tunable rectification behavior in 2D GeSe/MoSe(2) heterojunction diode. *Sci Rep*, 2021. **11**(1): p. 3688. DOI: 10.1038/s41598-021-83187-z

[115] Z. Zou, J. Liang, X. Zhang, C. Ma, P. Xu, X. Yang, et al., Liquid-metal-assisted growth of vertical GaSe/MoS(2) p-n heterojunctions for sensitive self-driven photodetectors. *ACS Nano*, 2021. **15**(6): p. 10039–10047. DOI: 10.1021/acsnano.1c01643

[116] H. Liu, X. Zhu, X. Sun, C. Zhu, W. Huang, X. Zhang, et al., Self-powered broad-band photodetectors based on vertically stacked WSe(2)/Bi(2)Te(3) p-n heterojunctions. *ACS Nano*, 2019. **13**(11): p. 13573–13580. DOI: 10.1021/acsnano.9b07563

[117] C. Kang, M. Ahsan Iqbal, S. Zhang, X. Weng, Y. Sun, L. Qi, et al., Cu(3) (HHTP)(2) c-MOF/ZnO ultrafast ultraviolet photodetector for wearable optoelectronics. *Chemistry*, 2022. **28**(64): p. e202201705. DOI: 10.1002/chem.202201705

[118] L. Guo, X. Liu, L. Gao, X. Wang, L. Zhao, W. Zhang, et al., Ferro-pyro-phototronic effect in monocrystalline 2D ferroelectric perovskite for high-sensitive, self-powered, and stable ultraviolet photodetector. *ACS Nano*, 2022. **16**(1): p. 1280–1290. DOI: 10.1021/acsnano.1c09119

[119] W. Song, J. Chen, Z. Li, and X. Fang, Self-powered MXene/GaN van der Waals heterojunction ultraviolet photodiodes with superhigh efficiency and stable current outputs. *Adv Mater*, 2021. **33**(27): p. e2101059. DOI: 10.1002/adma.202101059

[120] D. Li, R. Li, D. Zhou, F. Zeng, X. Qin, W. Yan, et al., High-performance self-powered photodetector with broadened spectrum absorption based on black phosphorus/Cs2SnI4 heterostructure. *Applied Surface Science*, 2023. **609**. DOI: 10.1016/j.apsusc.2022.155032

[121] J. Chen, Z. Zhang, Y. Ma, J. Feng, X. Xie, X. Wang, et al., High-performance self-powered ultraviolet to near-infrared photodetector based on WS2/InSe van der Waals heterostructure. *Nano Research*, 2022. DOI: 10.1007/s12274-022-5323-1

[122] H. Wang, Y. Gui, C. Dong, S. Altaleb, B.M. Nouri, M. Thomaschewski, et al., Self-powered broadband photodetector based on MoS2/Sb2Te3 heterojunctions: A promising approach for highly sensitive detection *Nanophotonics*, 2022. **11**(22): p. 5113–5119.

[123] Z.-Y. Yan, S. Li, Z. Liu, W.-J. Liu, F. Qiao, P.-G. Li, et al., Ti3C2/ϵ-Ga2O3 schottky self-powered solar-blind photodetector with robust responsivity. *IEEE Journal of Selected Topics in Quantum Electronics*, 2022. **28**(2, Optical Detectors): p. 1–8. DOI: 10.1109/jstqe.2021.3124824

[124] C. Thota, G. Murali, R. Dhanalakshmi, M. Reddeppa, N.H. Bak, G. Nagaraju, et al., 2D MXene/ 1D GaN van der Waals heterojunction for self-powered UV photodetector. *Applied Physics Letters*, 2023. **122**(3). DOI: 10.1063/5.0132756

[125] S. Yang, C. Wang, C. Ataca, Y. Li, H. Chen, H. Cai, et al., Self-driven photodetector and ambipolar transistor in atomically thin GaTe-MoS2 p-n vdW heterostructure. *ACS Appl Mater Interfaces*, 2016. **8**(4): p. 2533–2539. DOI: 10.1021/acsami.5b10001

[126] W. Jin, Y. Ye, L. Gan, B. Yu, P. Wu, Y. Dai, et al., Self-powered high performance photodetectors based on CdSe nanobelt/graphene Schottky junctions. *Journal of Materials Chemistry*, 2012. **22**(7). DOI: 10.1039/c2jm15913a

[127] Z. Gao, W. Jin, Y. Zhou, Y. Dai, B. Yu, C. Liu, et al., Self-powered flexible and transparent photovoltaic detectors based on CdSe nanobelt/graphene Schottky junctions. *Nanoscale*, 2013. **5**(12): p. 5576–55811. DOI: 10.1039/c3nr34335a

[128] C. Xie, L. Zeng, Z. Zhang, Y.H. Tsang, L. Luo, and J.H. Lee, High-performance broadband heterojunction photodetectors based on multilayered PtSe(2) directly grown on a Si substrate. *Nanoscale*, 2018. **10**(32): p. 15285–15293. DOI: 10.1039/c8nr04004d

[129] Y. Wu, X. Yan, X. Zhang, and X. Ren, A monolayer graphene/GaAs nanowire array Schottky junction self-powered photodetector. *Applied Physics Letters*, 2016. **109**(18). DOI: 10.1063/1.4966899

[130] L. Duan, F. He, Y. Tian, B. Sun, J. Fan, X. Yu, et al., Fabrication of self-powered fast-response ultraviolet photodetectors based on graphene/ZnO:Al Nanorod-array-film structure with stable schottky barrier. *ACS Appl Mater Interfaces*, 2017. **9**(9): p. 8161–8168. DOI: 10.1021/acsami.6b14305

Index

Page numerals in *italics* refer to figures and those in **bold** refer to tables.

F

M

S